Studies in Economic Theory

Editors

Charalambos D. Aliprantis
Purdue University
Department of Economics
West Lafayette, IN 47907-1310
USA

Nicholas C. Yannelis
Department of Economics
University of Illinois
Champaign, IL 61820
USA

Springer
Berlin
Heidelberg
New York
Barcelona
Hong Kong
London
Milan
Paris
Singapore
Tokyo

Titles in the Series

M. Ali Khan and Nicholas C. Yannelis (Eds.)
Equilibrium Theory in Infinite Dimensional Spaces

Charalambos D. Aliprantis, Kim C. Border and
Wilhelmus A. J. Luxemburg (Eds.)
Positive Operators, Riesz Spaces, and Economics

Donald G. Saari
Geometry of Voting

Charalambos D. Aliprantis and Kim C. Border
Infinite Dimensional Analysis

Jean-Pierre Aubin
Dynamic Economic Theory

Mordecai Kurz (Ed.)
Endogenous Economic Fluctuations

Jean-François Laslier
Tournament Solutions and Majority Voting

Ahmet Alkan, Charalambos D. Aliprantis,
Nicholas C. Yannelis (Eds.)
Theory and Applications

James C. Moore
Mathematical Methods for Economic Theory 1

James C. Moore
Mathematical Methods for Economic Theory 2

Mukul Majumdar, Tapan Mitra, Kazuo Nishimura
Optimization and Chaos

Mukul Majumdar
Tapan Mitra
Kazuo Nishimura

Optimization and Chaos

With 13 Figures

Springer

Prof. Mukul Majumdar
Prof. Tapan Mitra
Cornell University
Department of Economics
418 Uris Hall
Ithaca, NY 14853-7601
USA
mkm5@cornell.edu

Prof. Kazuo Nishimura
Kyoto University
Institute for Economic Research
Kyoto 606
Japan

ISSN 1431-8849
ISBN 3-540-67030-0 Springer-Verlag Berlin Heidelberg New York

Library of Congress Cataloging-in-Publication Data
Die Deutsche Bibliothek – CIP-Einheitsaufnahme
Optimization and chaos / Mukul Majumdar, Tapan Mitra, Kazuo Nishimura. – Berlin; Heidelberg; New York; Barcelona; Hong Kong; London; Milan; Paris; Singapore; Tokyo: Springer, 2000
(Studies in economic theory; 11)
ISBN 3-540-67030-0

Springer-Verlag is a company in the BertelsmannSpringer publishing group.

Printed in Germany

Hardcover design: Erich Kirchner, Heidelberg

SPIN 10733605 42/2202-5 4 3 2 1 0 – Printed on acid-free paper

Preface

This collection of essays brings together some articles on dynamic optimization models that exhibit chaotic behavior. Chapters 3, 4, 5, 6, 7, and 9 appeared in a Symposium on Chaotic Dynamical Systems in *Economic Theory* (Volume 4, Number 5, 1994). Also, Chapters 10,11, and 12 appeared in the *Journal of Economic Theory*. We would like to thank the authors, and Academic Press for permission to reprint. We are grateful to Professor C.D. Aliprantis for suggesting the idea of a book structured around the *Economic Theory* Symposium, and without the support and patience of Dr. Mueller this project could not have been completed.

We would like to thank Ms. Amy Gowan who cheerfully performed the arduous task of typing the manuscript. Thanks are also due to Xiao Qing Yu, Tridip Ray and Malabika Majumdar for their help at various stages in the preparation of the manuscript.

For a course on dynamic optimization addressed to students with a good background in economic theory and real analysis, one can assign Chapter 2 as a partial introduction to the basic techniques. Chapters 3 and 4 can be assigned to provide examples of simple optmization models generating complicated behavior. The role of discounting in generating complexity can be taken up through Chapters 10-12. Many of the major problems of intertemporal allocation theory cannot be discussed in the framework of Chapter 2: short run shocks, indivisibilities, increasing returns, structural breaks for the "late-comers in development", export-led strategies of develoment, sustained growth through human capital accumulation, just to indicate a few. Appropriate readings can be assigned for some of these topics. The course can be made more challenging by introducing the standard model of discounted dynamic programming under uncertainty with a single decision making agent [using any of the standard texts], and then extending the coverage to models involving a small number of interacting decision makers. The excellent review of Dutta and Sundaram (1998) is particuarly useful in the last context.

Parts of the book can also be used for a course on chaotic models in economic theory. In such a course, Chapters 1 and 2 will provide some of the core material. The instructor can move on to Chapters 3-6, to provide examples of bifurcation, cycling and chaos. The question of "robustness" can be dealt with by drawing from Chapters 3,7 and 9. Finally, parts of the last four chapters can be used to clarify the role of discounting.

Progress of a science, we are told, is made through funerals. It is our hope that some of the concepts and techniques will continue to be useful in the analysis of models that will engage researchers in the future.

List of Contributors

Robert Becker, Indiana University
Ciprian Foias, Indiana University
Mukul Majumdar, Cornell University
Tapan Mitra, Cornell University
Kazuo Nishimura, Kyoto University
Gerhard Sorger, University of London
Makoto Yano, Keio University

Contents

1

Dynamical Systems: A Tutorial

Mukul Majumdar and Tapan Mitra

1. Introduction

A dynamical system is described by a pair (X, f) where X is a nonempty set (called the *state space*) and f is a function (called the *law of motion*) from X into X. Thus, if x_t is the state of the system in period t, then

$$x_{t+1} = f(x_t) \tag{1.1}$$

is the state of the system in period $t+1$.

In this chapter we always assume that the state space X *is a (nonempty) metric space (the metric is denoted by* d*) and* the law of motion f *is a continuous function from* X *into* X.

Once the initial state x (i.e., the state in period 0) is specified, we write $f^0(x) \equiv x$, $f^1(x) = f(x)$, and for every positive integer $j \geq 1$

$$f^{j+1}(x) = f(f^j(x)) \tag{1.2}$$

We refer to f^j as the *j-th iterate* of f. From any initial x, the *trajectory* from x is the sequence $\tau(x) = \{f^j(x)_{j=0}^{\infty}\}$. The *orbit* from x is the set $\gamma(x) = \{y : y = f^j(x) \text{ for some } j \geq 0\}$. The long run or asymptotic behavior of a trajectory $\tau(x)$ is described by its limit set $\omega(x)$ defined as the set of all limit points of $\tau(x)$.

Example 1.1: Let $X = \Re$, $f(x) = -x$. Verify that:
$\tau(1) = (1, -1, 1, -1, 1...)$
$\gamma(1) = \{1, -1\}$
$\tau(0) = (0, 0, 0...)$
$\gamma(0) = \{0\}$

Example 1.2: $X = \Re$, $f(x) = x + 1$. Here, $f^j(x) = x + j$ for all $j \geq 1$.

2. Fixed Points and Periodic Points

A point $x \in X$ is a *fixed point* if $x = f(x)$. A fixed point is often called a *steady state* or a *rest point* or an *equilibrium* of the dynamical system.

Exercise 2.1: Show that in Example 1.1, the only fixed point of f is the point 0. In Example 1.2, the function does not have a fixed point.

Example 2.1: Let $X = \Re$, $f(x) = x^2$. If x is any fixed point of f, it must satisfy the equation $x = x^2$. This means that the only fixed points of f are $0, 1$. More generally, consider the class of functions $f_c(x) = x^2 + c$, where c is a real number. For $c > 1/4$, f_c does not have any fixed point; for $c = 1/4$, f_c has a unique fixed point $x = 1/2$; for $c < 1/4$, f_c has a pair of fixed points.

A point $x \in X$ is a *periodic point* of *period* $k \geq 2$, if $f^k(x) = x$ and $f^j(x) \neq x$ for $1 \leq j < k$. Thus, to prove that x is a periodic point of period, say, 3, one must prove that x is a fixed point of f^3 and that it is *not* a fixed point of f and f^2. *Some writers consider a fixed point as a periodic point of period one.*

Example 2.2: Consider $X = \Re$, $f(x) = x^2 - 1$. The fixed points are $[\sqrt{5} + 1]/2$ and $[\sqrt{5} - 1]/2$. Note that $f(0) = -1$; $f(-1) = 0$. Hence, both 0 and -1 are periodic points of period 2.

$\tau(0) = (0, -1, 0, -1, ...)$

$\tau(-1) = (-1, 0, -1, 0, ...)$

$\gamma(-1) = \{-1, 0\}$

$\gamma(0) = \{0, -1\}$

More generally, consider $f_c(x) = x^2 + c$.

Here, $f_c^2(x) = (x^2 + c)^2 + c = x^4 + 2x^2c + c^2 + c$. Hence, to compute periodic points of period 2 one has to solve the equation:

$$x^4 + 2x^2c + c^2 + c = x \tag{2.1}$$

Observe that if f_c does have any fixed points, then these will also solve the equation.

Exercise 2.2: Consider the dynamical system in Example 1.2. Show that there is no periodic point.

Exercise 2.3: Let $X = [0, 1]$. Consider the "tent map" defined by:

$$f(x) = \begin{cases} 2x & \text{for } x \in [0, 1/2] \\ 2(1-x) & \text{for } x \in [1/2, 1] \end{cases}$$

Note that f has two fixed points '0' and '2/3'. It is tedious to write out the functional form of f^2:

$$f^2(x) = \begin{cases} 4x & \text{for } x \in [0, 1/4] \\ 2(1-2x) & \text{for } x \in [1/4, 1/2] \\ 2(2x-1) & \text{for } x \in [1/2, 3/4] \\ 4(1-x) & \text{for } x \in [3/4, 1] \end{cases}$$

Verify the following:

(i) There are two periodic points of period 2, namely '2/5'and '4/5'.

(ii) There are three periodic points of period 3, namely '2/9', '4/9', '8/9'. It follows from a well-known theorem (see below) that there are periodic points of *all* periods

Let $\wp(X)$ be the set of all periodic points of X. We write $\aleph(X)$ to denote the set of non-periodic points.

We now note some useful results on the existence of fixed points of f.

Proposition 2.1: *Let $X = \Re$, and f be continuous. If there is a (nondegenerate) closed interval $I = [a, b]$ such that*
(i) $f(I) \subset I$ *or* (ii) $f(I) \supset I$
then there is a fixed point of f in I

Proof (sketch): (i) If $f(I) \subset I$, then $f(a) \geq a$, and $f(b) \leq b$. If $f(a) = a$ or $f(b) = b$, the conclusion is immediate. Otherwise, $f(a) > a$ and $f(b) < b$. This means that the function $g(x) = f(x) - x$ is positive at a and negative at b. Using the intermediate value theorem, $g(x^*) = 0$ for some x^* in (a, b). Then $f(x^*) = x^*$. (ii) If $f(I) \supset I$, by the Weierestrass' theorem, there are points

x_m and x_M in I such that $f(x_m) \leq f(x) \leq f(x_M)$ for all x in I. Write $f(x_m) = m$ and $f(x_M) = M$. Again, by the intermediate value theorem, $f(I) = [m, M]$. Since $f(I)$ is assumed to contain I, $m \leq a \leq b \leq M$. In other words,

$$f(x_m) = m \leq a \leq x_m,$$

and

$$f(x_M) = M \geq b \geq x_M.$$

The proof can now be completed by an argument similar to that in case (i).

Exercise 2.4: Let $X = [a, b]$, and f is a continuous function from X into X. Suppose that for all x in (a, b), the derivative $f'(x)$ exists and $|f'(x)| < 1$ Then f has a unique fixed point in X.

Proposition 2.2: *Let X be a nonempty compact convex subset of $\Re^n$, and f be continuous. Then there is a fixed point of f.*

A function $f : X \to X$ is a *contraction* if there is some β, $0 < \beta < 1$, such that for all $x, y \in X$, $x \neq y$, one has

$$d(f(x),\ f(y)) < \beta d(x, y)$$

If f is a contraction, f is continuous on X.

Proposition 2.3: *Let (X, d) be a (nonempty) complete metric space, and $f : X \to X$ a contraction. Then f has a unique fixed point $x^* \in X$. Moreover, for any x in X, the trajectory $\tau(x) = \{f^j(x)_{j=0}^{\infty}\}$ converges to x^*.*

Proof (sketch): Choose an arbitrary $x \in X$. Consider the trajectory $\tau(x) = (x_t)$ from x, where

$$x_{t+1} = f(x_t). \tag{2.2}$$

Note that $d(x_2, x_1) = d(f(x_1),\ f(x)) < \beta d(x_1, x)$; hence, for any $t \geq 1$,

$$d(x_{t+1}, x_t) < \beta^t d(x_1, x) \tag{2.3}$$

We note that

$$\begin{aligned} d(x_{t+2}, x_t) &\le d(x_{t+2}, x_{t+1}) + d(x_{t+1}, x_t) \\ &< \beta^{t+1} d(x_1, x) + \beta^t d(x_1, x) \\ &= \beta^t (1+\beta) d(x_1, x) \end{aligned}$$

It follows that, for any integer $k \ge 1$,

$$d(x_{t+k}, x_t) < [\beta^t/(1-\beta)] d(x_1, x)$$

and this implies that (x_t) is a Cauchy sequence. Since X is assumed to be complete, $\lim_{t\to\infty} x_t = x^*$ exists. By continuity of f, and (2.2),

$$f(x^*) = x^*$$

If there are two distinct fixed points x^* and x^{**} of f, we see that there is a contradiction:

$$0 < d(x^*, x^{**}) = d(f(x^*), f(x^{**})) < \beta d(x^*, x^{**}) \tag{2.4}$$

where $0 < \beta < 1$ Q.E.D.

A fixed point x^* of f is *locally stable* (locally attracting) if there is an open set U containing x^* such that for all $x \in U$, the trajectory $\tau(x)$ converges to x^*.

A fixed point x^* is *repelling* if there is an open set U containing x^* such that for any $x \in U$, $x \neq x^*$, there is some $k \ge 1$, $f^k(x) \notin U$.

Consider a dynamical system (X, f) where X is a (nondegenerate) closed interval $[a, b]$, and f is continuous on $[a, b]$. Suppose that f is also continuously differentiable on (a, b). A fixed point $x^* \in (a, b)$ is *hyperbolic* if $|f'(x^*)| \neq 1$.

Proposition 2.4: *Let $X = [a, b]$ and f be continuous on $[a, b]$ and continuously differentiable on (a, b).*

(a) *If $x^* \in (a, b)$ is a hyperbolic fixed point of f and $|f'(x^*)| < 1$, then x^* is locally stable.*

(b) *If $x^* \in (a, b)$ is a hyperbolic fixed point with $|f'(x^*)| > 1$, then x^* is repelling.*

Proof: (a) There is some $\mu > 0$ such that $|f'(x)| < \beta < 1$ for all x in $I = [x^* - \mu, x^* + \mu]$. By the mean value theorem,

$$|x - x^*| = |f(x) - f(x^*) \leq \beta|x - x^*| < \beta\mu < \mu.$$

Hence, f maps I into I and, by the mean value theorem is a contraction on I. The result follows from Proposition 2.3.
(b) Exercise. Q.E.D.

We can define "a hyperbolic periodic point of period k" and define locally attracting and repelling periodic points accordingly.

Example 2.3: Going back to Example 2.2, consider once again the map $f(x) = x^2 - 1$. f^2 has four fixed points: the fixed points of f and $0, -1$. Both 0 and -1 are periodic points of period 2. Since

$$f^2(x) = x^4 - 2x^2$$

we see that the derivative of f^2 with respect to x, denoted by $[f^2(x)]'$, is given by

$$[f^2(x)]' = 4x^3 - 4x$$

Now, $[f^2(x)]'_{x=0} = [f^2(x)]'_{x=1} = 0$. Hence, both 0 and 1 are attracting fixed points of f^2.

Let x_0 be a periodic point of periods 2 and $x_1 = f(x_0)$. By definition $x_0 = f(x_1) = f^2(x_0)$ and $x_1 = f(x_0) = f^2(x_1)$. Now if f is differentiable, by the chain rule,

$$[f^2(x_0)]' = f'(x_1) \bullet f'(x_0)$$

More generally, suppose that x_0 is a periodic point of period k and its orbit is denoted by $\{x_0, x_1, ..., x_{k-1}\}$. Then, $[f^k(x_0)]' = f'(x_{k-1})...f'(x_0)$. It follows that $[f^k(x_0)]' = [f^k(x_1)]' = ...$ $[f^k(x_{k-1})]'$. We can now extend Proposition 2.4 appropriately.

Exercise 2.5: By using the graphs, if necessary, verify that the fixed and periodic points of the tent map in Exercise 2.3 are repelling.

While the contraction property of f ensures that the trajectories enter any neighborhood of the fixed point, there are examples of simple nonlinear dynamical systems in which trajectories "wander around" the state space. We shall now examine this feature more formally.

Proposition 2.5: *Let* $\{A_j\}_{j=0}^{\infty}$ *be a sequence of nonempty compact sets in* X *such that*

$$f(A_j) \supseteq A_{j+1}. \tag{2.5}$$

Then there is a nonempty compact set $Q \subset A_0$ *such that for all* $x \in Q$,

$$f^j(x) \in A_j \text{ for all } j. \tag{2.6}$$

Proof: Write $Q_0 = A_0$. Let g_j be the restriction of f to A_j. Define $Q_1 = (g_0)^{-1}(A_1)$. Since Q_1 is a closed subset of a compact set A_0, it is itself compact. It is nonempty. Construct in this manner a sequence of nonempty compact sets $Q_2 = (g_0)^{-1}(g_1)^{-1}(A_2), ..., Q_{j+1} = (g_0)^{-1}...(g_j)^{-1}(A_{j+1})$. Then the sequence of nonempty compact sets $\{Q_j\}$ is "nested": $Q_{j+1} \subset Q_j$. Hence, $Q = \bigcap_{j=0}^{\infty} Q_j$ is nonempty and has the desired property by construction. Q.E.D.

Let us reflect on some of the implications of Proposition 2.5 by using the "tent map" of Exercise 2.3 for the sake of concreteness.

Exercise 2.3 (continued): Note that for the tent map both the intervals $H = [0, 1/2]$ and $T = [1/2, 1]$ have the property that $f(H) = f(T) = [0, 1]$. Let Ω be the uncountable set of all infinite sequences with two symbols H and T.

Choose an *arbitrary* element s of Ω This chosen s can be interpreted as an arrangement of a sequence $\{A_i\}$ of nonempty compact sets, where each A_i is either H or T. Let us stress informally that s can be viewed as a record of the outcomes of an infinite sequence of coin-tossings: the order in which H and T appear is not 'controlled' in any way whatsoever. Proposition 2.5 asserts that there is some x in $[0,1]$ from which the trajectory $\tau(x) = (f^j(x))$ satisfies $f^j(x) \in A_j$. See Saari (1991) for an elaboration of this theme.

3. Li-Yorke Chaos and the Sarkovskii Theorem

Here we take the state space X as a (nondegenerate) interval in the real line, (and f a continuous function from X into X). An interval is compact if it is closed and bounded. A subinterval of an interval I is an interval contained in I. Since f is continuous $f(I)$ is an interval. If I is a compact interval, so is $f(I)$. Suppose that a dynamical system (X, f) has a periodic point of period k. Can we conclude that it also has a periodic point of some other period $k' \neq k$? It is useful to look at a simple example first.

Example 3.1: Suppose that (X, f) has a periodic point of period $k (\geqq 2)$. Then it has a fixed point, (i.e., a periodic point of period one). To see this, consider the orbit γ of the periodic point of period k, and let us write

$$\gamma = \{x^{(1)}, ..., x^{(k)}\}$$

where $x^{(1)} < x^{(2)} < ... < x^{(k)}$. Both $f(x^{(1)})$ and $f(x^{(k)})$ must be in γ. This means that

$$f(x^{(1)}) = x^{(i)} \text{ for some } i > 1$$

and

$$f(x^{(k)}) = x^{(j)} \text{ for some } j < k.$$

Hence, $f(x^{(1)}) - x^{(1)} > 0$ *and* $f(x^{(k)}) - x^{(k)} < 0$. By the intermediate value theorem, there is some x in X such that $f(x) = x$.

We shall now state the Li-Yorke theorem and provide a brief sketch of the proof.

Proposition 3.1: *Let I be an interval and $f : I \to I$ be continuous. Assume that there is some point a in I for which there are points $b = f(a)$, $c = f(b)$ and $d = f(c)$ satisfying:*

$$d \leq a < b < c \ (or,\ d \geq a > b > c). \tag{3.1}$$

Then:

[T.1] *For every positive integer* $k = 1, 2, \ldots$ *there is a periodic point of period* k *in* I.

[T.2] (i) *There is an uncountable* S *of* $\aleph(X)$ *such that for all* p, q *in* S, $p \neq q$,

$$\limsup_{n \to \infty} |f^n(p) - f^n(q)| > 0; \tag{3.2}$$

$$\liminf_{n \to \infty} |f^n(p) - f^n(q)| = 0. \tag{3.3}$$

(ii) *If* $p \in S$, *and* $q \in \wp(X)$

$$\limsup_{n \to \infty} |f^n(p) - f^n(q)| > 0$$

Proof of [T.1] [sketch]:

Step 1: Let G be a real-valued continuous function on an interval I. For any compact subinterval I_1 of $G(I)$ there is a compact subinterval Q of I such that $G(Q) = I_1$.

Proof of Step 1: One can figure out the subinterval Q directly as follows. Let $I_1 = [G(x), G(y)]$ where x, y are in I. Assume that $x < y$. Let r be the last point of $[x, y]$ such that $G(r) = G(x)$; let s be the first point after r such that $G(s) = G(y)$. Then $Q = [r, s]$ is mapped onto I_1 under G. The case $x > y$ is similar. Q.E.D.

Step 2: Let I be an interval and $f : I \to I$ be continuous. Suppose that $(I_n)_{n=0}^{\infty}$ is a sequence of compact subintervals of I and, for all n,

$$I_{n+1} \subset f(I_n). \tag{3.4}$$

Then there is a sequence of compact subintervals (Q_n) of I such that, for all n,

$$Q_{n+1} \subset Q_n \subset Q_0 = I_0 \tag{3.5}$$

and

$$f^n(Q_n) = I_n. \tag{3.6}$$

Hence, there is

$$x \in \bigcap_n Q_n \text{ such that } f^n(x) \in I_n \text{ for all } n \tag{3.7}$$

Proof of Step 2: The construction of the sequence Q_n proceeds "inductively" as follows: Define $Q_0 = I_0$. Recall that f^0 is defined as the identity mapping, so $f^0(Q_0) = I_0$ and $I_1 \subset f(I_0)$. If Q_{n-1} is defined as a compact subinterval such that $f^{n-1}(Q_{n-1}) = I_{n-1}$, then $I_n \subset f(I_{n-1}) = f^n(Q_{n-1})$. Use Step 1, with $G = f^n$ on Q_{n-1} in order to get a compact subinterval Q_n of Q_{n-1} such that $f^n(Q_n) = I_n$. This completes the induction argument (establishing (3.5) and (3.6)). Compactness of Q_n leads to (3.7).

Now we prove [T.1]. Assume that $d \leq a < b < c$ (the other case $d \geq a > b > c$ is treated similarly).

Write $K = [a, b]$ and $L = [b, c]$.
Let k be any positive integer.
For $k > 1$, define a sequence of intervals (I_n) as follows:
$I_n = L$ for $n = 0, 1, 2, ...k - 2$; $I_{k-1} = K$, and $I_{n+k} = I_n$ for $n = 0, 1, 2.....$

For $k = 1$, let $I_n = L$ for all n.

Let Q_n be the intervals in Step 2. Notice that $Q_k \subset Q_0 = I_0$ and $f^k(Q_k) = I_k = I_0$. Hence, Proposition 2.1 applied to f^k gives us a fixed point p_k of f^k in Q_k. Now, p_k cannot have period less than k; otherwise, we need to have $f^{k-1}(p_k) = b$, contrary to $f^{k+1}(p_k) \in L$. Q.E.D.

Proof of [T.2]: Let $\mathcal{M}$ be the set of sequences $M = \{M_n\}_{n=1}^{\infty}$ of intervals with

A1: $M_n = K$ or $M_n \subset L$, and $f(M_n) \supset M_{n+1}$
if $M_n = K$ then
A2: n is the square of an integer and M_{n+1}, $M_{n+2} \subset L$

Of course if n is the square of an integer, then $n+1$ and $n+2$ are not, so the last requirement in (A.2) is redundant. For $M \in \mathcal{M}$ let $P(M, n)$ denote the number of i's in $\{1, ..., n\}$ for which $M_i = K$. For each $r \in (3/4, 1)$ choose $M^r = \{M_n^r\}_{n=1}^{\infty}$ to be a sequence in $\mathcal{M}$ such that

A3: $\lim_{n\to\infty} P(M^r, n^2)/n = r$.

Let $\mathcal{M}_0 = \{M^r : r \in (3/4, 1)\} \subset \mathcal{M}$. Then $\mathcal{M}_0$ is uncountable since $M^{r_1} \neq M^{r_2}$ for $r_1 \neq r_2$. For each $M^r \in \mathcal{M}_0$, by Step 2, there exists a point x_r, with $f^n(x_r) \in M^r_n$ for all n.
Let $S = \{x_r : r \in (3/4, 1)\}$. Then S is also uncountable. For $x \in S$, let $P(x, n)$ denote the number of i's in $\{1, ..., n\}$ for which $f^i(x) \in K$. We can never have $f^k(x_r) = b$, because then x_r would eventually have period 3, contrary to (A.2). Consequently $P(x_r, n) = P(M^r, n)$ for all n, and so

$$\rho(x_r) = \lim_{n\to\infty} P(X_r, n^2) = r$$

for all r. We claim that

A4: for $p, q \in S$, with $p \neq q$, there exist finitely many n's such that $f^n(p) \in K$ and $f^n(q) \in L$ or vice versa.

We may assume $\rho(p) > \rho(q)$. Then $P(p, n) - P(q, n) \to x$, and so there must be infinitely many n's such that $f^n(p) \in K$ and $f^n(q) \in L$.

Since $f^2(b) = d \leqq a$ and f^2 is continuous, there exists $\delta > 0$ such that $f^2(x) < (b + d)/2$ for all $x \in [b - \delta, b] \subset K$. If $p \in S$ and $f^n(p) \in K$, then (A.2) implies $f^{n+1}(p) \in L$ and $f^{n+2}(p) \in L$. Therefore $f^n(p) < b - \delta$. If $f^n(q) \in L$, then $f^n(q) \geqq b$ so

$$|f^n(p) - f^n(q)| > \delta.$$

By claim (A.4), for any p, $q \in S$, $p \neq q$, it follows

$$\limsup_{n\to\infty} |f^n(p) - f^n(q)| \geqq \delta > 0$$

Hence (3.1) is proved. This technique may be similarly used to prove that [T.2(ii)] is satisfied.

Proof of (3.3): Since $f(b) = c$, $f(c) = d$, $d \leqq a$, we may choose intervals $[b^n, c^n]$, $n = 0, 1, 2, ...$,such that

(a) $[b, c] = [b^0, c^0] \supset [b^1, c^1] \supset ... \supset [b^n, c^n] \supset ...$,

(b) $f(x) \in (b^n, c^n)$ for all $x \in (b^{n+1}, c^{n+1})$,
(c) $f(b^{n+1}) = c^n$, $f(c^{n+1}) = b^n$.

Let $A = \bigcap_{n=0}^{\infty} [b^n, c^n]$, $b^* = \inf A$ and $c^* = \sup A$, then $f(b^*) = c^*$ and $f(c^*) = b^*$, because of (c).

In order to prove (3.3) we must be more specific in our choice of the sequences M^r. In addition to our previous requirements on $M \in \mathcal{M}$, we assume that if $M_k = K$ for both $k = n^2$ and $(n+1)^2$ then $M_k = [b^{2n-(2j-1)}, b^*]$ for $k = n^2 + (2j-1)$, $M_k = [c^*, c^{2n-2j}]$ for $k = n^2 + 2j$ where $j = 1, ..., n$. For the remaining k's which are not squares of integers, we assume $M_k = L$.

It is easy to check that these requirements are consistent with (A1) and (A2), and that we can still choose M^r so as to satisfy (A3). From the fact that $\rho(x)$ may be thought of as the limit of the fraction of n's for which $f^{n^2}(x) \in K$, it follows that for any r^*, $r \in (3/4, 1)$ there exist infinitely many n such that $M_k^r = M_k^{r^*} = K$ for both $k = n^2$ and $(n+1)^2$. To show (3.3), let $x_r \in S$ and $x_{r^*} \in S$. Since $b^n \to b^*$, $c^n \to c^*$ as $n \to \infty$, for any $\varepsilon > 0$ there exists N with $|b^n - b^*| < \varepsilon/2$, $|c^n - c^*| < \varepsilon/2$ for all $n > N$. Then, for any n with $n > N$ and $M_k^r = M_k^{r^*} = K$ for both $k = n^2$ and $(n+1)^2$, we have

$$f^{n^2+1}(x_r) \in M_k^r = [b^{2n-1}, b^*]$$

with $k = n^2 + 1$ and $f^{n^2+1}(x_r)$ and $f^{n^2+1}(x_{r^*})$ both belong to $[b^{2n-1}, b^*]$. Therefore, $|f^{n^2+1}(x_r) - f^{n^2+1}(x_{r^*})| \leqq \varepsilon$. Since there are infinitely many n with this property, $\liminf_{n\to\infty} |f^n(x_r) - f^n(x_{r^*})| = 0$. Q.E.D.

We shall now state Sarkovskii's theorem on periodic points. Consider the following Sarkovskii ordering of the positive integers:

$$\begin{aligned} &3 \rhd 5 \rhd 7 ... \rhd 2.3 \rhd 2.5 ... \rhd 2^2 3 \rhd 2^2 5 \rhd ... \\ &\rhd 2^3.3 \rhd 2^3.5 ... \rhd 2^3 \rhd 2^2 \rhd 1 \end{aligned} \tag{SO}$$

In other words, first list all the odd integers beginning with 3; next list 2 times the odds, 2^2 times the odds, etc. Finally, list all the powers of 2 in decreasing order.

Proposition 3.2: *Let $X = \Re$, f be a continuous function from X into X. Supppose that f has a periodc point of period k. If $k \rhd l$ in the Sarkovskii ordering* (SO), *then f has a periodic point of period l.*

Proof: See Block, Guckenheimer, Misiurewicz, and Young (1980).

It follows that if f has only finitely many periodic points, then they all necessarily have periods which are powers of two.

Exercise 3.1: Show that if f has a periodic point of period 4 it has a periodic point of period 2. [Hint: Suppose that $\{x_1, x_2, x_3, x_4\}$ is the orbit of a periodic point of period 4. Let

$$x_1 < x_2 < x_3 < x_4$$

Choose a point a between x_2 and x_3. By considering all possible transitions among the points, one can show that there are two cases. The first case occurs if both $f(x_1) > a$ and $f(x_2) < a$. Then one must have $f(x_3) < a$ and $f(x_4) < a$. Let $I_0 = [x_1, x_2]$ and $I_1 = [x_3, x_4]$. Show that $f(I_1) \supset I_0$ and $f(I_0) \supset I_1$. It follows that f^2 has a fixed point, and it is in fact a periodic point of period 2.

The other case occurs when either one of x_1 or x_2 is mapped to the right of a but the other is not. For definiteness, suppose $f(x_1) > a$ and $f(x_2) < a$. Consequently, one has $f(x_2) = x_1$. Let $I_0 = [x_2, x_3]$ and $I_1 = [x_1, x_2]$. Then we have $f(I_0) \supset I_1$ and $f(I_1) \supset I_0 \cup I_1$. This is the same situation that occurs in the Li-Yorke theorem. Hence, there is a periodic point of period 2 (and, in fact, a periodic point of any period).

Example 3.1: If f has a periodic point of period $2^n (n \geq 3)$, it has a periodic point of period 2^m where $n > m$. Let $l = 2^{n-2}$ and consider $G(x) = f^l(x)$. A periodic point of period 2^n for f is a periodic point of period 4 of G. From Exercise 3.1 we know then that G has a periodic point of period 2. But this means that f has a periodic point of period 2^{n-1}.

Exercise 3.2: Let X be an interval and $f : X \to X$ be continuous and satisfy the following property: there are two subintervals A, B of X where $B = f(A)$ such that
(i) $A \cap B = \phi$
(ii) $f(B) \supset A \cup B$
Note that $f(B) \supset B$ implies that f has a fixed point. Also

$$f^2(A) \equiv f(f(A)) = f(B) \supset A$$

Hence, f^2 has a fixed point in A, i.e., there is some $y \in A$ such that $y = f^2(y)$. But $y \neq f(y)$ by Condition (i). Hence, f is a periodic point of period 2.
Show that (X, f) has periodic point of all periods. For a detailed discussion of such maps see Diamond (1976) and Day (1994).

4. Some Related Concepts

4.1 Devaney's Definition of Chaos

In his widely used text, Devaney (1986) introduced a definition of a chaotic dynamical system that captures several interesting properties that we first discuss.

a. Topological Transitivity

(X, f) is *topologically transitive* if for any pair of nonempty open sets U and V, there exists $k \geq 1$ such that $f^k(U) \cap V \neq \phi$.

Proposition 4.1: *If there is some x such that $\gamma(x)$, the orbit from x, is dense in X, then (X, f) is topologically transitive.*

Proposition 4.2: *Let X be a (nonempty) compact metric space. Assume that (X, f) is topologically transitive. Then there is some $x \in X$ such that the orbit $\gamma(x)$ from x is dense in X.*

Proof: Since X is compact, it has a countable base of open sets, i.e., there is a family $\{V_n\}$ of open sets in X with the property that if M is any open subset of X, there is some $V_n \subset M$.

Corresponding to each V_n, define the set O_n as follows:

$$O_n = \{x \in X : f^j(x) \in V_n, \quad \text{for some } j \geq 0\}$$

O_n is open, by continuity of f. By topological transitivity it is also dense in X. By Baire category theorem, the intersection $O = \cap_n O_n$ is also open and dense in X. Take any $x \in O$, and consider the orbit $\gamma(x)$ from x. Take any y in X and any open M containing y. Then M contains some V_n. Since x belongs to the corresponding O_n, there is some element of $\gamma(x)$ in V_n. Hence, $\gamma(x)$ is dense in X. Q.E.D.

b. Sensitivity to Initial Condition

A dynamical system (X, f) has *sensitive dependence on initial condition* if there is $\partial > 0$ such that, for any $x \in X$ and any neighborhood N of x there exist y and an integer $j \geq 0$ with the property $|f^j(x) - f^j(y)| > \partial$.

Devaney asserted that if a dynamical system "possesses sensitive dependence on initial condition, then for all practical purposes, the dynamics defy numerical computation. Small errors in computation which are introduced by round-off may become magnified upon iteration. The results of numerical computation of an orbit, no matter how accurate, may bear no resemblance whatsoever with the real orbit".

For a dynamical system (X, f) where the state space X is a subset of the real line, there is another concept of sensitive dependence explored by Guckenheimer (1979). Let us state it formally here for future reference. A dynamical system (X, f) has *Guckenheimer dependence on initial conditions* if there is a set of $Y \subset X$ of positive Lebesgue measure and an $\varepsilon > 0$ such that given any $x \in Y$ and any neighborhood U of x, there is $y \in Y$ and $n \geq 0$ such that $|f^n(x) - f^{(n)}(y)| > \varepsilon$.

c. A Chaotic Dynamical System

A dynamical system (X, f) is *chaotic in the sense of Devaney* if:

(i) (X, f) is topologically transitive.
(ii) (X, f) has sensitive dependence on initial condition.
(iii) The set of periodic points $\wp(X)$ of X is dense in X.

Exercise 4.1: Let X be a metric space (with an infinite number of elements) and $f : X \to X$ a continuous function. One can show that the properties (i) and (iii) imply (ii). See Banks et. al. (1992) for details.

Example 4.1: Let $X = [0, 1]$ and $f(x) = 4x(1 - x)$ is chaotic in Devaney's sense.

Example 4.2: Consider the space of all sequences of '0' and '1'; formally

$$\mathcal{S}_2 = \{s = (s_t) : s_t = 0 \text{ or } 1\}.$$

On $\mathcal{S}_2$ define a metric d by

$$d(s, s') = \sum_{t=0}^{\infty} |s_t - s'_t|/2^t$$

Note the following property:

R.4.1: Suppose that s and s' are in $\mathcal{S}_2$ and $s_t = s'_t$ all $t = 0, 1, \ldots n$. Then $d(s, s') \leq 1/2^n$; on the other hand, if $d(s, s') < 1/2^n$, then $s_t = s'_t$ for all $t = 0, \ldots n$.

Proof: Suppose that $s_t = s'_t$ for $t = 0, \ldots, n$. Then

$$d(s, s') = \sum_{t=0}^{n} \frac{|s_t - s'_t|}{2^t} + \sum_{t=n+1}^{\infty} \frac{|s_t - s'_t|}{2^t} \leq \frac{1}{2^n}$$

On the other hand, if $s_t \neq s'_t$ for k some $k \leq n$, then we must have

$$d(s, s') \geqq \frac{1}{2^k} \geq \frac{1}{2^n}$$

Consequently, $d(s, s') < 1/2^n$ implies $s_t = s'_t$, $t = 0, ..., n$. Q.E.D.

Define the "shift map" $\sigma : \mathcal{S}_2 \to \mathcal{S}_2$ as;

$$\sigma(s_0, s_1, ...) = (s_1, s_2, ...).$$

R.4.2: σ is continuous.

Proof: Let $\varepsilon > 0$ be given and $s = (s_0, s_1, ...)$. Choose n such that $1/2^n < \varepsilon$. Let $\delta = (1/2^{n+1})$. If $m = (m_0, m_1....)$ satisfies $d(s, m) < \delta$, then by R.4.1, we must have $s_t = m_t$ for $0 \leq t \leq n+1$. Hence, $d(\sigma(s), \sigma(m)) \leq 1/2^n < \varepsilon$, again by R.4.1. Q.E.D.

R.4.3: a) The set of periodic points of σ is dense in $\mathcal{S}_2$.
b) There is a dense orbit for σ in $\mathcal{S}_2$.

Proof: a) In order to prove that the set of periodic points is dense in $\mathcal{S}_2$, we have to show that if $s = (s_0, s_1, ..., s_t, ...)$ is an arbitrary element of $\mathcal{S}_2$, there is a sequence τ_n of periodic points in $\mathcal{S}_2$ that converges to s (in the metric d). This sequence is constructed as follows:

$$\tau_0 = (s_0, s_0, s_0, s_0, ...)$$
$$\tau_1 = (s_0, s_1; s_0, s_1; ...)$$
$$\bullet$$
$$\bullet$$
$$\tau_n = (s_0, s_1, ..., s_n; s_0, ..., s_n; , ...)$$

Then, by R.4.1., $d(s, \tau_n) \leq 1/2^n$.

(b) Consider

$$s^* = (0, 1| \qquad 0, 0; 0, 1; 1, 0; 1, 1| \qquad 0, 0, 0; 0, 0, 1.....)$$

The sequence s^* is constructed by listing all blocks of 0's and 1's of length n, then length $n+1$, etc. Clearly, some iterate of applied to s^* yields a sequence that agrees with any given sequence in an arbitrarily large number of places. Hence (b) follows (from an application of R.4.1). Q.E.D.

5. Ergodic Chaos

Let $(X, \Im, \mu)$ be a probability space: here X is a set, $\Im$ is a σ field of subsets of X, and μ is a probability measure on $\Im$. The elements of $\Im$ are *measurable subsets* of X.

Example 5.1: X is a metric space; $\Im$ is the Borel σ field of X; μ is any probability measure on X.

Example 5.2: $X = [0,1]$; $\Im$ is the Borel σ field of X; m is the Lebesgue measure. Recall that for any subinterval (c,d) of $[0,1]$, $m\{(c,d)\} = d - c$. Also, m is nonatomic, i.e., $m\{x\} = 0$ for any $x \in X$.

A mapping (transformation, function) f from X into X is *measurable* if, for any $S \in \Im$, $f^{-1}(S) \in \Im$.

Example 5.3: Let X be any metric space, $\Im$ its Borel σ field and $f : X \to X$ be continuous. Then f is measurable.

In what follows, any mapping is understood to be a measurable mapping.

A measurable mapping f is *measure preserving* and μ is *invariant* if $\mu(A) = \mu(f^{-1}(A))$.

If f is a measurable mapping on a probability space $(X, \Im, \mu)$ and E is any measurable subset of X, a point $x \in E$ is *recurrent* (with respect to E and f) if $f^n(x) \in E$ for at least one positive integer n.

Proposition 5.1: *If f is measure preserving, and $E \in \Im$, then (μ) almost every point of E is recurrent.*

Proof: If not then the set F of those points of E that never return to E is a set of positive measure. The set F is measurable since

$$F = E \cap f^{-1}(X - E) \cap f^{-2}(X - E) \cap \ldots$$

If $x \in F$, then none of the points $f(x)$, $f^2(x)$, $f^3(x)$, ... belongs to F, or, in other words, F is disjoint from $f^{-n}(F)$ for all positive n. It follows that F, $f^{-1}F$, $f^{-2}F$ are all pairwise disjoint, since

$$f^{-n}F \cap f^{-(n+k)}F = f^{-n}(F \cap f^{-k}F).$$

Since f is measure preserving, and the measure of X is one, this is a contradiction. Q.E.D.

The recurrence theorem implies a stronger version. Not only is it true that for almost every $x \in E$, at least one term of the sequence $f(x)$, $f^2(x)$,... belongs to E; in fact, for almost every x in E, there are infinitely many values of n such that $f^n(x) \in E$. The idea of the proof is to apply the recurrence theorem to each iterate of f. If F_n is the set of those points of E that never return to E under the action of f^n, then by the recurrence theorem $\mu(F_n) = 0$. If $x \in E - (F_1 \cup F_2 \cup ...)$, then $f^n(x) \in E$, for some positive n since $x \in E - F_1$. Since $x \in E - F_n$, it follows that $f^{kn}(x) \in E$ for some positive k. The strengthened version of the recurrence theorem follows.

We now turn to the concept of ergodicity. A set $E \in \Im$ is invariant under f if and only if $f^{-1}(E) = E$; this means that "$x \in E$" if and only if "$f(x) \in E$". A measurable function f is *ergodic* if and only if it has trivial invariant sets, i.e., $E \in \Im$ and $f^{-1}(E) = E$ implies that either $\mu(E) = 0$ or $\mu(E) = 1$. In this case we shall refer to μ as an *ergodic measure* (for f).

A major result in the theory of dynamical systems is the celebrated "ergodic theorem".

Proposition 5.2: *Let $(X, \Im, \mu)$ be a probability space and f be measure preserving and ergodic. Let $g(\bullet)$ be an integrable function. Then*

$$\lim_{n \to \infty} 1/n \sum_{i=0}^{n-1} g(f^i(x)) = \int_X g d\mu \quad \textit{for almost all } x \in X \qquad (5.1)$$

For an extended discussion, see Halmos (1956).

An immediate implicaton of the ergodic theorem is the following: for any measurable E, and for μ-almost every x, $\mu(E)$ is the fraction of time that the trajectory from x spends in E.

Consider the case when X is a metric space, and $\Im$ is its Borel σ field. A probability measure μ is *strictly positive* if the μ measure of any nonempty open set is positive.

Proposition 5.3: *Let X be a separable metric space; $\Im$ its Borel σ field, and μ be a strictly positive probability measure. If f is any measure preserving, ergodic mapping from X into X, then for μ almost every $x \in X$, the orbit $\gamma(x)$ is dense in X.*

Proof: The orbit of x is not dense if and only if there is a nonempty basic open set G such that x belongs to the intersection of all $X - f^n G$. Since this intersection is an invariant set disjoint from G and $\mu(G) > 0$, it follows that it has μ measure zero. If x does not belong to any of this countable class of sets of measure zero (one for each basic open set), then x has a dense orbit. Q.E.D.

We now introduce the notion of the *support* of a measure.

Proposition 5.4: *Let X be a separable metric space, $\Im$ its Borel σ field and μ a proability measure on x. Then there exists a unique closed set C_μ satisfying* (i) $\mu(C_\mu) = 1$; (ii) *if D is any closed set such that $\mu(D) = 1$, then $C_\mu \subset D$. Moreover,* (iii) *C_μ is the set of all points $x \in X$ having the property that $\mu(U) > 0$ for each open set U containing x.*

Proof: See Parthasarathy (1967, p. 28).

The closed set C_μ in Proposition 5.3 is called the *support* of μ. Note that if the support of μ is the entire set X, then μ is necessarily strictly positive.

Exercise 5.1: Consider $X = [0, 1]$, and let m be the Lebesgue measure on the Borel σ field of x. Then m is necessarily strictly positive. Indeed, any probability measure equivalent to m is also strictly positive. Recall that μ and m are equivalent if "$\mu(E) = 0$ if and only if $m(E) = 0$".

Exercise 5.2: Let $X = [0,1]$. Suppose that a probability measure v is given by its density function $i(\cdot)$.

$$v(E) = \int_E i(x)m(dx)$$

If the density function $i : [0,1] \to [0,1]$ is strictly positive, i.e., $i(x) > 0$ for all $x \in [0,1]$, then v is strictly positive.

In what follows in this section, the state space X of the dynamical system (X, f) is taken to be a closed interval $[a, b]$.

A dynamical system (X, f) is said to exhibit *ergodic chaos* if there is an ergodic, invariant measure v which is absolutely continuous with respect to m, the Lebesgue measure.

An *attractor* for f is a closed subset F of X such that $\omega(x) = F$ for all x in a set of positive Lebesgue measure. We now state two results on the existence of ergodic chaos.

Proposition 5.5: *Let (X, f) be a dynamical system with $X = [a, b]$. Suppose that there is some $y \in (a, b)$ and f restricted to both (a, y) and (y, b) is*

(i) *strictly monotone;*

(ii) *twice continuously differentiable; and*

(iii) *there is $\lambda > 1$ such that $f'(x) \geq \lambda > 1$ for μ-almost every x in (a, y) and (y, b).*

Then there is a unique ergodic invariant measure μ which is absolutely continuous with respect to m, and the support of m is the attractor for m-almost every $x \in [a, b]$.

Proposition 5.6: *Let (X, f) be a dynamical system with $X = [a, b]$. Suppose that there is $x^* \in (a, b)$ such that*

(i) *f is twice continuously differentiable on $[a, b]$*

(ii) *$f'(x) > 0$ for $x < x^*$; $f'(x^*) = 0$; $f''(x^*) < 0$; $f'(x) < 0$ for $x > x^*$.*

(iii) *$f(x) > x$ for all $x \in (a, x^*)$; $f(x^*) \in (x^*, b)$; $\mathcal{S}f(x) \equiv [f'''(x)/f'(x)] - 3/2[f''(x)/f'(x)]^2 < 0$ for all $x \in X$ except $x = x^*$.*

(iv) *There exists* $K \geq 2$ *such that* $y = f^K(x^*)$ *is an unstable fixed point of* f.

Then f *exhibits ergodic chaos.*

For a more detailed discussion see Day and Pianigiani (1991) and Misiurewicz (1981).

6. The Quadratic Family

Let $X = [0,1]$ and $I = [1,4]$. The *quadratic family of maps* is then defined by

$$f_\mu(x) = \mu x(1-x) \text{ for } (x,\mu) \in X \times I$$

We interpret x as the *variable* and μ as the *parameter* of the map h.

A few observations about the quadratic family are useful at this point. Note that, for each parameter specification $\mu \in I$, the state space is the same. Thus we can conveniently examine a family of dynamical systems $(X, f_\mu(x)$ parametrized by μ.

For each $\mu \in I$, f_μ has exactly one *critical point* (i.e., a point where $f'_\mu(x) = 0$, and this critical point (equal to 0.5) is independent of the parameter μ.

6.1. Stable Periodic Orbits

Even though there may be an infinite number of periodic orbits for a given dynamical system (as in the Li-Yorke theorem), a striking result due to Julia and Singer, informs us that there can be *at most one stable* periodic orbit.

Proposition 6.1: *Let* $X = [0,1]$, $I = [1,4]$; *given some* $\overline{\mu} \in I$, *define* $f_{\overline{\mu}}(x) = \overline{\mu}x(1-x)$ *for* $x \in X$. *Then there can be at most one stable periodic orbit. Furthermore, if there is a stable periodic orbit, then* $\omega(0.5)$, *the limit set of* $x^* = 0.5$, *must coincide with this orbit.*

Suppose, now, that we have a stable periodic orbit. This means that the asymptotic behavior (limit sets) of trajectories from all initial states "near" this periodic orbit must coincide with the periodic orbit. But what about the asymptotic behavior of trajectories from other initial states? If one is interested in the behavior of a "typical" trajectory, a remarkable result, due to Misiurewicz (1983), settles this question. Recall that m denotes the Lebesgue measure on (the Borel σ field of) $[0, 1]$.

Proposition 6.2: *Let $X = [0, 1]$, $I = [1, 4]$; given some $\overline{\mu} \in I$, define $f_{\overline{\mu}}(x) = \overline{\mu}x(1 - x)$ for $x \in X$. Suppose there is a stable periodic orbit. Then for m almost every $x \in [0, 1]$, $\omega(x)$ coincides with this orbit.*

Combining the above two results, we have the following scenario. Suppose we do have a stable periodic orbit. Then there are no other stable periodic orbits. Furthermore, the (unique) stable periodic orbit "attracts" the trajectories from almost every initial state. Thus we can make the qualitative prediction that the asymptotic behavior of the "typical" trajectory will be just like the given stable periodic orbit.

Proposition 6.3: *Let $X = [0, 1]$, $I = [1, 4]$; given some $\overline{\mu} \in I$, define $f_{\overline{\mu}}(x) = \overline{\mu}x(1 - x)$ for $x \in X$. Suppose there is a stable periodic orbit. Then the dynamical system does not have Guckenheimer dependence on initial conditions.*

It is important to note that the above scenario (existence of a stable periodic orbit) is by no means inconsistent with condition (3.1) of the Li-Yorke theorem (and hence with its implications). Let us elaborate on this point following Devaney (1986) and Day and Pianigiani (1991). Consider $\mu = 3.839$, and for this μ, simply write $f(x) = \mu x(1-x)$ for $x \in X$. Choosing $x^* = 0.1498$, it can be checked then that there is $0 < \varepsilon < 0.0001$ such that $f^3(x)$ maps the interval $U \equiv [x^* - \varepsilon, x^* + \varepsilon]$ into itself, and $|[f^3(x)]'| < 1$ for all $x \in U$. Hence, there is $\widehat{x} \in U$ such that $f^3(\widehat{x}) = \widehat{x}$, and $|[f^3(\widehat{x})]'| < 1$. Thus, $\widehat{x}$ is a periodic point of period 3, and it can be checked (by choice of the range of ε) that $f^3(\widehat{x}) = \widehat{x} < f(\widehat{x}) < f^2(\widehat{x})$ so that condition (3.1) of Proposition 3.1 is satisfied. Also, $\widehat{x}$ is a periodic point of period 3 which is *stable*, so that Proposition 3.1 is also applicable. Then we may conclude that the set $\mathcal{S}$ of "chaotic" inital states in Proposition 6.3 must be of Lebesgue measure zero. In other words, Li-Yorke chaos exists but is *not* "observed" when $\mu = 3.839$.

7. Comparative Statics and Dynamics

7.1 Bifurcation Theory

Fixed and periodic points formally capture the intuitive idea of a *stationary* state of a dynamical system. In his *Foundations*, Samuelson (1947) noted that (p.513):

"*Stationary* is a descriptive term characterizing the behavior of an economic variable over time; it usually implies constancy, but is occasionally generlized to include behavior periodically repetitive over time."

Bifurcation theory deals with the question of changes in the stationary state of a dynamical system with respect to variations of a parameter that affects the law of motion. Perhaps it is best to introduce the main ideas through an explicit example.

Example 2.1(continued): Consider a class of dynamical systems with a common state space $X = \Re$, and with the laws of motion given by

$$f(x) = x^2 + c \tag{7.1}$$

where 'c' (a real number) is a parameter. Our task is to study the implications of changes in the parameter c. In order to understand stationary states, we have to explore the nature of fixed points of f_c and its iterates f_c^j. To begin with consider the dependence of the fixed points of f_c on the value of c. The fixed points of f_c are obtained by solving

$$f_c(x) = x \tag{7.2}$$

for a given value of c; i.e., solving

$$x^2 - x + c = 0 \tag{7.3}$$

The exact values are given by

$$\begin{aligned} p_+(c) &= (1+\sqrt{1-4c})/2 \\ p_-(c) &= (1-\sqrt{1-4c})/2 \end{aligned} \tag{7.4}$$

Observe that $p_+(c)$ and $p_-(c)$ are real if and only if $1-4c \geq 0$ or $c \leq 1/4$. Thus, when $c > 1/4$, we can say that the dynamical system (with the state space $X = \Re$) has no fixed point when $c = 1/4$, we have

$$p_+(1/4) = p_-(1/4) = 1/2 \tag{7.5}$$

Now when $c < 1/4$, both $p_+(c)$ and $p_-(c)$ are real and distinct (and $p_+(c) > p_-(c)$).

Thus, as the parameter values decreases through 1/4, the dynamical system undergoes a *bifurcation* from the situation of no fixed piont through a unique fixed point 'splittig into' two.

We can also explore the 'local' dynamics of the system from initial states close to the fixed point(s). Since $f_c'(x) = 2x$ (does not depnd on c), we see that $f_{1/4}'(1/2) = 1$ (so the fixed point 1/2 is 'neutral' when $c = 1/4$). Now from (7.4), we see that:

$$\begin{aligned} f_c'(p_+(c)) &= 1+\sqrt{1-4c} \\ f_c' p_-(c)) &= 1-\sqrt{1-4c} \end{aligned} \tag{7.6}$$

For $c < 1/4$, $f_c'(p_+(c) > 1$ so that the fixed point is repelling.

Now consider $p_-(c)$; of course, $f_c'(p_-(c)) = 1$ when $c = 1/4$. When $c < 1/4$ but sufficiently close to 1/4, $f_c'(p_-(c)) < 1$, so that the fixed point p_- is locally stable or attracting. It will continue to be attracting as long as $|f_c'(p_-(c))| < 1$, i.e.,

$$-1 < f_c'(p_-(c)) < 1$$

or

$$-1 < 1-\sqrt{1-4c} < 1.$$

It follows that $p_-(c)$ is locally stable for all c satisfying

$$-3/4 < c < 1/4$$

when $c = -3/4$, $p_-(c)$ is again 'neutral' (i.e., $|f_c'(p_c)| = 1$) and for $c < -3/4$, $|f_c'(p_-(c))| > 1$ so that $p_-(c)$, too, becomes repelling.

Using a graphical analysis [See Devaney (1992, p. 54-55)] one can show the following:

(i) for $c \leq 1/4$, if the initial state $x > p_+(c)$ or $x < -p_+(c)$ then the trajectory from x tend to infinity;

(ii) for $-3/4 < c < 1/4$ all the trajectories starting from $x \in (-p_+(c), p_+(c))$ tend to the attracting fixed point $p_-(c)$.

As the parameter c decreases through $-3/4$, the fixed point $p_-(c)$ loses it stability property; but more is true. We see a 'period doubling bifurcation': a pair of periodic points is "born". To examine this, consider the equation $f_c^2(x) = x$, i.e.,

$$x^4 + 2cx^2 - x + c^2 + c = 0 \tag{7.7}$$

using the fact that both p_+ and p_- are solutions to (7.7), we see that there are two other roots given by

$$\begin{aligned} q_+(c) &= (-1 + \sqrt{-4c - 3})/2 \\ q_-(c) &= (-1 - \sqrt{-4c - 3})/2 \end{aligned} \tag{7.8}$$

Clearly, $q_+(c)$ and $q_-(c)$ are real if an only if $c \leq -3/4$.

Of course, when $c = -3/4$, $q_+(c) = q_-(c) = -1/2 = p_-(c)$. Furthermore, for $-5/4 < c < -3/4$, the periodic points are locally stable.

To summarize:

a change in the parameter may affect the number as well as the local stability properties of fixed and periodic points of a family of dynamical systems.

The above indicates that bifurcations occur near non-hyperbolic fixed and periodic points. This is indeed the only place where bifurcations of fixed points occur, as the follwoing result demonstrates.

Proposition 7.1: *Let f_λ be a one-parameter family of functions and suppose that $f_{\lambda_0}(x_0) = x_0$ and $f'_{\lambda_0}(x_0) \neq 1$. Then there are intervals I about x_0 and N about λ_0 and a smooth function $p : N \to I$ such that $p(\lambda_0) = x_0$ and $f_\lambda(p(\lambda)) = p(\lambda)$. Moreover, f_λ has no other fixed points in I.*

Proof: Consider the function defined by $G(x, \lambda) = f_\lambda(x) - x$. By hypothesis, $G(x_0, \lambda_0) = 0$ and

$$\frac{\partial G}{\partial x}(x_0, \lambda_0) = f'_{\lambda_0}(x_0) - 1 \neq 0$$

By the Implicit Function Theorem, there are intervals I about x_0 and N about λ_0, and a smooth function $p : N \to I$ such that $p(\lambda_0) = x_0$ and $G(p(\lambda), \lambda) = 0$ for all $\lambda \in N$. Moreover, $G(x, \lambda) \neq 0$ unless $x = p(\lambda)$. This concludes the proof.

Remark: For theoretical simplicity, it is often convenient to assume that the fixed point set of f_λ is statioanry as λ is varied. The previous result allows us to make this assumption. Suppose that f_λ and $f_\lambda(p(\lambda)) = p(\lambda)$ are as in Proposition 7.1. Consider the new function

$$g_\lambda(z) = f_\lambda(z + p(\lambda)) - p(\lambda).$$

Clearly, $g_\lambda(0) = f_\lambda(p(\lambda)) - p(\lambda) = 0$ for all λ, so 0 is always fixed. Moroever, g_λ is topologically conjugate [Devaney (1986, p. 47)] to f_λ via the simple map $h_\lambda(x) = x - p(x)$. Hence the dynamics of f_λ and g_λ agree, but g_λ is simpler to handle since fixed point remains stationary at 0 as λ varies.

The above proposition (as well as all that follow) obviously hold for periodic points by replacing f with f^n.

We now turn to the general setting of bifurcation theory.

Proposition 7.2 (The saddle-node bifurcation): *Suppose that*

(i) $f_{\lambda_0}(0) = 0$
(ii) $f'_{\lambda_0}(0) = 1$
(iii) $f''_{\lambda_0}(0) \neq 0$
(iv) $\frac{\partial f_\lambda}{\partial \lambda}\big|_{\lambda=\lambda_0} \neq 0$

Then there exists an interval I about 0 and a smooth function $p : I \to R$ such that

$$f_{p(x)}(x) = x.$$

Moreover, $p'(0) = 0$ and $p''(0) \neq 0$.

Remark: The signs of $f''_\lambda(0)$ and

$$\left.\frac{\partial f_\lambda}{\partial \lambda}\right|_{\lambda=\lambda_0}$$

determine the "direction" of the bifurcation.

Proof: Define $G(x, \lambda) = f_\lambda(x) - x$. Note that $G(x, \lambda) = 0$ implies that f_λ has a fixed point at x. We will apply the Implicit Function Theorem to G.

Note that $G(0, \lambda_0) = 0$ and that

$$\frac{\partial G}{\partial \lambda}(0, \lambda_0) = \left.\frac{\partial f_\lambda}{\partial \lambda}\right|_{\lambda=\lambda_0(0)} \neq 0.$$

Hence there exists a smooth function $p(x)$ satisfying $G(x, p(x)) = 0$. From the chain rule, we have

$$\frac{\partial G}{\partial x} + \frac{\partial G}{\partial \lambda} p'(x) = 0.$$

Therefore

$$p'(x) = \frac{-\frac{\partial G}{\partial x}}{\frac{\partial G}{\partial \lambda}}.$$

Verify that $p'(0) = 0$ and $p''(0) \neq 0$ [see Devaney (1986)]. Q.E.D.

Proposition 7.3 (Period-doubling bifurcation): *Suppose*

(i) $f_\lambda(0) = 0$ *for all* λ *in an interval about* λ_0

(ii) $f'_{\lambda_0}(0) = -1$

(iii) $f'''_{\lambda_0}(0) \neq 0$

(iv) $\left.\frac{\partial (f_\lambda^2)'}{\partial \lambda}\right|_{\lambda=\lambda_0}(0) \neq 0$

Then there is an interval I *about* 0 *and a function* $p : I \to R$ *such that*

$$f_{p(x)}(x) \neq x$$

but

$$f^2_{p(x)}(x) = x.$$

Proof: For this proof we define $G(x,\lambda) = f_\lambda^2(x) - x$. Also set

$$H(x,\lambda) = \begin{cases} \frac{G(x,\lambda)}{x} & x \neq 0 \\ \frac{\partial G}{\partial x}(0,\lambda) & x = 0. \end{cases}$$

One checks easily that H is smooth and satisfies

$$\frac{\partial H}{\partial x}(0,\lambda_0) = \frac{\partial^2 G}{\partial x^2}(0,\lambda_0)$$

$$\frac{\partial^2 H}{\partial x^2}(0,\lambda_0) = \frac{\partial^3 G}{\partial x^3}(0,\lambda_0).$$

We now apply the Implicit Function Theorem to H. Note that

$$\begin{aligned} H(0,\lambda_0) &= \frac{\partial G}{\partial x}(0,\lambda_0) \\ &= (f_{\lambda_0}^2)'(0) - 1 \\ &= f_{\lambda_0}'(0) \bullet f_\lambda'(0) - 1 \\ &= 0. \end{aligned}$$

We have by assumption that

$$\begin{aligned} \frac{\partial H}{\partial \lambda}(0,\lambda_0) &= \left.\frac{\partial}{\partial \lambda}\right|_{\lambda=\lambda_0} (f_\lambda^2)'(0) - 1 \\ &= \frac{\partial (f_\lambda^2)'}{\partial \lambda}(0) \\ &\neq 0. \end{aligned}$$

Hence there is a smooth function $p(x)$ defined on a neighborhood of 0 and satisfying $p(0) = \lambda_0$ and $H(x,p(x)) = 0$. In particular,

$$\frac{1}{x} G(x,p(x)) = 0$$

or $x \neq 0$ and it follows that x is a period two point for $f_{p(x)}$.

As above, we compute

$$p'(0) = \frac{-\frac{\partial H}{\partial x}(0,\lambda_0)}{\frac{\partial H}{\partial \lambda}(0,\lambda_0)} = 0 \text{ and } p''(0) \neq 0 \qquad \text{Q.E.D.}$$

7.2 Robustness

Let $X = [a, b]$ and suppose that a continuous function f satisfies the Li-Yorke condition (3.1) with strict inequality throughout, i.e., suppose that there are points a, b, c, d such that

$$\begin{aligned} d &= f(c) < a < b = f(a) < c = f(b) \\ & \text{or} \\ d &= (f(c)) > a > b = f(a) > c = f(b) \end{aligned} \tag{7.9}$$

Consider $C(X)$ the space of all bounded continuous real valued functions on X. Let $||f|| = \max_{x \in X} |f(x)|$. The remarkable conclusions of Theorem 3.1 then hold with respect to the dynamical system (X, f). But the complexity is "robust" in the following sense.

Proposition 7.4: *Suppose that $X = [a, b]$ and f satisfies (7.9). In addition, suppose that $a < r(X, f) < R(X, f) < b$, where $R(X, f)$ and $r(X, f)$ are respectively the maximum and the minimum of f on the interval X. Then there exists some $\varepsilon > 0$ such that for all g with $||g - f|| < \varepsilon$, the conclusions of Proposition 3.1 hold with g in place of f.*

Proof: see Bala and Majumdar [1992, Corollary 3.1].

Finally, consider the parameteric family $\mathcal{F}$ of maps $f_\mu : [0, 1] \times [3, 4] \to [0, 1]$ defined as:

$$\mathcal{F} = \{f_\mu(x) = \mu x(1 - x), x \in [0, 1], \mu \in [2, 4]$$

The following fundamental theorem of Jakobson (1981) throws light on the issue of robust ergodic chaos.

Proposition 7.5: *There is a set $M \subset [3, 4]$ with $m(M) > 0$ such that if $X = [0, 1]$, $\mu \in M$, and $f_\mu \equiv \mu x(1 - x)$, then (X, f_μ) exhibits ergodic chaos.*

2

Introduction to Dynamic Optimization Theory

Tapan Mitra

1. Introduction

Dynamic optimization models and methods are currently in use in a number of different areas in economics, to address a wide variety of issues. The purpose of this chapter is to provide an introduction to the subject of dynamic optimization theory which should be particularly useful in economic applications. The current section provides a brief history of our subject, to put this survey in proper perspective. It then provides an overview of the topics covered in the various sections of this review.

The contribution by Frank Ramsey (1928) on the optimum rate of savings of a nation is generally regarded as the paper which introduced the use of dynamic optimization methods in addressing economic problems. Ramsey analyzed a continuous-time dynamic optimization model, and developed a modification of the standard calculus of variations method to deal with the problem of existence of an optimum savings rate, when all generations (current and future) over an infinite horizon are to be treated equally in the objective function.

Almost thirty years were to pass before there was a renewed interest in the problem of optimum savings as formulated by Ramsey. The method of optimal control formulated by Pontryagin et.al. (1962), had a significant impact on this line of research. Ramsey's problem was reformulated and studied in depth in terms of the continuous-time one-sector neoclassical model by Cass (1965), Koopmans (1965) and Samuelson (1965), and in the continuous-time two-sector neoclassical model by Srinivasan (1964) and Uzawa (1964).

The 1950s had seen the successful application of the more basic methods of real and convex analysis in general equilibrium theory by Arrow-Debreu (1954), McKenzie (1954), Gale(1955) and Nikaido(1956), replacing the earlier calculus treatments of the problem of existence of equilibrium and its Pareto-Optimality. In his seminal work on efficient allocation of resources over time in an infinite-horizon framework, Malinvaud (1953) recognized that problems of intertemporal allocation could be addressed by employing similar methods, thereby bringing the treatment of capital theory closer to that of the theory of general equilibrium. In his justly celebrated survey of the state of economic science, Koopmans (1957) reviewed these developments and advocated the use of these methods more generally in all branches of economic theory. The impact of this line of thinking on the theory of dynamic optimization took some time to materialize. The standard problem of dynamic optimization was formulated both as a discrete-time problem, and in alternative versions of the so-called reduced form model, by Radner (1967a), using dynamic programming methods, and by Gale (1967) and McKenzie (1968), using the methods of duality theory. Gale's paper appeared along with the papers by McFadden (1967) and Radner (1967b) in a symposium of the Review of Economic Studies, which had a substantial impact on the methods subsequently used in dynamic optimization theory.

The papers by Gale (1967) and McKenzie (1968) were concerned, true to the spirit of the Ramsey exercise, with objective functions in which future utilities were not discounted. Infinite-horizon programs were compared by means of some version of the "overtaking criterion" proposed by Atsumi (1965) and Weizsacker (1965). The central problem was seen as the problem of existence of an optimal program, and its dynamic behavior (convergence to the unique stationary optimal stock) was a by-product, obtained en route to solving this central problem.

The 1950s and early sixties had given rise to a literature on finite-horizon pure capital accumulation oriented dynamic optimization exercises, where optimality was defined in terms of only the state of the economy at the end of the horizon. Samuelson (1949) had conjectured that programs, optimal according to this criterion, would stay close (for most of the planning horizon) to the balanced growth path which had the largest growth factor, associated with a von Neumann equilibrium (1945). This was made more explicit in Dorfman, Samuelson and Solow (1958) and came to be known as the "turnpike conjecture". Definitive proofs of this conjecture were obtained in alternative frameworks by Morishima(1961), Radner(1961) and McKenzie(1963a). The central idea contained in Radner's paper was that optimal programs which stayed away (uniformly) from the von Neumann equilibrium would suffer a (uniform) value loss, and McKenzie (1963b, 1963c, 1971) recognized that this was the key concept to generalizations of the turnpike property.

When the interest of the profession shifted from purely capital accumulation oriented models to consumption oriented optimal growth models of the Ramsey-type, this key concept remained. Both Atsumi (1965) and McKenzie (1968) recognized that this idea could be used to advantage in studying asymptotic properties of optimal programs, with the golden-rule equilibrium (associated with a program yielding maximum sustainable utility) replacing the notion of the von Neumann equilibrium.

With a shift in emphasis of many economies away from planning at the national level, there was a corresponding change in interpretation of dynamic optimization problems of the Ramsey type. The problem being solved was previously viewed as a normative problem the "social planner" ought to solve; it was now viewed as a descriptive problem that a typical representative agent (more precisely, an infinitely-lived dynasty of the typical agent) solves. The Ramsey objection to discounting future utilities as "ethically indefensible" on the part of the social planner was no longer relevant. If the representative agent did discount the future, the optimization problem would have to reflect this. Thus, the central problem to be solved in describing the agent's behavior would be a discounted dynamic optimization problem of the Ramsey-type.

This reformulation of the focus of the subject had two significant consequences. The issue of the existence of an optimal program, which had occupied center-stage for undiscounted dynamic optimization models, became a relatively unimportant aspect of the theory for discounted models, since it was a relatively straightforward exercise, under discounting, to establish the existence of an optimal program. In contrast, description of dynamic behavior of optimal programs became considerably more difficult.

Examples due to Kurz (1968) for continuous-time models, and Sutherland (1970) for discrete-time models indicated that there could be multiple stationary optimal states. Further, even if the stationary optimal state was unique, optimal programs starting from other initial states need not converge to it over time. Then, Samuelson (1973) presented an example due to Weitzman, which showed that optimal programs could cycle around a unique stationary state independent of the magnitude of the discount factor, and these cycles were not "boundary phenomena". While this destroyed any hope of a general turnpike theorem for discounted models, Samuelson conjectured that with (differential) strict concavity of the utility function, a turnpike property for optimal programs would continue to hold for high discount factors. This led to a considerable literature on the discounted turnpike problem. In alternative frameworks, Samuelson's conjecture was shown to be valid by Brock and Scheinkman (1976), Cass and Shell (1976), Rockafellar (1976) and Scheinkman (1976). Building on these papers, which appeared in a symposium of the Journal of Economic Theory, the theory was refined in the contributions of Araujo and Scheinkman (1977), Bewley (1980) and McKenzie (1982,1983), among others. A definitive survey of these developments are presented in McKenzie (1986).

A natural question that arose from this literature was how to describe dynamic optimal behavior when the discount factor was not close to unity. We will not go into the details of the history of how this question was tackled. It suffices to note that Chapters 3 - 13 of this book are devoted to various facets of this question, demonstrating the possibility and robustness of cyclical and chaotic optimal behavior in a variety of discounted dynamic optimization exercises. It should also be clear from these chapters that some important aspects of this question still remain unanswered.

We turn now to an overview of Sections 2-9 of this chapter. There are two aspects of the review of dynamic optimization theory that should be pointed out at this stage. First, to provide an introduction to the key ideas of this theory, the setting for this review is deliberately one-dimensional (the state space is a compact subset of the real line) although many of the topics covered here have their multidimensional counterparts in the literature. We provide some guidance to this literature in the bibliographic notes of Section 10. Second, we limit our coverage of dynamic optimal behavior to turnpike theory for high discount factors, leaving to the reader the task of sampling Chapters 3-13 of the book for the nature of dynamic optimal behavior when the discount factor is not close to unity.

Section 2 presents the standard reduced-form dynamic optimization model, as well as five examples of dynamic optimization exercises in economics which can be reduced to this standard form. The existence of optimal programs in this standard reduced form is given a complete self-contained treatment.

Two well-known characterizations of optimality have figured prominently in the literature of dynamic optimization. The first uses a primal approach, and characterizes optimality in terms of the existence a value function satisfying the functional equation of dynamic programming (often referred to as Bellman's optimality principle). This approach is developed in Section 3, where basic properties of the value and policy functions are derived.

The second characterization (usually referred to as the price characterization of optimality) is based on a dual approach, and is developed for convex structures, where separation theorems for convex sets play a crucial role. Here an optimal program is characterized in terms of existence of a sequence of dual variables or shadow prices, in terms of which (generalized) profit is maximized at the program at each date compared to any alternative activity available at that date and, in addition, an appropriate transversality condition is satisfied asymptotically. This material is developed in Section 4, where we establish the basic price characterization result , and study some of its implications.

Section 5 is devoted to sensitivity analysis (also known as comparative dynamics), where we examine how optimal policies (and therefore programs) change with changes in the parameters of the dynamic optimization model. Two aspects of such changes are studied in some detail : continuity (Section 5.1) and monotonicity (Section 5.2, 5.3). The former topic is based on the concept of uniform continuity, and the latter on the concept of supermodularity.

Section 6 presents the main result on the existence of a (non-trivial) stationary optimal stock, using the Kakutani fixed point theorem. The related concepts of a discounted golden-rule stock and a modified golden-rule are also introduced and discussed here.

While Sections 2-6 are developed using the methods of real and convex analysis, with differentiability assumptions playing no role, Sections 7-9 use the assumption of smooth preferences. Section 7 develops two basic properties of the dynamic optimization models under appropriate smoothness assumptions. First, if the utility function is continuously differentiable (in the interior of the transition possibility set), then the value function is continuously differentiable (in the interior of the state space), and the derivative of the value function is equal to the derivative of the utility function with respect to the initial state at the optimal point. Second, if the utility function is twice continuously differentiable (in the interior of the transition possibility set), then the policy function is differentiable at the stationary optimal stock, and the magnitude of the derivative of the policy function at the stationary optimal stock is equal to smaller (in absolute value) of the characteristic roots associated with the Ramsey-Euler equation at the stationary optimal stock.

Section 8 examines the issue of uniqueness of the stationary optimal stock. A simple example of non-uniqueness is constructed to show how easily this phenomenon can arise. Then, it is shown that there is a bound on the discount rate (depending on the transition possibility set and the utility function) such that for all higher discount rates, there is a unique (non-trivial) stationary optimal stock.

Section 9 is concerned with global asymptotic stability of the non-trivial stationary optimal stock. First, we present the example of Weitzman, as reported in Samuelson (1973), to show that even if the stationary optimal stock is unique, it is not globally asymptotically stable. Then, we pursue Samuelson's idea that if the Hessian of the utility function is negative definite at the golden-rule, then there is a bound on the discount factor (depending on the transition possibility set and the utility function), such that for all higher discount factors, there can be no cyclical optimal programs. This, in turn, yields global asymptotic stability of the non-trivial stationary optimal stock in our one-dimensional state space framework.

2. Dynamic Optimization Problems

We will describe the reduced-form dynamic optimization problem in Section 2.1 below. This is by now the standard form used to describe dynamic optimzation problems arising in economics. The reduced-form model is widely used because it has essentially a simple mathematical structure; one keeps track of the transition over time of only the state variable, from one state to another. It is also an extremely flexible model to interpret, and a wide variety of dynamic optimization problems in economics can be "reduced" to this form. We illustrate this by discussing five examples in Section 2.2.

2.1 The Reduced Form Model

A *state space* $X = [0, b] \subset \Re$ is given, where $b > 0$. Time is measured in discrete periods $t \in \{0, 1, 2, ...\} \equiv \mathbf{N}$. At each time t the state of the economic system is described by a number $x_t \in X$.

The typical dynamic optimization problem we will be concerned with seeks to maximize the objective functional

$$\sum_{t=0}^{\infty} \delta^t u(x_t, x_{t+1}) \tag{2.1}$$

over the set of all sequences $(x_t)_0^\infty$ satisfying the constraints:

$$(x_t, x_{t+1}) \in \Omega \qquad t \in \{0, 1, 2, ...,\} \tag{2.2}$$

$$x_0 = x \tag{2.3}$$

In the description of the above problem, δ is to be interpreted as a *discount factor*, u as a (reduced form) *utility function*, Ω as the *transition possiblity set*, and $x \in X$ as the *initial state.*

The following basic assumptions will be maintained throughout:

A1: $\Omega \subset X \times X$ *is a closed and convex set containing* $(0, 0)$.

A2: $u : \Omega \to \Re$ *is a bounded, concave, and upper semicontinuous function.*

A3: $\delta \in (0, 1)$.

Given (Ω, u, δ) and $x \in X$, we refer to a sequence $(x_t)_0^\infty$ satisfying (2.2) and (2.3) as a *program* (starting) from (the initial state) x. A solution to problem (2.1)-(2.3) is referred to as an *optimal program* from x.

Proposition 2.1: *If* (Ω, u, δ) *satisfies* A1-A3, *then there exists a solution to problem* (2.1)-(2.3) *for every initial state* $x \in X$.

Proof: Since u is bounded, we can find $B > 0$ such that $|u(x, y)| \leq B$ for all $(x, y) \in \Omega$. Thus, for every sequence $(x_t)_0^\infty$ satisfying (2.2),

$$\sum_{t=0}^{\infty} \delta^t u(x_t, x_{t+1}) \tag{2.4}$$

is an absolutely convergent series, and hence a convergent series, with

$$\sum_{t=0}^{\infty} \delta^t u(x_t, x_{t+1}) \leq B/(1 - \delta) \tag{2.5}$$

Let $S(x) = \sup\{\sum_{t=0}^{\infty} \delta^t u(x_t, x_{t+1}) : (x_t)_0^\infty$ is a sequence satisfying (2.2) and (2.3)$\}$. We will now show that this supremum is actually attained by some $(x_t)_0^\infty$ satisfying (2.2) and (2.3).

By definition of $S(x)$, there is a sequence of programs $(x_t^n)_{t=0}^\infty$, $n = 1, 2, 3, ...$, starting from the initial state x, such that

$$\sum_{t=0}^{\infty} \delta^t u(x_t^n, x_{t+1}^n) \geq S(x) - (1/n) \tag{2.6}$$

By the Cantor diagonal process, we can find a subsequence n' (of n) such that for each $t \in \{0, 1, 2, ...\}$, we have a number x_t^0, such that

$$x_t^{n\prime} \to x_t^0 \qquad \text{as } n' \to \infty \tag{2.7}$$

Since Ω is closed, $(x_t^0, x_{t+1}^0) \in \Omega$ for $t \in \{0, 1, 2, ...\}$, and $x_0^0 = x_0^{n\prime} = x$. Thus, $(x_t^0)_{t=0}^{\infty}$ is a program starting from x.

We claim that

$$\sum_{t=0}^{\infty} \delta^t u(x_t^0, x_{t+1}^0) = S(x). \tag{2.8}$$

If the claim were not true, then we could find $\varepsilon > 0$, such that

$$\sum_{t=0}^{\infty} \delta^t u(x_t^0, x_{t+1}^0) \leq S(x) - \varepsilon \tag{2.9}$$

Pick T large enough so that $B\delta^{T+1}/(1 - \delta) \leq (\varepsilon/4)$. For $t \in \{0, 1, ..., T\}$, we can use (2.7) and the upper semicontinuity of u to get

$$\limsup_{n' \to \infty} u(x_t^{n\prime}, x_{t+1}^{n\prime}) \leq u(x_t^0, x_{t+1}^0) \tag{2.10}$$

We can now pick an interger $N > (4/\varepsilon)$, such that for $t \in \{0, 1, ..., T\}$, we have

$$u(x_t^{n\prime}, x_{t+1}^{n\prime}) \leq u(x_t^0, x_{t+1}^0) + [\varepsilon(1 - \delta)/4]$$

whenever $n' \geq N$. Then, for $n' \geq N$, we can obtain the following string of inequalities:

$$\begin{aligned}
\sum_{t=0}^{\infty} \delta^t u(x_t^0, x_{t+1}^0) &\geq \sum_{t=0}^{T} \delta^t u(x_t^0, x_{t+1}^0) - (\varepsilon/4) \\
&\geq \sum_{t=0}^{T} \delta^t u(x_t^{n\prime}, x_{t+1}^{n\prime}) - (\varepsilon/2) \\
&\geq \sum_{t=0}^{\infty} \delta^t u(x_t^{n\prime}, x_{t+1}^{n\prime}) - 3(\varepsilon/4) \\
&\geq S(x) - (1/n') - 3(\varepsilon/4) \\
&> S(x) - \varepsilon
\end{aligned}$$

which contradicts (2.9) and establishes our claim (2.8). This means, of course, that $(x_t^0)_0^\infty$ is an optimal solution to (2.1)-(2.3). Q.E.D.

2.2 Examples

We now discuss five examples, arising in various contexts of capital theory, which can be treated as special cases of the reduced-form model of Section 2.1. For each example, described in terms of its "primitives", we show how it can be converted to its reduced form.

Example 2.1. (The One-Sector Model of Optimal Growth)

Let $G : \Re_+^2 \to \Re_+$ be a concave, non-decreasing, continuous and constant returns to scale *production function*, indicating the net output level $G(K, L)$ producible with inputs of capital (K) and labor (L). Labor is assumed to grow exogenously at a rate $n \geq 0$; that is $L_t = L_0(1+n)^t$ for $t \in \mathbf{N}$ and $L_0 > 0$. Capital depreciates at a constant rate d where $0 < d \leq 1$. The basic growth equation is

$$G(X_t, L_t) = C_{t+1} + K_{t+1} - (1-d)K_t$$

where $K_t \geq 0$ is the capital stock at time t, and $C_{t+1} \geq 0$ is the consumption at time $(t+1)$. Dividing through by L_{t+1}, and denoting $[K_t/L_t]$ by x_t and (C_{t+1}/L_{t+1}) by c_{t+1} for $t \in N$, we have

$$[G(x_t, 1)/(1+n)] = c_{t+1} + x_{t+1} - [(1-d)/(1+n)]x_t$$

Then letting $g(x) = G(x, 1)/(1+n)$ for $x \geq 0$, and $f(x) = g(x) + [(1-d)/(1+n)]x$ for $x \geq 0$, we obtain

$$c_{t+1} = f(x_t) - x_{t+1} \text{ for } t \in \mathbf{N}$$

A continuous, concave and increasing *welfare function*, $w : \Re_+ \to \Re$, evaluates the (period) welfare, $w(c)$, corresponding to a per capita consumption level, $c \geq 0$. The *discount factor*, $\delta \in (0,1)$, indicates the weight placed by the planner (representative agent) on tomorrow's welfare relative to today's welfare in the intertemporal objective function.

The optimal growth problem can be written as:

$$
(P1)\begin{cases}
\text{Max } \sum_{t=0}^{\infty} \delta^t w(c_{t+1}) \\
\text{subject to } c_{t+1} = f(x_t) - x_{t+1}, \text{ and } (x_t, c_{t+1}) \geq 0 \\
\text{for } t \in \mathbf{N} \\
x_0 = x > 0
\end{cases}
$$

We now indicate how to convert this example to its reduced form. First, we have to choose an appropriate state space. Let us assume that $\lim_{x \to \infty} [g(x)/x] = 0$. Then, we can find a number $b > 0$, such that $f(x) < x$ for $x > b$. In this case, if the initial capital-labor ratio, x_0, is in $[0, b]$, then all future capital-labor ratios, x_{t+1} (for $t \in \mathbf{N}$) must lie in $[0, b]$. Further, it is easy to verify that a solution to (P1) must have $x_t \in [0, b]$ for some finite t. Thus, it is legitimate to define the state space to be $X \equiv [0, b]$, since all the long-run dynamics is confined to this set.

Now, defining $\Omega = \{(x, z) \in \Re^2_+ : z \leq f(x)\}$, and $u : \Omega \to \Re$ by $u(x, z) = w(f(x) - z)$, a solution to problem (2.1)-(2.3) for (Ω, u, δ) corresponds precisely to a solution to problem (P1), and vice versa.

Example 2.2 (Two-Sector Model of Economic Growth)

The two-sector model of optimal economic growth, originally discussed by Uzawa (1964) and Srinivasan (1964), is a generalization of the one-sector model, discussed in Example 2.2 above. The severe restriction imposed in the one-sector model, that consumption can be traded off against investment on a one-to-one basis, is relaxed in the two-sector model, and it is principally this aspect that makes the two-sector model a considerably richer framework in studying economic growth problems than its one-sector predecessor.

Production uses two inputs, capital and labor. Given the total amounts of capital and labor available to the economy, the inputs are allocated to two "sectors" of production, the consumption good sector, and the investment good sector. Output of the former sector is consumed (and cannot be used for investment purposes), and output of the latter sector is used to augment the capital stock of the economy (and cannot be consumed). Thus, the consumption-investment decision amounts to a decision regarding allocation of capital and labor between the two production sectors.

The discounted sum of outputs of the consumption good sector is to be maximized to arrive at the appropriate sectoral allocation of inputs in each period. [A welfare function on the output of the consumption good sector can be used, instead of the output of the consumption good sector itself, in the objective function, but it is usual to assimilate this welfare function, when it is increasing and concave, in the "production function" of the consumption good sector.]

Formally, the model is specified by (F, G, d, δ), where

(a) the *production function* in the *consumption good sector*, $F : \Re^2_+ \to \Re_+$, satisfies:

(i) F is continuous and homogeneous of degree one on $\Re^2_+$.

(ii) F is non-decreasing on $\Re^2_+$.

(iii) F is concave on $\Re^2_+$.

(b) the *production function* in the *investment good sector*, $G : \Re^2_+ \to \Re_+$, satisfies:

(i) G is continuous and homogeneous of degree one on $\Re^2_+$.

(ii) G is non-decreasing on $\Re^2_+$.

(iii) G is concave on $\Re^2_+$.

(iv) $\lim_{K\to\infty} [G(K, 1)/K] = 0$

(c) the *depreciation factor*, d, satisfies $0 < d \leq 1$;

(d) the *discount factor*, δ, satisfies $0 < \delta < 1$.

The optimal growth problem can be written as:

$$(P2)\begin{cases} \text{Max } \sum_{t=0}^{\infty} \delta^t c_{t+1} \\ \text{subject to } c_{t+1} = F(k_t, n_t) \text{ for } t \in \mathbf{N} \\ x_{t+1} = G(x_t - k_t, 1 - n_t) + (1 - d)x_t \text{ for } t \in \mathbf{N} \\ 0 \leq k_t \leq x_t,\ 0 \leq n_t \leq 1, \text{ for } t \in \mathbf{N} \\ x_0 = x > 0 \end{cases}$$

Here x_t is the total capital available at date t, which is allocated between the consumption good sector (k_t) and the investment good sector ($x_t - k_t$). Labor is exogenously available at a constant amount (normalized to unity), which is allocated between the consumption good sector (n_t) and the investment good sector ($1 - n_t$). Note that an exogenously growing labor force (at a constant growth rate) can be accommodated easily by interpreting k_t and x_t as per worker capital stocks, and reinterpreting the depreciation factor, d.

As in the one-sector model, we can find $b > 0$, such that for $x \in [0, b]$, we have $G(x, 1) + (1 - d)x$ in $[0, b]$, and furthermore for any solution to (P2), $x_t \in [0, b]$ for some t. Thus, it is appropriate to define $X = [0, b]$ as the state space.

To convert the problem to its reduced-form, we can define the transition possibility set, Ω, as:

$$\Omega = \{(x, z) \text{ in } \Re^2_+ : z \leq G(x, 1) + (1 - d)x\}$$

and the utility function, u, as

$$\begin{aligned} u(x, z) &= \text{Max } \; F(k, n) \\ \text{Subject to } 0 &\leq k \leq x, \; 0 \leq n \leq 1 \\ z &\leq G(x - k, 1 - n) + (1 - d)x \end{aligned}$$

Then a solution to (P2) corresponds exactly to a solution (2.1)-(2.3) for (Ω, u, δ), and vice versa.

Example 2.3 (Optimal Exploitation of a Fishery)

The economic exploitation of a fishery figures prominently in the literature on the management of renewable resources (see, for example, Clark (1976) for references to the relevant literature). The following is an idealized description (following Dasgupta (1982)) of the relevant optimization problem.

The resource stock (biomass of the fish species) at the end of period t is denoted by x_t. The biological reproduction function (also called the stock recruitment function) yields the biomass, y_{t+1}, in period $(t + 1)$ given by

$$y_{t+1} = f(x_t)$$

If c_{t+1} is harvested in period $(t+1)$ then the stock at the end of time period $(t+1)$ is x_{t+1} given by

$$x_{t+1} = f(x_t) - c_{t+1}$$

It is usual to assume that $f(0) = 0$, f is continuous, non-decreasing and concave, with $\lim_{x \to 0}[f(x)/x] > 1$ and $\lim_{x \to \infty}[f(x)/x] = 0$ [although there are important variations, especially dealing with non-concave functions, f].

We can find $b > 0$ such that whenever x_t is in $[0, b]$, x_{t+1} is in $[0, b]$ also. Thus, $[0, b]$ is a legitimate choice for the state space, X.

Harvesting the fishery is not costless. Specifically, the harvest, c_{t+1}, depends on the resource biomass available for exploitation, y_{t+1}, and the labor effort, e_{t+1}:

$$c_{t+1} = G(y_{t+1}, e_{t+1})$$

with G a non-decreasing, concave function in both its variables (increasing in both arguments when (y, e) is stricty positive), such that given any $y \geq 0$, $G(y, e) \leq y$ for all e. We express the effort, e_{t+1}, required to havest c_{t+1} when the biomass available for harvest is y_{t+1} as:

$$e_{t+1} = H(c_{t+1}, y_{t+1})$$

with H increasing in its first argument and decreasing in its second. This is the cost of exploiting the fishery. The benefit obtained depends on the harvest level, c_{t+1}:

$$B_{t+1} = B(c_{t+1})$$

where B is an increasing, concave function. [For a competitive fishery, B is linear, with $B(c_{t+1}) = pc_{t+1}$, where $p > 0$ is the price (assumed constant over time) per unit of the fish.] The return in period $(t+1)$ is then the benefit minus the cost:

$$w(c_{t+1}, x_t) = B(c_{t+1}) - H(c_{t+1}, f(x_t))$$

The optimization problem can then be written as

$$
(P3)\begin{cases}
\text{Max } \sum_{t=0}^{\infty} \delta^t w(c_{t+1}, x_t) \\
\text{Subject to } c_{t+1} = f(x_t) - x_{t+1} \text{ for } t \in \mathbf{N} \\
c_{t+1} \geq 0, x_{t+1} \geq 0 \text{ for } t \in \mathbf{N} \\
x_0 = x > 0
\end{cases}
$$

To convert the problem to its reduced form, we define the transition possibility set, Ω, as:

$$\Omega = \{(x, z) \text{ in } \Re_+^2 : z \leq f(x)\}$$

and the utility function, u, as:

$$u(x, z) = w(f(x) - z, x)$$

Then, a solution to (P3) corresponds exactly to a solution to (2.1)-(2.3) for (Ω, u, δ).

Example 2.4 (A Three-Sector Model with an Aging Process)

This example is due to Weitzman, as reported in Samuelson (1973). It involves a pure aging process in one of the sectors, which produces wine, and therefore might be called an "Austrian" capital-theoretic model.

The representative agent has one unit of labor. She can allocate it between two sectors: one of these sectors produces bread, the other produces grape juice. It takes one unit of labor per unit of bread production, and one unit of labor per unit of grape juice production. A third sector produces wine, and its production process requires one unit of grape juice per unit of wine, and one period of time (for the juice to ferment).

As a consumer, the agent derives satisfaction from the consumption of bread and wine, but not from grape juice.

Let x_t denote the allocation of labor in period t to grape-juice production, with $x_0 > 0$ (initial allocation) as given. Then $(1 - x_t)$ of labor is allocated to production of bread. Thus, in period $(t+1)$, where $t \geq 0$, we have bread production of $(1 - x_{t+1})$, grape-juice production of x_{t+1}, and wine produciton of x_t.

Assume that the consumer's felicity function is

$$f(b, w) = b^\alpha w^\beta, \text{ with } (\alpha, \beta) >> 0, \ \alpha + \beta \leq 1$$

where b and w are the amounts of bread and wine consumed.

The agent's optimization problem can be written as:

$$(P4) \begin{cases} \text{Max } \sum_{t=0}^{\infty} \delta^t f(b_{t+1}, w_{t+1}) \\ \text{Subject to } b_{t+1} = 1 - x_{t+1} \text{ for } t \in \mathbf{N} \\ 0 \leq x_t \leq 1 \text{ for } t \in \mathbf{N} \\ x_0 = x > 0 \end{cases}$$

To convert the problem to its reduced form, we can define the state space $X = [0, 1]$, the transition possibility set, Ω, to be:

$$\Omega = \{(x, z) \in X^2\}$$

and the utility function, u, to be:

$$u(x, z) = f(1 - x_{t+1}, x_t) = (1 - x_{t+1})^\alpha x_t^\beta$$

Then an optimal solution to (P4) corresponds exactly to an optimal solution to (2.1)-(2.3) for (Ω, u, δ).

Example 2.5 (A Model of Forest Management)

This example is a simplified version of the forest management model, developed in Mitra and Wan (1985, 1986), based on the pioneering work in this area by Faustmann (1849). Our exposition of the example is based on the presentation of Wan (1987).

An agent has a plot of land (normalized to unity), which is good for growing a certain type of tree. The growth process for the tree is as follows: after saplings are planted, the tree grows in timber content for two years, the timber content being a $\in (0, 1)$ after one year, and 1 after two years (per unit of land). After two years, the tree decays and the timber becomes worthless.

The agent derives a return from selling the timber, the timber being assumed of homogeneous quality, whether it is from one or two year old trees. Planting and cutting trees are assumed costless activities.

Let x_t denote the amount of land with two-year old trees at the end of period t. Then, $(1-x_t)$ is the amount of land with one-year old trees at the end of period t.

At the end of period t, all two year old trees are cut down and sold as timber. This clears up x_t of land for reforestation. In contrast one-year old trees might or might not be cut down. The one-year old trees which are *not* cut down at the end of period t become the two-year old trees at the end of period $(t+1)$. Thus, the land cleared up by cutting down one-year old trees at the end of period t is $[(1-x_t)-x_{t+1}]$. The total timber content of all trees cut down at the end of period t is, therefore, given by

$$c_t = x_t + a[(1-x_t) - x_{t+1}]$$

The agent's optimization problem can be written as:

$$(P5)\left\{\begin{array}{c} \text{Max } \sum_{t=0}^{\infty} \delta^t R(c_t) \\ \text{Subject to } c_t = x_t + a[(1-x_t) - x_{t+1}] \text{ for } t \in \mathbf{N} \\ x_{t+1} \leq (1-x_t) \text{ for } t \in \mathbf{N} \\ 0 \leq x_t \leq 1, \text{ for } t \in \mathbf{N} \\ x_0 = x \geq 0 \end{array}\right.$$

To convert this problem to its reduced form, we can define the state space to be $X = [0,1]$, the transition possibility set, Ω, to be:

$$\Omega = \{(x,z) \in X^2 : z \leq (1-x)\}$$

and the utility function, u, to be:

$$u(x,z) = R(a + (1-a)x - az)$$

Then, an optimal solution to (P5) corresponds to an optimal solution to (2.1)-(2.3) for (Ω, u, δ).

3. Dynamic Programming

Dynamic optimization problems of the type described in Section 2 can be studied conveniently by the method of dynamic programming. In this section, we provide the basic results that can be established for the class of problems defined by (2.1)-(2.3) by using the dynamic programming approach.

We can associate with each dynamic optimization problem two functions, called the (optimal) *value function* and the (optimal) *policy function* [or, more generally, the (optimal) policy correspondence]. The value function defines the maximized value of the objective function (2.1), given the constraints (2.2), (2.3), corresponding to each initial state, x. If there is a unique optimal solution $(x_t^*)_0^\infty$ to the problem (2.1)-(2.3) for each initial state $x \in X$, the policy function defines the first period optimal state, x_1^*, corresponding to each initial state, x. [More generally, the policy correspondence describes the set of states that it is optimal to go to in period 1, corresponding to each initial state, x.]

Given the stationary recursive nature of the optimization problem, the policy function, in fact, generates the entire optimal solution, starting from any initial state. The value function helps us to study the properties of the policy function, given the connection between the two through the functional equation of dynamic programming. Thus, the value and policy functions are extremely useful objects of study in describing dynamic optimal behavior.

3.1 The Value Function

Given the existence of an optimal solution to problem (2.1)-(2.3) fore every $x \in X$ [Proposition 2.1], we can define the *value function*, $V : X \to \Re$ as follows:

$$V(x) = \sum_{t=0}^{\infty} \delta^t u(x_t^*, x_{t+1}^*) \tag{3.1}$$

where $(x_t^*)_0^\infty$ is an optimal solution to (2.1)-(2.3) corresponding to the initial state $x \in X$.

The following result summarizes the basic properties of the value function.

Proposition 3.1: (i) *The value function, V, is a concave and continuous function on X.* (ii) *V satisfies the following functional equation of dynamic programming*

$$V(x) = \max_{y \in \Omega_x} \{u(x, y) + \delta V(y)\} \tag{3.2}$$

for all $x \in X$ [where $\Omega_x = \{y \in X : (x, y) \in \Omega\}$]. (iii) *$V$ is the only continuous function on X which satisfies (3.2).* (iv) *$(x_t)_0^\infty$ is an optimal solution to (2.1)-(2.3) if and only if*

$$V(x_t) = u(x_t, x_{t+1}) + \delta V(x_{t+1}) \text{ for } t \in \mathbf{N} \tag{3.3}$$

Proof: (i) Let x, $x' \in X$ and let $0 < \lambda < 1$. Let $(x_t)_0^\infty$ and $(x'_t)_0^\infty$ be optimal programs from x and x' respectively. Define $x'' = \lambda x + (1 - \lambda)x'$, and let (x''_t) be an optimal program from x''. By convexity of Ω, the sequence $(\lambda x_t + (1 - \lambda)x'_t)_0^\infty$ is a program from x''. Thus, we have

$$\begin{aligned} V(x'') &= \sum_{t=0}^{\infty} \delta^t u(x''_t, x''_{t+1}) \\ &\geq \sum_{t=0}^{\infty} \delta^t u(\lambda x_t + (1 - \lambda)x'_t, \lambda x_{t+1} + (1 - \lambda)x'_{t+1}) \\ &\geq \sum_{t=0}^{\infty} \delta^t [\lambda u(x_t, x_{t+1}) + (1 - \lambda)u(x'_t, x'_{t+1})] \\ &= \lambda V(x) + (1 - \lambda)V(x') \end{aligned}$$

the second inequality following from the concavity of u on Ω. Thus, V is concave on X.

In order to establish the continuity of V on X, we first establish its upper semicontinuity. This can be done by following essentially the method used in the proof of Proposition 2.1.

If V were not upper semicontinuous on X, we can find $x^n \in X$ for $n = 1, 2, 3, \ldots$, with $x^n \to x^0$ and $V(x^n) \to V$ as $n \to \infty$, with $V > V(x^0)$. Let us denote by (x_t^n) an optimal program from x^n for $n = 0, 1, 2, \ldots$.

Denoting $[V - V(x^0)]$ by ε, we can find T large enough so that $B\delta^{T+1}/(1-\delta) \le (\varepsilon/4)$.

Clearly, we can find a subsequence n' (of n) such that for $t \in \{0, 1, ..., T\}$

$$x_t^{n\prime} \to x_t^0 \text{ as } n' \to \infty.$$

Then, using the upper semicontinuity of u, we can find N such that for $t \in \{0, 1, ..., T\}$

$$u(x_t^{n\prime}, x_{t+1}^{n\prime}) \le u(x_t^0, x_{t+1}^0) + [\varepsilon(1-\delta)/4]$$

whenever $n' \ge N$. Thus, for $n' \ge N$, we obtain the following string of inequalities

$$\begin{aligned} V(x^0) &= \sum_{t=0}^{\infty} \delta^t u(x_t^0, x_{t+1}^0) \\ &\ge \sum_{t=0}^{T} \delta^t u(x_t^0, x_{t+1}^0) - (\varepsilon/4) \\ &\ge \sum_{t=0}^{T} \delta^t u(x_t^{n\prime}, x_{t+1}^{n\prime}) - (\varepsilon/2) \\ &\ge \sum_{t=0}^{\infty} \delta^t u(x_t^{n\prime}, x_{t+1}^{n\prime}) - 3(\varepsilon/4) \\ &= V(x^{n\prime}) - 3(\varepsilon/4) \end{aligned}$$

Since $V(x^{n\prime}) \to V$, we have $V(x^0) \ge V - 3(\varepsilon/4) > V - \varepsilon = V(x^0)$, a contradiction. Thus, V is upper semicontinuous on X.

V is concave on X, and hence continuous on *int* $X = (0, b)$. To show that V is continuous at 0, let x^n be a sequence of points in X $(n = 1, 2, 3, ...)$ converging to 0. Then $V(x^n) = V((1 - (x^n/b))\ 0 + (x^n/b)b) \ge [1 - (x^n/b)]V(0) + (x^n/b)V(b)$. Letting $n \to \infty$, $\liminf_{n\to\infty} V(x^n) \ge V(0)$. On the other hand, since V is upper semicontinuous on X, $\limsup_{n\to\infty} V(x^n) \le V(0)$. Thus, $\lim_{n\to\infty} V(x^n)$ exists and equals $V(0)$. The continuity of V at b is established similarly.

(ii) Let $y \in \Omega_x$, and let $(y_t)_0^\infty$ be an optimal program from y. Then $(x, y_0, y_1, ...)$ is a program from x, and hence, by definition of V,

$$\begin{aligned} V(x) &\geq u(x, y_0) + \sum_{t=1}^{\infty} \delta^t u(y_{t-1}, y_t) \\ &= u(x, y_0) + \delta \sum_{t=1}^{\infty} \delta^{t-1} u(y_{t-1}, y_t) \\ &= u(x, y_0) + \delta \sum_{t=0}^{\infty} \delta^t u(y_t, y_{t+1}) \\ &= u(x, y_0) + \delta V(y) \end{aligned}$$

So, we have established that

$$V(x) \geq u(x, y) + \delta V(y) \text{ for all } y \in \Omega_x \tag{3.4}$$

Next, let $(x_t)_0^\infty$ be an optimal program from x, and note that

$$\begin{aligned} V(x) &= u(x_0, x_1) + \delta \left[\sum_{t=1}^{\infty} \delta^{t-1} u(x_t, x_{t+1}) \right] \\ &= u(x_0, x_1) + \delta \left[\sum_{t=0}^{\infty} \delta^t u(x_{t+1}, x_{t+2}) \right] \\ &\leq u(x_0, x_1) + \delta V(x_1) \end{aligned}$$

Using (3.4), we then have

$$V(x) = u(x_0, x_1) + \delta V(x_1) \tag{3.5}$$

Now, (3.4) and (3.5) establish (3.2).

(iii) Let W be a continuous function on X satisfying (3.2). Let $D \equiv \max_{x \in X} |V(x) - W(x)|$. Given any $x \in X$, denote by y a solution to the maximization problem on the right-hand side of (3.2). Then, $V(x) = u(x, y) + \delta V(y)$, and $W(x) \geq u(x, y) + \delta W(y)$, so that

$$[V(x) - W(x)] \leq \delta[V(y) - W(y)] \leq \delta D \tag{3.6}$$

Denote by y' a solution to the maximization problem on the right-hand side of (3.2), when V is replaced by W. Then, $W(x) = u(x, y') + \delta W(y')$, and $V(x) \geq u(x, y') + \delta V(y')$, so that

$$[W(x) - V(x)] \leq \delta[W(y') - V(y')] \leq \delta D \tag{3.7}$$

Using (3.6) and (3.7) we get

$$|V(x) - W(x)| \leq \delta D$$

and since $x \in X$ was arbitrary,

$$D \equiv \max_{x \in X} |V(x) - W(x)| \leq \delta D$$

Thus, $D = 0$, and so $W = V$ on X.

(iv) If $(x_t)_0^\infty$ satisfies (3.3), then we get for any $T \geq 1$,

$$V(x) = \sum_{t=0}^{T} \delta^t u(x_t, x_{t+1}) + \delta^{T+1} V(x_{T+1})$$

Since $|V(x)| \leq B/(1-\delta)$ for all $x \in X$ and $\delta^{T+1} \to 0$ as $T \to \infty$, we have

$$V(x) = \sum_{t=0}^{\infty} \delta^t u(x_t, x_{t+1})$$

Then, by the definition of V, $(x_t)_0^\infty$ is an optimal program (from x).

To establish the converse implication, let $(x_t^*)_0^\infty$ be an optimal solution to (2.1)-(2.3). Then, for each $T \geq 1$, $(x_T^*, x_{T+1}^*, ...)$ is an optimal solution to (2.1)-(2.3) for $x = x_T^*$. For, if $(y_t)_0^\infty$ is a program from $y_0 = x_T^*$ such that

$$\sum_{t=0}^{\infty} \delta^t u(y_t, y_{t+1}) > \sum_{t=T}^{\infty} \delta^{t-T} u(x_t^*, x_{t+1}^*)$$

then the sequence $(x_0^*, ..., x_T^*, y_1, y_2, ...)$ has a discounted sum of utilities

$$\sum_{t=0}^{T-1} \delta^t u(x_t^*, x_{t+1}^*) + \sum_{t=T}^{\infty} \delta^t u(y_t, y_{t+1}) > \sum_{t=0}^{\infty} \delta^t u(x_t^*, x_{t+1}^*)$$

which contradicts the optimality of $(x_t^*)_0^\infty$ from $x = x_0^*$. Now, using the result (3.5) in (ii) above, we have $V(x_t^*) = u(x_t^*, x_{t+1}^*) + \delta V(x_{t+1}^*)$ for $t \in \mathbf{N}$.

3.2 The Policy Function

An optimal solution to the problem (2.1)-(2.3) need to be unique. For some states x_t, there might be more than one successor state x_{t+1} satisfying (3.3). To ensure that an optimal solution to (2.1)-(2.3) is unique for every $x \in X$, we use a strict concavity assumption on u.

A4: u is strictly concave in its second argument

If A4 holds, then given any $x \in X$, there is a unique solution to the maximization problem on the right-hand side of (3.2). For if y and $y'(y' \neq y)$ in Ω_x both solved this problem, then $V(x) = u(x,y) + \delta V(y)$ and $V(x) = u(x,y') + \delta V(y')$. However, $(x, 0.5y + 0.5y') \in \Omega$, and $u(x, 0.5y + 0.5y') + \delta V(0.5y + 0.5y') > 0.5u(x,y) + 0.5u(x,y') + \delta[0.5V(y) + 0.5V(y')] = V(x)$, a contradiction to (3.2).

For each x, denote the unique state $y \in \Omega_x$ which solves the maximization problem on the right hand side of (3.2) by $h(x)$. We will call $h : X \to X$ the *policy function.*

The following result summarizes the basic properties of the policy function.

Proposition 3.2: (i) *The policy function $h : X \to X$ is continuous on X.*

(ii) *For all $(x,y) \in \Omega$ with $y \neq h(x)$ we have*

$$u(x,y) + \delta V(y) < V(x) = u(x, h(x)) + \delta V(h(x)) \tag{3.8}$$

(iii) *$(x_t)_0^\infty$ is an optimal solution to (2.1)-(2.3) if and only if*

$$x_{t+1} = h(x_t) \text{ for } t \in \mathbf{N} \tag{3.9}$$

Proof: (i) Let $(x^n)(n = 1, 2, 3, ...)$ be a sequence of points in X converging to x^0. Then for $n = 1, 2, 3, ...$

$$V(x^n) = u(x^n, h(x^n)) + \delta V(h(x^n)) \tag{3.10}$$

Let $(y^{n'})$ be an arbitrary convergent subsequence of $(h(x^n))$ converging to y^0. We claim that $y^0 = h(x^0)$. For the subsequence n', using (3.10), the upper semicontinuity of u and the continuity of V,

$$\begin{aligned} V(x^0) &= \limsup_{n'\to\infty} u(x^{n'}, y^{n'}) + \delta V(y^0) \\ &\leq u(x^0, y^0) + \delta V(y^0) \end{aligned}$$

But, by (3.2), $V(x^0) \geq u(x^0, y^0) + \delta V(y^0)$, so that $V(x^0) = u(x^0, y^0) + \delta V(y^0)$. This means that $y^0 = h(x^0)$, establishing our claim. Thus, $h(x^n) \to h(x^0)$ as $n \to \infty$, establishing continuity of h.

(ii) We have $V(x) = u(x, h(x)) + \delta V(h(x))$ from (3.2) and the definition of h. For all $y \in \Omega_x$, we have $u(x, y) + \delta V(y) \leq V(x)$ by (3.2). Further, if equality holds in the previous weak inequality, then $y = h(x)$. Thus if $y \neq h(x)$, the strict inequality must hold. This establishes (3.8).

(iii) If $(x_t)_0^\infty$ is an optimal solution to (2.1)-(2.3), then by Proposition 3.1 (iv), for every $t \in \mathbf{N}$,

$$V(x_t) = u(x_t, x_{t+1}) + \delta V(x_{t+1})$$

so that by Proposition 3.2 (ii), $x_{t+1} = h(x_t)$.

Conversely, if (3.9) holds, then by Proposition 3.2 (ii), we have for $t \in \mathbf{N}$

$$V(x_t) = u(x_t, x_{t+1}) + \delta V(x_{t+1})$$

so that $(x_t)_0^\infty$ is an optimal solution to (2.1)-(2.3) by Proposition 3.1 (iv).

4. Duality Theory

Optimal programs can be characterized in terms of dual variables (or shadow prices). At the shadow prices supporting an optimal program, there is no activity which yields a higher "generalized profit" at any date (value of utility plus value of terminal stocks minus value of initial stocks at that date) than the activity chosen along the optimal program at that date. This support property is referred to as the *competitive condition*, and the shadow prices as *competitive prices*.

In addition, optimal programs must satisfy the condition that the aymptotic value of the stocks (at the shadow prices) must be zero, which is referred to as the *transversality condition*.

The two conditions taken together are also sufficient to ensure optimality, so that they provide a complete characterization of optimal programs, generally called a *price characterization of optimality*.

It was observed by Weitzman (1973) that the shadow prices appearing in the competitive condition also provide a support for the value function, which was introduced in Section 3. Given the competitive condition, the validty of the transversality condition is equivalent to the property that the competitive prices support the value function.

This connection between dynamic programming and duality theory forms the basis of Weitzman's approach to the price characterization of optimal programs, which we will describe below.

It is convenient (although not essential) to develop the theory under an additional "monotonicity assumption", which ensures that the relevant competitive prices are non-negative at each date.

A5: If $(x, z) \in \Omega$, and $x' \in X$, $x' \geq x$, $0 \leq z' \leq z$, then $(x', z') \in \Omega$ and $u(x', z') \geq u(x, z)$.

Proposition 4.1: *If $(x_t)_0^\infty$ is a program from $x \in X$, and there is a sequence $(p_t)_0^\infty$ such that $p_t \in \Re_+$ for $t \in \mathbf{N}$, and the following conditions hold:*

$$\text{(i) } \delta^t u(x_t, x_{t+1}) + p_{t+1}x_{t+1} - p_t x_t \geq \delta^t u(x, z) + p_{t+1}z - p_t x$$
$$\textit{for all } (x, z) \in \Omega,\ t \in \mathbf{N} \tag{4.1}$$

$$\text{(ii) } \lim_{t \to \infty} p_t x_t = 0 \tag{4.2}$$

then $(x_t)_0^\infty$ is an optimal program from x.

Proof: Let $(x'_t)_0^\infty$ be any program from x. Using (4.1), we have for $t \geq 0$,

$$\delta^t[u(x'_t, x'_{t+1}) - u(x_t, x_{t+1})] \leq (p_{t+1}x_{t+1} - p_t x_t) - (p_{t+1}x'_{t+1} - p'_t x'_t)$$

Summing this from $t = 0$ to $t = T$

$$\begin{aligned}\sum_{t=0}^{T} \delta^t[u(x'_t, x'_{t+1}) - u(x_t, x_{t+1})] &\leq (p_{T+1}x_{T+1} - p_0 x_0) - \\ &\quad (p_{T+1}x'_{T+1} - p_0 x'_0) \\ &= (p_{T+1}x_{T+1} - p_{T+1}x'_{T+1}) \\ &\leq p_{T+1}x_{T+1}\end{aligned}$$

Letting $T \to \infty$, we get

$$\sum_{t=0}^{\infty} \delta^t u(x'_t, x'_{t+1}) - \sum_{t=0}^{\infty} \delta^t u(x_t, x_{t+1}) \leq \lim_{T\to\infty} p_{T+1}x_{T+1} = 0$$

which proves that $(x_t)_0^\infty$ is optimal from x.

Remark 4.1: (i) Condition (4.2) can be replaced by

$$\liminf_{t\to\infty} p_t x_t = 0 \tag{4.3}$$

in Proposition 4.1.

(ii) Condition 4.2 is called the *transversality condition.* Its significance was first noted by Malinvaud (1953) in his study of intertemporal efficiency. Condition 4.1 is referred to as the *competitive condition*; the sequence $(p_t)_0^\infty$ of dual variables are called *competitive prices.* The result of Proposition 4.1 is then paraphrased as: "A competitive program satisfying the transversality condition is optimal."

In contrast to Proposition 4.1, the converse result (Proposition 4.2 below) exploits the convex structure of the dynamic optimization model.

Proposition 4.2: *Suppose $(x_t)_0^\infty$ is an optimal program from $x > 0$. Suppose also that there is some $(\widehat{x}, \widehat{z})$ in Ω with $\widehat{z} > 0$.*

Then, there is a sequence $(p_t)_0^\infty$ *such that* $p_t \in \Re_+$ *for* $t \in \mathbf{N}$ *and the following conditions hold:*

$$\text{(i) } \delta^t u(x_t, x_{t+1}) + p_{t+1}x_{t+1} - p_t x_t \geq \delta^t u(x, z) + p_{t+1}z - p_t x \text{ for all } (x, z) \in \Omega \text{ and } t \in \mathbf{N} \tag{4.4}$$

$$\text{(ii) } \delta^t V(x_t) - p_t x_t \geq \delta^t V(x) - p_t x \text{ for all } x \in X \text{ and } t \in \mathbf{N} \tag{4.5}$$

$$\text{(iii) } \lim_{t\to\infty} p_t x_t = 0 \tag{4.6}$$

In order to establish Proposition 4.2, we first state and prove two Lemmas.

Lemma 4.1: *If* $x \in X$ *and* $x > 0$ *, then there is* $p \in \Re_+$ *such that*

$$V(x) - px \geq V(y) - py \;\; \text{for all } y \in X \tag{4.7}$$

Proof: If $x \in (0, b]$, and we define p to be the left-hand derivative of V at x, then (4.7) holds. [Note that the left-hand derivative of V is well defined for all $x \in (0, b)$ by concavity of V, and also at $x = b$ by A5]. A5 ensures that p is non-negative.

Lemma 4.2: *Suppose* (x_t) *is an optimal program from* $\overline{x} \in X$*, and suppose also that there is* $(\widehat{x}, \widehat{z})$ *in* Ω *with* $\widehat{z} > 0$*. If for some* $t \in \mathbf{N}$ *there is* $p_t \in \Re_+$ *satisfying*

$$\delta^t V(x_t) - p_t x_t \geq \delta^t V(x) - p_t x \text{ for all } x \in X \tag{4.8}$$

then there is $p_{t+1} \in \Re_+$ *satisfying*

$$\delta^{t+1} V(x_{t+1}) - p_{t+1}x_{t+1} \geq \delta^{t+1} V(x) - p_{t+1}x \;\; \text{for all } x \in X \tag{4.9}$$

and, furthermore,

$$\begin{aligned} \delta^t u(x_t, x_{t+1}) + p_{t+1}x_{t+1} - p_t x_t \;&\geq\; \delta^t u(x, y) \\ + p_{t+1}y - p_t x \text{ for all } (x, y) \;&\in\; \Omega \end{aligned} \tag{4.10}$$

Proof: Using Proposition 3.1, we have $V(x_t) = u(x_t, x_{t+1}) + \delta V(x_{t+1})$, and for all $(x, y) \in \Omega$, $V(x) \geq u(x, y) + \delta V(y)$. Using these in (4.8), we get

$$\begin{aligned}\theta_{t+1} &\equiv \delta^t u(x_t, x_{t+1}) + \delta^{t+1} V(x_{t+1}) - p_t x_t \\ &\geq \delta^t u(x, y) + \delta^{t+1} V(y) - p_t x \text{ for all } (x, y) \in \Omega\end{aligned}$$

Thus, we have

$$\theta_{t+1} - \delta^t u(x, y) + p_t x \geq \delta^{t+1} V(y) \text{ for all } (x, y) \in \Omega \tag{4.11}$$

Define two sets A and B as follows:

$$\begin{aligned}\mathrm{A} &= \{(w, y) \in \Re^2 : (x, y) \in \Omega \text{ for some } x \in X \\ \text{and } w &> \theta_{t+1} - \delta^t u(x, y) + p_t x\}\end{aligned}$$

$$\mathrm{B} = \{(w, y) \in \Re^2 : y \in X \text{ and } w \leq \delta^{t+1} V(y)\}$$

Clearly, A and B are non-empty and convex (since u is concave on Ω and V is concave on X). Also, by (4.11), A and B are disjoint. Hence, by the Minkowski separation theorem, there is (μ, v) in $\Re^2$ with $(\mu, v) \neq 0$ and $\alpha \in \Re$ such that

$$\mu w + v y \geq \alpha \text{ for all } (w, y) \in A \tag{4.12}$$

$$\mu w + v y \leq \alpha \text{ for all } (w, y) \in B \tag{4.13}$$

Using (4.13), we have $\mu \geq 0$. Define $q_{t+1} = (-v)$. Then, using (4.12) and (4.13),

$$\begin{aligned}\mu[\theta_{t+1} - \delta^t u(x, y) + p_t x] - q_{t+1} y &\geq \\ \mu \delta^{t+1} V(y') - q_{t+1} y' \text{ for all } (x, y) &\in \Omega \\ \text{and all } y' &\in X\end{aligned} \tag{4.14}$$

Substituting $x = x_t$ and $y = x_{t+1}$ in (4.14),

$$\begin{aligned}\mu \delta^{t+1} V(x_{t+1}) - q_{t+1} x_{t+1} &\geq \mu \delta^{t+1} V(y') - q_{t+1} y' \\ \text{for all } y' &\in X\end{aligned} \tag{4.15}$$

Substituting $y' = x_{t+1}$ in (4.14),

$$\begin{aligned}\mu[\delta^t u(x_t, x_{t+1}) - p_t x_t] + q_{t+1}x_{t+1} &\geq \\ \mu[\delta^t u(x, y) - p_t x] + q_{t+1}y \text{ for all } (x, y) &\in \Omega \end{aligned} \tag{4.16}$$

We claim now that $\mu \neq 0$. For if $\mu = 0$, then by (4.15), we have $q_{t+1}x_{t+1} \leq q_{t+1}y'$ for all y' in X, while by (4.16), $q_{t+1}x_{t+1} \geq q_{t+1}y$ for all y such that $(x, y) \in \Omega$ for some $x \in X$. Thus

$$q_{t+1}x_{t+1} = q_{t+1}y \text{ for all } y \text{ such that } (x, y) \in \Omega \text{ for some } x \in X \tag{4.17}$$

Since $(\widehat{x}, \widehat{z}) \in \Omega$ with $\widehat{z} > 0$ and $(\widehat{x}, 0) \in \Omega$ by A5, we have $q_{t+1}x_{t+1} = q_{t+1}\widehat{z} = q_{t+1}0 = 0$. Thus, $q_{t+1} = (-v) = 0$, and so $(\mu, v) = 0$, a contradiction. Thus, $\mu \neq 0$, and since $\mu \geq 0$, we have $\mu > 0$. Define $p_{t+1} = (q_{t+1}/\mu)$ and use (4.15), (4.16) to get

$$\begin{aligned}\delta^{t+1}V(x_{t+1}) - p_{t+1}x_{t+1} &\geq \delta^{t+1}V(y') - p_{t+1}y' \\ \text{for all } y' &\in X \end{aligned} \tag{4.18}$$

$$\begin{aligned}\delta^t u(x_t, x_{t+1}) - p_t x_t + p_{t+1}x_{t+1} &\geq \\ \delta^t u(x, y) - p_t x + p_{t+1}y \text{ for all } (x, y) &\in \Omega \end{aligned} \tag{4.19}$$

It remains to show that $p_{t+1} \geq 0$. If $x_{t+1} \in [0, b)$, then by choosing $y' \in X$, $y' > x_{t+1}$ in (4.18), we get $p_{t+1} \geq 0$ by the (weak) monotonicity of V on X (which follows from A5). If $x_{t+1} = b$, then by choosing $(x, y) = (x_t, y)$ with $y \in [0, b)$ in (4.19), we get $p_{t+1} \geq 0$ by A5.

We can now provide the proof of Proposition 4.2.

Proof (of Proposition 4.2): Using Lemma 4.1, there is $p_0 \in \Re_+$ satisfying

$$V(x_0) - p_0 x_0 \geq V(x) - p_0 x \text{ for all } x \in X$$

Using Lemma 4.2, there is a sequence $(p_t)_0^\infty$ with $p_t \in \Re_+$ for $t \in \mathbf{N}$ satisfying (4.4) and (4.5)

Using $x = 0$ in (4.5), we get

$$\delta^t[V(x_t) - V(0)] \geq p_t x_t \text{ for } t \in \mathbf{N}$$

Since $|V(x)|$ is bounded on X by $B/(1-\delta)$, and $\delta^t \to 0$ as $t \to \infty$, and $p_t x_t \geq 0$ for $t \in \mathbf{N}$, we have $p_t x_t \to 0$ as $t \to \infty$, establishing (4.6).

5. Sensitivity Analysis

Sensitivity analysis is concerned with studying the nature of the change in the (optimal) policy function, given a change in the parameters of the dynamic optimization problem.

In the present context, the parameters of the dynamic optimization problem (2.1)-(2.3) are (Ω, u, δ) and the initial state, x.

Our treatment of sensitivity analysis will confine itself to the case where the transition possibility set, Ω (and the state space, X) are fixed. We will then study the effect of changes in (u, δ) and the initial state, x, on the nature of the policy function.

We will be concerned with two types of sensitivity questions:

(i) does h change *continuously* with a change in the parameters?

(ii) does h change *monotonically* with a change in the parameters?

We have already seen (in Section 3) that $h(x)$ changes continuously with respect to changes in the initial state x. Thus, in the present section, we will address question (i) above with respect to variations in (u, δ).

It is difficult to address (even formulate) question (ii) above with respect to variations in the utility function, u. Thus, we will address question (ii) with respect to variations in δ and the initial state, x.

5.1 Continuity

The principal result of this subsection is that the policy function is uniformly continuous on the space of all (u, δ), endowed with the metric of uniform convergence. Our exposition is based on Mitra and Sorger (1999).

To proceed, given the space $\mathcal{M}$ of (Ω, u, δ) satisfying A1-A4, define the metric

$$\begin{aligned}\rho((\Omega, u, \delta), (\Omega, u', \delta')) &= \max\{|u(x,y) - u'(x,y)| \\ &: (x,y) \in \Omega\} + |\delta - \delta'|\end{aligned}$$

Proposition 5.1: *For every positive integer n, let $u_n : \Omega \to \Re$ be a utility function and let δ_n be a discount factor, such that $(\Omega, u_n, \delta_n) \in M$. Further, let $u : \Omega \to \Re$ be a utility function and δ be a discount factor with $(\Omega, u, \delta) \in M$, such that*

$$\rho((\Omega, u_n, \delta_n), (\Omega, u, \delta)) \to 0 \quad as\ n \to \infty$$

Denote by V_n and h_n [V and h] the value and policy functions for (Ω, u_n, δ_n) [(Ω, u, δ)]. Then V_n converges uniformly to V and h_n converges uniformly to h.

Proof: Since u_n converges uniformly to u, it follows that the sequence $(u_n)_{n=1}^{+\infty}$ is uniformly bounded. Therefore we can find real numbers m and M such that $m \leq u(x,y) \leq M$ and $m \leq u_n(x,y) \leq M$ for all $(x,y) \in \Omega$ and all $n \in \{1, 2, ...\}$. Without loss of generality we may assume that $m = 0$.

Step 1: We start by proving that V_n converges uniformly to V. Let $\varepsilon > 0$ be given. There exists $T \geq 1$ such that

$$\frac{M}{1-\delta}\left(\frac{1+\delta}{2}\right)^{T+1} \leq \frac{\varepsilon}{4}.$$

Because of the convergence of the sequences $(\delta_n)_{n=1}^{+\infty}$ and $(u_n)_{n=1}^{+\infty}$ we can find an integer N such that for all $n \geq N$ the following three properties hold:

$$\begin{aligned}\sup\{|u_n(x,y) - u(x,y)| &: (x,y) \in \Omega\} < \varepsilon(1-\delta)/8, \\ \delta_n &\leq (1+\delta)/2, \\ \max\{\left|\delta_n^t - \delta^t\right| &: t \in \{1, 2, ..., T\}\} \leq \varepsilon/(4MT).\end{aligned}$$

Consider an arbitrary state $x \in X$ and fix an arbitrary $n \geq N$. Let $(x_t)_{t=0}^{+\infty}$ be an optimal path from initial state x for (Ω, u, δ).

Note that this path is also feasible for the model (Ω, u_n, δ_n). The following chain of inequalities holds because of the above properties:

$$\begin{aligned}
V_n(x) &\geq \sum_{t=0}^{+\infty} \delta_n^t u_n(x_t, x_{t+1}) \\
&\geq \sum_{t=0}^{T} \delta_n^t u_n(x_t, x_{t+1}) \\
&= \sum_{t=0}^{T} (\delta_n^t - \delta^t) u_n(x_t, x_{t+1}) + \sum_{t=0}^{T} \delta^t u_n(x_t, x_{t+1}) \\
&\geq -(\varepsilon/4) + \sum_{t=0}^{T} \delta^t u_n(x_t, x_{t+1}) \\
&\geq -(\varepsilon/4) + \sum_{t=0}^{T} \delta^t u(x_t, x_{t+1}) - (\varepsilon/8) \\
&> \sum_{t=0}^{+\infty} \delta^t u(x_t, x_{t+1}) - \sum_{t=T+1}^{+\infty} \delta^t u(x_t, x_{t+1}) - (\varepsilon/2) \\
&\geq V(x) - (\varepsilon/4) - (\varepsilon/2) \\
&> V(x) - \varepsilon.
\end{aligned}$$

Using an analogous calculation one can also show that $V(x) > V_n(x) - \varepsilon$. Since both $n \geq N$ and $x \in X$ have been chosen arbitrarily, we have shown that V_n converges uniformly to V.

Step 2: Now we prove that h_n converges uniformly to h. To this end first note that h and h_n are continuous functions for all n and that X is compact. If h_n would not converge uniformly to h then, by Royden (1988, p. 162, Exercise 40e) it would be possible to find $x_0 \in X$, $\theta > 0$, and a sequence $(x_n)_{n=1}^{+\infty}$ such that $\lim_{n \to +\infty} x_n = x_0$ and $|h_n(x_n) - h(x_0)| \geq \theta$ for all n. By compactness of X we may assume without loss of generality that $z_0 = \lim_{n \to +\infty} h_n(x_n)$ exists. Since $z_0 \neq h(x_0)$ one can find $\varepsilon > 0$ such that

$$V(x_0) \geq u(x_0, z_0) + \delta V(z_0) + \varepsilon.$$

Because of uniform convergence of u_n to u and V_n to V, convergence of δ_n to δ, as well as continuity of u and V, one can find an intetger N such that for all $n \geq N$ and all $x \in X$ the following properties hold:

$$\begin{aligned}|V_n(x) - V(x)| &< \varepsilon/8,\\ |V(x_n) - V(x_0)| &< \varepsilon/8,\\ |u(x_n, h_n(x_n)) - u(x_0, z_0)| &< \varepsilon/8,\\ |V(h_n(x_n)) - V(z_0)| &< \varepsilon/8,\\ |\delta_n - \delta| &< \varepsilon(1-\delta)/(8M).\end{aligned}$$

From these conditions we obtain

$$\begin{aligned}V_n(x_n) &= [V_n(x_n) - V(x_n)] + [V(x_n) - V(x_0)] + V(x_0)\\ &\geq -(\varepsilon/8) - (\varepsilon/8) + V(x_0)\\ &\geq u(x_0, z_0) + \delta V(z_0) + (3\varepsilon/4)\\ &\geq u(x_n, h_n(x_n)) + \delta V(h_n(x_n)) + (\varepsilon/2)\\ &= u(x_n, h_n(x_n)) + \delta V(h_n(x_n)) + (\varepsilon/2) + (\delta - \delta_n)\\ &\quad V(h_n(x_n))\\ &\geq u(x_n, h_n(x_n)) + \delta_n V(h_n(x_n)) + (3\varepsilon/8)\\ &\geq u(x_n, h_n(x_n)) + \delta_n V_n(h_n(x_n)) + (\varepsilon/4)\\ &= V_n(x_n) + (\varepsilon/4).\end{aligned}$$

Clearly, this is a contradiction and the result is proved. Q.E.D.

5.2 Monotonicity with Respect to the Initial State

The concept that is crucial to establishing monotonicity properties of the policy funciton is known as *supermodularity*, and was introduced into the optimization theory literature by Topkis (1978).

Let A be a subset of $\Re^2$ and f a function from A to $\Re$. Then f is *supermodular* on A if whenever (a, b) and (a', b') belong to A with $(a', b') \geq (a, b)$, we have

$$f(a, b) + f(a', b') \geq f(a, b') + f(a', b) \tag{5.1}$$

provided (a, b') and (a', b) belong to A.

If A is a rectangular region, then whenever (a, b) and (a', b') belong to A, we have (a, b') and (a', b) also in A. Further, for such a region if f is continuous on A and C^2 on *int* A, then

$$f_{12}(a, b) \geq 0 \text{ for all } (a, b) \in int\ A \tag{5.2}$$

is equivalent to the condition that f is supermodular on A. [For this result, see Ross (1983), Benhabib-Nishimura (1985).]

Let (Ω, u, δ) be a dynamic optimization model. The principal result on the montonicity of its policy function, h, with respect to the initial condition (x), is that if u is supermodular on Ω, then h is monotone non-decreasing in x. [This result is based on the analysis in Topkis (1978), Benhabib and Nishimura (1985).]

Proposition 5.2: *Let (Ω, u, δ) be a dynamic optimization model, with value function V and policy function h. If u is supermodular on Ω, then h is montone non-decreasing on X.*

Proof: Let x, x' belong to X with $x' > x$. Denote $h(x)$ by z and $h(x')$ by z'. We want to show that $z' \geq z$. Suppose, on the contrary that $z' < z$.

Since $(x, z) \in \Omega$ and $z' \in X$ with $z' \leq z$, we have $(x, z') \in \Omega$. Since $(x, z') \in \Omega$ and $x' \in X$ with $x' \geq x$, we have $(x', z) \in \Omega$. Using the definition of supermodularity, and $(x', z) \geq (x, z')$, we have

$$u(x, z') + u(x', z) \geq u(x, z) + u(x', z') \tag{5.3}$$

Since $z = h(x)$ and $z' = h(x')$, we have

$$V(x) = u(x, z) + \delta V(z) \tag{5.4}$$

and

$$V(x') = u(x', z') + \delta V(z') \tag{5.5}$$

Since $(x, z') \in \Omega$ and $(x', z) \in \Omega$, and $z' \neq z$, we have $z' \neq h(x)$, $z \neq h(x')$, so

$$V(x) > u(x, z') + \delta V(z') \tag{5.6}$$

and

$$V(x') > u(x', z) + \delta V(z) \tag{5.7}$$

Adding (5.4) and (5.5),

$$V(x) + V(x') = u(x, z) + u(x'z') + \delta V(z) + \delta V(z') \tag{5.8}$$

Adding (5.6) and (5.7)

$$V(x) + V(x') > u(x, z') + u(x', z) + \delta V(z') + \delta V(z) \tag{5.9}$$

Using (5.8) and (5.9),

$$u(x, z) + u(x', z') > u(x, z') + u(x', z)$$

which contradicts (5.3) and establishes the result. Q.E.D.

Remark 5.1: It follows from the above result that, in the framework of Proposition 5.2, if (x_t) is an optimal program starting from $x \in X$, then either (i) $x_{t+1} \geq x_t$ for all $t \in \mathbf{N}$ or (ii) $x_{t+1} \leq x_t$ for all $t \in \mathbf{N}$.

5.3 Monotonicity with Respect to the Discount Factor

Let (Ω, u) be given with u supermodular on Ω. Then, for every specification of the discount factor, δ, and initial state x, we can solve the dynamic optimization problem (2.1)-(2.3) and obtain the value $V(\delta, x)$ and policy $h(\delta, x)$.

It turns out that (i) V is supermodular as a function of the two variables (δ, x), and (ii) it follows from this fact that h is monotone non-decreasing in δ.

The second step of this two-step result is fairly straightfoward, using arguments similar to those employed in Section 5.2. The first step is more involved because one cannot use the functional equation of dynamic programming directly to obtain this property of V (since V appears on both sides of the equation). Thus, the method employed is an iterative procedure, in which a sequence of functions, V_n, are shown to be supermodular, and to be converging to the value function, V (Proposition 5.3). In order to make this iterative procedure work, we need to make sure that the property of supermodularity of the maximand is preserved by the maximum value in a maximization problem, a general result due to Topkis (1978). We provide a proof of this result as well (see Lemma 5.1 below) in our special context.

Our analysis in this section is based on Dutta (1987) and Amir, Mirman and Perkins (1991), where similar results are obtained in the more special context of the one-sector model of optimal growth (Example 2.1 of Section 2). Our exposition follows Mitra and Nishimura (1999) closely.

Lemma 5.1: *Let (Ω, u) be given, with u supermodular on Ω. Let $G(\delta, x)$ be a supermodular function on $I \times X$, where $I = (0,1)$, which is continuous on $I \times X$ and non-decreasing and concave on X. Let H be a funciton on $I \times X$ defined by*

$$H(\delta, x) = \max_{z \in \Omega_x}[u(x,z) + \delta G(\delta, z)] \tag{5.10}$$

Then H is supermodular and continuous on $I \times X$, and non-decreasing and concave on X.

Proof: Given the information on G, the maximization problem on the right hand side of (5.1) has a solution. Thus H is well-defined. It is clearly non-decreasing on X.

Using the concavity of u on Ω and G on X, H is concave on X. The continuity of H on $I \times X$ can be established by using arguments similar to those employed in the proof of Proposition 3.1.

It remains to check the supermodularity of H on $I \times X$. Let (δ, x) and (δ', x') belong to $I \times X$ with $(\delta', x') \geq (\delta, x)$. Let $\tilde{z}$ [*resp.*$\tilde{z}'$] be a solution to the maximization problem (5.1) corresponding to (δ', x) [*resp.*(δ, x')].

There are two cases to consider: (i) $\tilde{z}' \geq \tilde{z}$ (ii) $\tilde{z} > \tilde{z}'$.

Case (i) $[\tilde{z}' \geq \tilde{z}]$: From the definitions of $\tilde{z}$ and $\tilde{z}'$ it follows that

$$H(\delta, x') = u(x', \tilde{z}') + \delta G(\delta, \tilde{z}') \tag{5.11}$$

and

$$H(\delta', x) = u(x, \tilde{z}) + \delta' G(\delta', \tilde{z}) \tag{5.12}$$

Also, considering the maximization problem (5.10) for (δ, x) and (δ', x') yield:

$$H(\delta, x) \geq u(x, \tilde{z}) + \delta G(\delta, \tilde{z}) \tag{5.13}$$

and

$$H(\delta', x') \geq u(x', \tilde{z}') + \delta' G(\delta', \tilde{z}') \tag{5.14}$$

Since $(\delta', \tilde{z}') \geq (\delta, \tilde{z})$ and G is supermodular, we get

$$G(\delta', \tilde{z}') + G(\delta, \tilde{z}) \geq G(\delta', \tilde{z}) + G(\delta, \tilde{z}')$$

which yields

$$G(\delta', \tilde{z}') - G(\delta', \tilde{z}) \geq G(\delta, \tilde{z}') - G(\delta, \tilde{z}) \geq 0 \tag{5.15}$$

using the fact that G is non-decreasing on X. Thus, we get

$$\begin{aligned} \delta'[G(\delta', \tilde{z}') - G(\delta', \tilde{z})] &\geq \delta'[G(\delta, \tilde{z}') - G(\delta, \tilde{z})] \\ &\geq \delta[G(\delta, \tilde{z}') - G(\delta, \tilde{z})] \end{aligned}$$

which in turn yields

$$\delta' G(\delta', \tilde{z}') + \delta G(\delta, \tilde{z}) \geq \delta' G(\delta', \tilde{z}) + \delta G(\delta, \tilde{z}') \tag{5.16}$$

Adding (5.11) and (5.12), and (5.13) and (5.14), and using (5.16), we get

$$H(\delta, x) + H(\delta', x') \geq H(\delta, x') + H(\delta', x)$$

which shows that H is supermodular.

Case (ii) $[\tilde{z} > \tilde{z}']$: Note that $(x, \tilde{z}) \in \Omega$ and $\tilde{z} > \tilde{z}'$, so that $(x, \tilde{z}') \in \Omega$. Also, $(x, \tilde{z}) \in \Omega$ and $x' \geq x$, so that $(x', \tilde{z}) \in \Omega$. Thus, considering the maximization problem (5.10) for (δ, x) and (δ', x') yield:

$$H(\delta, x) \geq u(x, \tilde{z}') + \delta G(\delta, \tilde{z}') \tag{5.17}$$

$$H(\delta', x') \geq u(x', \tilde{z}) + \delta' G(\delta', \tilde{z}) \tag{5.18}$$

Also, note that (5.11) and (5.12) continue to be valid. Adding (5.11) and (5.12), and (5.17) and (5.18), and using the supermodularity of u, we get

$$H(\delta, x) + H(\delta', x') \geq H(\delta, x') + H(\delta', x)$$

which establishes the supermodularity of H. Q.E.D.

Proposition 5.3: *Let (Ω, u) be given with u supermodular on Ω. Then (i) $V(\delta, x)$ is supermodular on $I \times X$, and (ii) $h(\delta, x)$ is monotone non-decreasing on I.*

Proof: Normalize $u(0,0) = 0$, and define:

$$V_0(\delta, x) = u(x, 0) \tag{5.19}$$

$$V_{n+1}(\delta, x) = \max_{z \in \Omega_x} [u(x, z) + \delta V_n(\delta, z)] \tag{5.20}$$

Note that $V_0(\delta, x)$ is continuous on $I \times X$ and non-decreasing and concave on X. Further, it is (trivially) supermodular. Thus, using Lemma 5.1, (5.20) defines a sequence $\{V_n(\delta, x)\}$ of functions on $I \times X$, such that V_n is continuous and supermodular on $I \times X$, and non-decreasing and concave on X.

Since u is bounded on X, we can find $B > 0$ such that $|u(x, z)| \leq B$ for all $(x, z) \in \Omega$. Then $|V_n(\delta, x)| \leq B/(1 - \delta)$ for each $\delta \in I$, as can be easily verified by induction.

Next, we show that $V_{n+1}(\delta, x) \geq V_n(\delta, x)$ for $n = 0, 1, 2, \ldots$. Notice that $V_1(\delta, x) \geq u(x, 0) + \delta V_0(\delta, 0) = u(x, 0) + \delta u(0, 0) =$

$u(x,0) = V_0(\delta,x)$. Assume now that $V_{n+1}(\delta,x) \geq V_n(\delta,x)$ for $n = 0,...,N$, where $N \geq 0$. We will now show that $V_{N+2}(\delta,x) \geq V_{N+1}(\delta,x)$. Let $\overline{z}$ be the solution to the problem:

$$\max_{z\in\Omega_x}[u(x,z) + \delta V_N(\delta,z)]$$

Then $V_{N+2}(\delta,x) = \max_{z\in\Omega_x}[u(x,z) + \delta V_{N+1}(\delta,z)]$
$\geq u(x,\overline{z}) + \delta V_{N+1}(\delta,\overline{z}) \geq u(x,\overline{z}) + \delta V_N(\delta,\overline{z}) = V_{N+1}(\delta,x)$, by definition of $\overline{z}$.

Fix any δ in I, and consider the sequence of functions $\{V_n(\delta,\bullet)\}_{n=0}^{\infty}$. Define $\overline{V}(\delta,x) = \lim_{n\to\infty} V_n(\delta,x)$ for each $x \in X$, noting that the limit is well defined. Then (using Theorem 7.13, p. 150 of Rudin (1976)) $V_n(\delta,\bullet)$ in fact converges uniformly to $\overline{V}(\delta,\bullet)$, and so $\overline{V}(\delta,\bullet)$ is continuous on X. Also, $\overline{V}(\delta,\bullet)$ is concave on X.

Fixing δ in I, denote by z_n the solution to the maximization problem on the right hand side of (5.20). Then, there is a subsequence of $\{z_n\}$ which converges to some $\overline{z}$. Using the fact that

$$V_{n+1}(\delta,x) = u(x,z_n) + \delta V_n(\delta,z_n)$$

we obtain

$$\overline{V}(\delta,x) \leq u(x,\overline{z}) + \delta\overline{V}(\delta,\overline{z}) \tag{5.21}$$

Also, for every $z \in \Omega_x$, we have $V_{n+1}(\delta,x) \geq u(x,z) + \delta V_n(\delta,z)$, so that

$$\overline{V}(\delta,x) \geq u(x,z) + \delta\overline{V}(\delta,z) \tag{5.22}$$

Using (5.21) and (5.22), we get

$$\overline{V}(\delta,x) = \max_{z\in\Omega_x}[u(x,z) + \delta\overline{V}(\delta,z)]$$

Since $\overline{V}(\delta,\bullet)$ is continuous on X, we must have $\overline{V}(\delta,\bullet) = V(\delta,\bullet)$ by Proposition 3.1.

Since V_n is supermodular on $I \times X$, and $V_n(\delta,x) \to V(\delta,x)$ as $n \to \infty$, for all (δ,x) in $I \times X$, V is supermodular on $I \times X$. This establishes (i).

To establish (ii), let $x \in X$, and δ, $\delta' \in I$ with $\delta' > \delta$. Denote $h(\delta, x)$ by z and $h(\delta', x)$ by z'. We want to show that $z' \geq z$. Suppose, on the contrary, that $z' < z$.

Since $z = h(\delta, x)$ and $z' = h(\delta', x)$, we get

$$V(\delta, x) = u(x, z) + \delta V(\delta, z) \tag{5.23}$$

and

$$V(\delta', x) = u(x, z') + \delta' V(\delta', z') \tag{5.24}$$

Also, we clearly have (since $z \neq z'$),

$$V(\delta, x) > u(x, z') + \delta V(\delta, z') \tag{5.25}$$

and

$$V(\delta', x) > u(x, z) + \delta' V(\delta', z) \tag{5.26}$$

Using (5.23)-(5.26), we have

$$\delta V(\delta, z') + \delta' V(\delta', z) \leq \delta V(\delta, z) + \delta' V(\delta', z')$$

which yields

$$\delta'[V(\delta', z) - V(\delta', z')] \leq \delta[V(\delta, z) - V(\delta, z')] \tag{5.27}$$

Since $z > z'$, $[V(\delta', z) - V(\delta', z')] \geq 0$, so that, using $\delta' > \delta$, we get

$$\delta'[V(\delta', z) - V(\delta', z')] \geq \delta[V(\delta', z) - V(\delta', z')] \tag{5.28}$$

Using (5.27) and (5.28) we get

$$V(\delta', z) - V(\delta', z') \leq V(\delta, z) - V(\delta, z')$$

which yields

$$V(\delta', z) + V(\delta, z') \leq V(\delta', z') + V(\delta, z) \tag{5.29}$$

But since $\delta' > \delta$ and $z > z'$, (5.29) violates the supermodularity of V. Q.E.D.

6. Existence of a Stationary Optimal Stock

The concept of a (non-trivial) stationary optimal stock plays a central role in the theory of optimal intertemporal allocation. Its primary significance derives from the fact that it is the rest point of the dynamical system (X, h), which is a global attractor for optimal programs from all positive initial stocks (usually referred to as the turnpike property) when the discount factor is close to unity.

We will discuss this stability property in Section 9 below. In the present section, we confine ourselves to discussing the existence of a stationary optimal stock. Our exposition is primarily based on Flynn (1980), McKenzie (1982) and Khan and Mitra (1986).

6.1 Existence of a Discounted Golden-Rule Stock

A *discounted golden-rule stock* k is an element of X such that
(i) $(k, k) \in \Omega$
(ii) $u(k, k) \geq u(x, y)$ for all $(x, y) \in \Omega$ such that $\delta y - x \geq \delta k - k$
(iii) $u(k, k) > u(0, 0)$

We establish the existence of a discounted golden-rule stock under a "δ-normality" condition. We say that (Ω, u, δ) is *δ-normal* if there exists $(\overline{x}, \overline{y}) \in \Omega$ such that $\delta\overline{y} \geq \overline{x}$ and $u(\overline{x}, \overline{y}) > u(0, 0)$.

Proposition 6.1: *If (Ω, u, δ) satisfies A1-A3 and A5, and is δ-normal, then there exists a discounted golden-rule stock.*

Proof: For z in X, define $\phi(z) = \{(x, y) \in \Omega : \delta y - x \geq \delta z - z\}$ and $\psi(z) = \{(x, y) \in \Omega : u(x, y) \geq u(x', y')$ for all $(x', y') \in \phi(z)\}$.

Note that for each $z \in X$, the set $\phi(z)$ is non-empty (since $(0, 0) \in \Omega$), compact (since Ω is closed) and convex (since Ω is convex). Thus, for each $z \in X$, the set $\psi(z)$ is non-empty (since u is upper semicontinuous on Ω) and convex (since u is concave).

Next, we show the upper hemicontinuity of ψ. Let z^* be an arbitrary point of X. Consider a sequence $z^n \in X$, with $z^n \to z^*$ as $n \to \infty$. Let $(x^n, y^n) \in \psi(z^n)$, and $(x^n, y^n) \to (\widehat{x}, \widehat{y})$. We want to show that $(\widehat{x}, \widehat{y}) \in \psi(z^*)$. Since Ω is closed, $(\widehat{x}, \widehat{y}) \in \phi(z^*)$. Suppose $(\widehat{x}, \widehat{y})$ is not in $\psi(z^*)$. Then there is some $(x^*, y^*) \in \psi(z^*)$ and $\varepsilon > 0$ such that $u(x^*, y^*) \geq u(\widehat{x}, \widehat{y}) + \varepsilon$.

Now, since u is an upper semicontinuous function, $\limsup_{n\to\infty} u(x^n, y^n) \leq u(\widehat{x}, \widehat{y})$. Thus, there is N_1 such that for $n \geq N_1$, $u(x^n, y^n) \leq u(\widehat{x}, \widehat{y}) + \varepsilon/3$. Consequently, for $n \geq N_1$,

$$u(x^*, y^*) \geq u(x^n, y^n) + 2\varepsilon/3 \tag{6.1}$$

Choose $0 < \lambda < 1$ such that $(1-\lambda)[u(0,0) - u(x^*, y^*)] \geq -\varepsilon/3$. We claim that there is an N_2 such that for $n \geq N_2$, $(\lambda x^*, \lambda y^*) \in \phi(z^n)$. To see this, observe that $(0,0) \in \Omega$ and convexity of Ω imply that $(\lambda x^*, \lambda y^*) \in \phi(\lambda z^*)$. Since $z^n \to z^*$, there is N_2 such that for $n > N_2$, $z^n \geq \lambda z^*$. Thus $\delta\lambda y^* - \lambda x^* \geq (\delta - 1)\lambda z^* \geq (\delta - 1)z^n$, establishing our claim.

Since $(x^n, y^n) \in \psi(z^n)$, we have for $n \geq N_2$,

$$\begin{aligned} u(x^n, y^n) &\geq u(\lambda x^*, \lambda y^*) \geq \lambda u(x^*, y^*) + (1-\lambda)u(0,0) \\ &= u(x^*, y^*) + (1-\lambda)[u(0,0) - u(x^*, y^*)] \\ &\geq u(x^*, y^*) - \varepsilon/3. \end{aligned}$$

Using this in (6.1) for $n \geq Max(N_1, N_2)$,

$$u(x^*, y^*) \geq u(x^n, y^n) + 2\varepsilon/3 \geq u(x^*, y^*) + \varepsilon/3,$$

which leads to a contradiction and completes the demonstration that ψ is upper hemicontinuous.

Define for $z \in X$, $Q(z) = \{x \in X : (x, y) \in \psi(z)\}$. We will show that this correspondence Q satisfies all the requirements of Kakutani's fixed-point theorem.

Clearly, Q is a non-empty, convex-valued correspondence from X to the subsets of X. We check that Q is upper hemicontinuous. To see this, take an arbitrary $z^* \in X$. Let $z^n \in X$, with $z^n \to z^*$ as $n \to \infty$. Let $x^n \in Q(z^n)$, and $x^n \to \widehat{x}$ as $n \to \infty$. We have to show that $\widehat{x} \in Q(z^*)$. Since $x^n \in Q(z^n)$, there is y^n such that $(x^n, y^n) \in \psi(z^n)$. This means $(x^n, y^n) \in \phi(z^n)$, and we can pick a subsequence $(x^{n'}, y^{n'})$ tending to $(\widehat{x}, \widehat{y}) \in \phi(z^*)$. Since ψ is upper hemicontinuous, $(\widehat{x}, \widehat{y}) \in \psi(z^*)$ and so $\widehat{x} \in Q(z^*)$.

Thus, all the conditions of Kakutani's fixed point theorem are fulfilled, and there exists $x^o \in Q(x^o)$. This means there is some y^o such that $(x^o, y^o) \in \psi(x^o)$. That is,

$$u(x^o, y^o) \geq u(x, y) \text{ for all } (x, y) \in \phi(x^o).$$

But $(x^o, y^o) \in \phi(x^o)$ implies $x^o \leq y^o$, and we obtain from A5 that $(x^o, x^o) \in \Omega$ and $u(x^o, x^o) \geq u(x^o, y^o) \geq u(x, y)$ for all $(x, y) \in \Omega$ with $\delta y - x \geq \delta x^o - x^o$. Given δ-normality, there is $(x', y') \in \phi(x^o)$ such that $u(x', y') > u(0, 0)$. Thus $u(x^o, x^o) > u(0, 0)$, and hence x^o is a discounted golden-rule stock.

6.2 Existence of a Non-Trivial Stationary Optimal Stock

An optimal program $(x(t))_0^\infty$ from k is a *stationary optimal program* if $x(t) = k$ for $t \in \mathbf{N}$. A *stationary optimal stock* k is an element of X such that $(k)_0^\infty$ is a stationary optimal program. It is called *non-trivial* of $u(k, k) > u(0, 0)$.

Using Proposition 6.1, we can establish the existence of a non-trivial stationary optimal stock.

Proposition 6.2: *If* (Ω, u, δ) *satisfies A1-A3 and A5, and is* δ*-normal, then there exists a non-trivial stationary optimal stock.*

Proof: By Proposition 6.1, there is a discounted golden-rule stock, k. Now, let $(x_t)_0^\infty$ be any program from k. We will show that

$$\sum_{t=0}^{\infty} \delta^t u(k, k) \geq \sum_{t=0}^{\infty} \delta^t u(x_t, x_{t+1}) \tag{6.2}$$

Let $X_T = \sum_{t=0}^{T-1} (1-\delta)\delta^t x_t/(1-\delta^T)$ and $Y_T = \sum_{t=0}^{T-1} (1-\delta)\delta^t x_{t+1}/(1-\delta^T)$ for $T = 1, 2, 3, \ldots$. Given convexity of Ω, we have (X_T, Y_T) in Ω for $T = 1, 2, 3, \ldots$. Since $x_t \in X$ for all $t \in \mathbf{N}$, the pair $(\overline{x}, \overline{y}) \equiv \lim_{T \to \infty} (X_T, Y_T)$ is well-defined and in Ω.

Now, by concavity of u, and the fact that $0 < \delta < 1$, Jensen's inequality yields $u(\overline{x}, \overline{y}) \geq \sum_{t=0}^{\infty} (1 - \delta)\delta^t u(x_t, x_{t+1})$. But $(\overline{x} - \delta\overline{y}) = (1 - \delta) \left[\sum_{t=0}^{\infty} \delta^t x_t - \sum_{t=0}^{\infty} \delta^{t+1} x_{t+1} \right] = (1 - \delta)k$. Since k is a discounted golden-rul stock, $u(k,k) \geq u(\overline{x},\overline{y})$, which in turns implies that

$$\sum_{t=0}^{\infty} \delta^t u(k,k) \geq \sum_{t=0}^{\infty} \delta^t u(\overline{x},\overline{y}) = u(\overline{x},\overline{y})/(1-\delta) \geq \sum_{t=0}^{\infty} \delta^t u(x_t, x_{t+1})$$

which establishes (6.2). Thus, k is a stationary optimal stock. Since k is a discounted golden-rule stock, we have $u(k,k) > u(0,0)$, and so k is a non-trivial stationary optimal stock.

6.3 The Existence of a Modified Golden-Rule

A *modified golden-rule* is a pair (k,p) such that
(i) $(k,k) \in \Omega$, $p \in \Re_+$
(ii) $u(x,y) + \delta py - px \leq u(k,k) + \delta pk - pk$ for all $(x,y) \in \Omega$
Given Proposition 6.1, we can establish the existence of a modified golden-rule by providing a price support to the discounted golden-rule stock.

Proposition 6.3: *If* (Ω, u, δ) *satisfies A1-A3 and A5, and is* δ*-normal, then there exists a modified golden-rule.*

Proof: By Proposition 6.1, there is a discounted golden-rule stock, k.

Define two sets A and B as follows:

$A = \{(a,b) \in \Re^2 : u(x,y) - u(k,k) \geq a,\ (\delta y - x) - (\delta k - k) \geq b$ for some $(x,y) \in \Omega\}$

$B = \{(a,b) \in \Re^2 : a > 0 \text{ and } b > 0\}$

Then A and B are non-empty, convex sets. By the definition of a discounted golden-rule stock, A and B are disjoint. Then, by a separation theorem (see Nikaido (1968), Theorem 3.5, p. 35), there is $(\mu, v) \in \Re^2_+$, $(\mu, v) \neq 0$ such that

$$\mu a + vb \leq 0 \text{ for all } (a,b) \in A$$

We claim that $\mu \neq 0$. For if $\mu = 0$, then we have $vb \leq 0$ for all $(a,b) \in A$. Define $a = u(0,0) - u(k,k)$ and $b = (k - \delta k)$. Since $(0,0) \in \Omega$, $(a,b) \in A$, and so $v(k - \delta k) \leq 0$. Since $u(k,k) > u(0,0)$, we have $k > 0$ and so $(k - \delta k) > 0$. Thus, $v \leq 0$, which means $v = 0$, contradicting the fact that $(\mu, v) \neq 0$. Thus, $\mu > 0$, and so

$$a + (v/\mu)b \leq 0 \text{ for all } (a,b) \in A.$$

Defining $p = (v/\mu)$, we have

$$u(x,y) + \delta py - px \leq u(k,k) + \delta pk - pk \text{ for all } (x,y) \in \Omega$$

Thus, (k, p) is a modified golden-rule.

7. Smooth Preferences

So far, we have developed the theory of dynamic optimization without assumptions on the smoothness of the reduced form utility function.

However, significant progress in the theory can be made (beyond the material covered in Sections 2-6) if one assumes that the utility function is smooth in the interior of the transition possibility set.

We will concentrate on two sets of ideas that emerge from this theory. First, if the utility function, u, is C^1 in the interior of Ω, then an interior optimal program satisfies the Ramsey-Euler equations. Further, the value function is continuously differentiable in the interior of the state space, and the derivative of the value function is the derivative of the utility function (with respect to its first argument) when evaluated at the optimal point (by the envelope theorem).

Going back to the material on duality theory covered in Section 5, we now see that the shadow prices supporting an optimal program are given exactly by the derivatives of the value function (at the appropriate optimal states). This leads to a restatement of the characterization of optimal programs in terms of the Ramsey-Euler equations and a transversality condition that is quite useful. In particular, any interior stationary (and, in fact, any periodic) optimal program can be characterized solely in terms of the Ramsey-Euler equations.

The second set of ideas is developed under the hypotheses that the utility function, u, is C^2 in the interior of Ω, and optimal programs are in the interior of Ω (when the initial state is in the interior of X). Now, the Ramsey-Euler equations involve continuously differentiable functions which can be linearized around an interior stationary optimal stock to study the local stability of optimal programs in terms of the characteristic roots associated with the Ramsey-Euler equation at the stationary optimal stock.

The interesting feature that emerges is that the dynamic behavior of optimal programs in governed by the smaller (in absolute value) of the characteristic roots associated with the Ramsey-Euler equation. This feature is made precise in the result that (i) the policy function is differentiable at the stationary optimal stock, and (ii) the derivative of the policy function at the stationary optimal stock is equal to the smaller (in absolute value) of the characteristic roots associated with the Ramsey-Euler equation at the stationary optimal stock.

7.1 Ramsey-Euler Equations and the Differentiability of the Value Function

Ramsey-Euler equations can be derived as first-order necessary conditions of the optimization problem (2.1)-(2.3). Thus, if (x_t^*) solves the optimization problem (2.1)-(2.3), then for each $t \in \mathbf{N}$, x_{t+1}^* must solve the problem:

$$(Q)\begin{cases} \text{Max } u(x_t^*, y) + \delta u(y, x_{t+2}^*) \\ \text{subject to } y \in \Omega_{x_t^*}, x_{t+2}^* \in \Omega_y \end{cases}$$

Introduce now the following smoothness assumption on u:

A6: u is continuously differentiable in the interior of Ω.

If the optimal program is in the interior of Ω for all t (that is, if $(x_t^*, x_{t+1}^*) \in \text{int } \Omega$ for all $t \in \mathbf{N}$), then the first-order conditions for (Q) are:

$$u_2(x_t^*, x_{t+1}^*) + \delta u_1(x_{t+1}^*, x_{t+2}^*) = 0 \text{ for } t \in \mathbf{N} \tag{7.1}$$

which are the Ramsey-Euler equations.

We summarize this finding in the following result.

Proposition 7.1: *Suppose the policy function, h, satisfies $(x, h(x)) \in$ int Ω for all $x \in$ int X, then*

$$u_2(x, h(x)) + \delta u_1(h(x), h^2(x)) = 0$$

for all $x \in$ int X.

If an optimal program (x_t) is interior (that is, $(x_t, x_{t+1}) \in \text{int } \Omega$ for all $t \in \mathbf{N}$), then the shadow prices (p_t) obtained in Proposition 4.2 must satisfy (using (4.4))

$$\delta^t u_1(x_t, x_{t+1}) = p_t$$

If V were differentiable, then we would also have from (4.5)

$$\delta^t V'(x_t) = p_t$$

Thus, when V is differentiable, we should have (for $x \in \text{int } X$),

$$V'(x) = u_1(x, h(x))$$

We now show that (under A7), V is continuously differentiable in int X, and the above formula, relating the derivatives of the value and utility functions, holds. This result is due to Benveniste and Scheinkman (1979).

Proposition 7.2: *Suppose the policy function, h, satisfies $(x, h(x)) \in$ int Ω for all $x \in$ int X. Then V is continuously differentiable on int X, and*

$$V'(x) = u_1(x, h(x)) \tag{7.2}$$

for all $x \in$ int X.

Proof: Pick any $x^0 \in \text{int } X$, and note that by assumption $(x^0, h(x^0)) \in \text{int } \Omega$. Then there is a neighborhood N of x^0 such that $(x, h(x^0)) \in \text{int } \Omega$ for all $x \in N$. Define W on N by

$$W(x) = u(x, h(x^0)) + \delta V(h(x^0)) \tag{7.3}$$

Then W is concave and continuously differentiable on N. Further, $W(x^0) = V(x^0)$, and for all $x \in N$, $W(x) = u(x, h(x^0)) + \delta V(h(x^0)) \leq V(x)$. Let p be any subgradient of V at x^0. Then

$$p(x - x^0) \geq V(x) - V(x^0) \geq W(x) - W(x^0)$$

Since W is concave and continuously differentiable on N, $p = W'(x^0)$. Thus, V has a unique subgradient at x^0, so it must be differentiable at x^0, with $p = V'(x^0)$. Since $x^0 \in \text{int } X$ was arbitrary, V is differentiable on int X. Since V is concave, V is in fact, continuously differentiable on int X.

Using (7.3) to evaluate the derivative of W at x^0, we get

$$W'(x^0) = u_1(x^0, h(x^0)) \tag{7.4}$$

Since $W'(x^0) = p = V'(x^0)$, we get

$$V'(x^0) = u_1(x^0, h(x^0)) \tag{7.5}$$

Since $x^0 \in \text{int } X$ was arbitrary, this establishes (7.2). Q.E.D.

Given the above results, it is now possible to restate the results on the price characterization of optimal programs somewhat differently as follows.

Corollary 7.1: *If $(x_t)_0^\infty$ is a program from $x \in X$, with $(x_t, x_{t+1}) \in$ int Ω for $t \in \mathbf{N}$, and the following conditions hold:*

$$\text{(i) } u_2(x_t, x_{t+1}) + \delta u_1(x_{t+1}, x_{t+2}) = 0 \tag{7.6}$$

$$\text{(ii) } \lim_{t \to \infty} \delta^t u_1(x_t, x_{t+1}) x_t = 0 \tag{7.7}$$

then $(x_t)_0^\infty$ is an optimal program from x.

Proof: Define $p_t = \delta^t u_1(x_t, x_{t+1})$ for $t \in \mathbf{N}$. Note that $p_t \in \Re_+$ for $t \in \mathbf{N}$ by A5. Then for all $(x, z) \in \Omega$ and $t \in \mathbf{N}$,

$$\begin{aligned} u(x, z) - u(x_t, x_{t+1}) \;\; &\leq \;\; u_1(x_t, x_{t+1})(x - x_t) \\ &\quad + u_2(x_t, x_{t+1})(z - x_{t+1}) \end{aligned}$$

Multiplying through by δ^t, and using (7.6) and the definition of p_t,

$$\begin{aligned} \delta^t[u(x,z) - u(x_t, x_{t+1})] \;\leq\; & p_t(x - x_t) \\ & - p_{t+1}(z - x_{t+1}) \end{aligned} \tag{7.8}$$

Transposing terms in (7.8) yields (4.1). Also, (7.7) and the definition of p_t yields (4.2). Now the result follows from Proposition 4.1.

Remark 7.1: (i) It is worth noting that unlike Proposition 4.1, the sufficiency result of Corollary 7.1 does exploit the convex structure of the dynamic optimization model.

(ii) It follows from Corollary 7.1 that if x is a state satisfying $(x,x) \in \text{int } \Omega$, and $u_2(x,x) + \delta u_1(x,x) = 0$,
then X is a stationary optimal stock, since the transversality condition (7.7) is automatically satisfied.

(iii) Generalizing the result in remark (ii) above, if $(x_0, x_1, ..., x_N)$ (with $N \geq 1$) is a vector such that $(x_t, x_{t+1}) \in \text{int } \Omega$, and
$u_2(x_t, x_{t+1}) + \delta u_1(x_{t+1}, x_{t+2}) = 0$ for $t = 0, ..., N-2$ (if $N \geq 2$), $u_2(x_{N-1}, x_N) + \delta u_1(x_N, x_0) = 0$, then $(x_0, x_1, ..., x_N)$ is a periodic optimal program.

A converse of Corollary 7.1 can be established, using Proposition 4.2.

Corollary 7.2: *Suppose $(x_t)_0^\infty$ is an optimal program from $x \in X$, such that $(x_t, x_{t+1}) \in$ int Ω for $t \in \mathbf{N}$. Then the following conditions hold:*

(i) $u_2(x_t, x_{t+1}) + \delta u_1(x_{t+1}, x_{t+2}) = 0$ *for* $t \in \mathbf{N}$

(ii) $\lim\limits_{t \to \infty} \delta^t u_1(x_t, x_{t+1}) x_t = 0$

(iii) $V'(x_t) = u_1(x_t, x_{t+1})$ *for* $t \in \mathbf{N}$

Proof: Using (4.4) and the fact that (x_t, x_{t+1}) is in int Ω for $t \in \mathbf{N}$, we have

$$\delta^t u_1(x_t, x_{t+1}) = p_t \tag{7.9}$$

$$\delta^t u_2(x_t, x_{t+1}) = -p_{t+1} \tag{7.10}$$

This yields (i). Using (4.6), (ii) follows from (7.9).

By the proof of Proposition 7.2, V is differentiable at x_t for each $t \in \mathbf{N}$. So, using (4.5),

$$\delta^t V'(x_t) = p_t \tag{7.11}$$

Thus, (iii) follows from (7.9) and (7.11).

7.2 Characteristic Roots of the Ramsey-Euler Equation and Differentiability of the Policy Function

The study of the differentiability of the policy function is one of the hard topics of dynamic optimization theory. For the present exposition, we will confine our attention to the differentiability of the policy function at a stationary optimal stock, which turns out to be the most useful aspect of this theory for applications.

The differentiability property of the policy function at an SOS can be used to establish that the derivative of the policy function equals the smaller (in absolute value) characteristic root associated with the Ramsey-Euler equation, at the SOS. (See Santos (1991) for the most general result).

We will approach the topic in reverse order. That is, we will establish a relation betwenn the Dini derivatives of the policy function and the characteristic roots associated with the Ramsey-Euler equation at the SOS. This will then be used to establish simultaneously the differentiability of the policy function at the SOS, and the equality of the derivative of the policy function to the smaller (in absolute value) characteristic root associated with the Ramsey-Euler equation. This approach is based on Mitra and Nishimura (1999).

We proceed with our analysis under the following condition:

C1: There is a stationary optimal stock, $k_\delta \in X$, such that $(k_\delta, k_\delta) \in \text{int } \Omega$, $u_{11}(k_\delta, k_\delta) < 0$, $u_{22}(k_\delta, k_\delta) < 0$ and $0 < (1 + \delta)|u_{12}(k_\delta, k_\delta)| < [-u_{22}(k_\delta, k_\delta)] + \delta[-u_{11}(k_\delta, k_\delta)]$.

We will carry out our analysis for the case $u_{12}(k_\delta, k_\delta) > 0$, noting that the analysis for the case $u_{12}(k_\delta, k_\delta) < 0$ is analogous (and is worked out explicitly in Mitra and Nishimura (1999)).

We write k for k_δ to simplify the notation. Since $(k, k) \in$ int Ω, we can find an open interval $N_1(k)$ containing k, such that $N_1(k)^2$ is in the interior of Ω, and for all (x, z) in $N_1(k)^2$, we have $u_{12}(x, z) > 0$. Now, we can find an open interval $N(k)$, containng k, with $N(k) \subset N_1(k)$, such that for all $x \in N(k)$, $h(x)$ and $h^2(x)$ belong to $N_1(k)$.

Then, for all x, x' in $N(k)$, with $x' > x$, we must have $h(x') > h(x)$. To see this, note that we have

$$u_2(x', h(x')) = -\delta V'(h(x'))$$

$$u_2(x, h(x)) = -\delta V'(h(x))$$

Thus, by the Mean-Value Theorem,

$$\begin{aligned} & \widetilde{u}_{21}(x' - x) + \widetilde{u}_{22}(h(x') - h(x)) \\ = \; & \delta[V'(h(x)) - V'(h(x'))] \end{aligned}$$

where $\widetilde{u}_{21}$ and $\widetilde{u}_{22}$ are evaluated at a point (y, z) in between $(x, h(x))$ and $(x'(h(x'))$, as given by the Mean-Value Theorem. If $h(x') \leq h(x)$, then $V'(h(x')) \geq V'(h(x))$, so that

$$\widetilde{u}_{21}(x' - x) + (-\widetilde{u}_{22})(h(x) - h(x')) \leq 0$$

But since $\widetilde{u}_{21} > 0$, $(-\widetilde{u}_{22}) > 0$, we have a contradiction.

We break up our analysis into several steps.

Step 1: We can write the characteristic equation [associated with the Ramsey-Euler equation at $x = k_\delta$] as follows:

$$u_{12}(k, k) + [u_{22}(k, k) + \delta u_{11}(k, k)]\lambda + \delta u_{12}(k, k)\lambda^2 = 0 \quad (7.12)$$

Denoting the roots of the characteristic equation by λ_1 and λ_2 (where λ_1 has the least absolute value), we can infer that k is "saddle-point stable"; that is

$$0 < \lambda_1 < 1 < \lambda_2 \quad (7.13)$$

To see this, observe first that by using C1, λ_1 and λ_2 are real, and

$$\lambda_1\lambda_2 = (1/\delta); (\lambda_1 + \lambda_2) = -\{[u_{22}(k,k) + \delta u_{11}(k,k)]/\delta u_{12}(k,k)\} \tag{7.14}$$

Thus, λ_1 and λ_2 are both positive, with $(\lambda_2)^2 \geq \lambda_1\lambda_2 = (1/\delta)$, so that $\lambda_2 > 1$.

Using (7.14) and C1, we obtain $(1-\lambda_1)(1-\lambda_2) = 1 - \lambda_1 - \lambda_2 + \lambda_1\lambda_2 = [1/\delta u_{12}(k,k)][(1+\delta)u_{12}(k,k) + u_{22}(k,k) + \delta u_{11}(k,k)] < 0$. Since $\lambda_2 > 1$, we must have $\lambda_1 < 1$. This verifies (7.13).

Step 2: If $x \in N(k)$, $x \neq \widehat{k}$, then

$$\widetilde{u}_{21} + [(-\widetilde{u}_{22}) + \delta(-\overline{u}_{11})]\{[h(x) - h(k)]/(k - x)\}$$

$$+\delta\overline{u}_{12}\{[h^2(k) - h^2(x)]/(k-x)\} = 0 \tag{7.15}$$

where $\widetilde{u}_{21}$ and $\widetilde{u}_{22}$ are evaluated at an appropriate convex combination of $(x, h(x))$ and (k,k), and $\overline{u}_{11}$ and $\overline{u}_{12}$ are evaluated at an appropriate convex combination of $(h(x), h^2(x))$ and (k,k) as given by the Mean-Value Theorem.

To see this, write the Ramsey-Euler equations:

$$\begin{aligned} u_2(k,k) + \delta u_1(k,k) &= 0 \\ u_2(x, h(x)) + \delta u_1(h(x), h^2(x)) &= 0 \end{aligned}$$

Use the Mean-Value Theorem to get

$$\widetilde{u}_{21}(k-x) + (-\widetilde{u}_{22})(h(x) - k) + \delta(-\overline{u}_{11})(h(x) - k)$$

$$+\delta\overline{u}_{12}[k - h^2(x)] = 0$$

Dividing by $(k - x) \neq 0$, we obtain (7.15).

Step 3: For $x \in N(k), n \neq k$, define

$$m = \min\left[\liminf_{x\to k^+}\left|\frac{h(x)-h(k)}{x-k}\right|, \liminf_{x\to k^-}\left|\frac{h(x)-h(k)}{x-k}\right|\right]$$

$$M = \max\left[\limsup_{x\to k^+}\left|\frac{h(x)-h(k)}{x-k}\right|, \limsup_{x\to k^-}\left|\frac{h(x)-h(k)}{x-k}\right|\right]$$

We claim that

(i) M is either λ_1 or λ_2; (ii)m is either λ_1 or λ_2.

We will establish only (i), since the proof of (ii) is similar. Let us define:

$$\begin{aligned} g(\lambda) &= u_{21}(k,k) + u_{22}(k,k)\lambda + \delta u_{11}(k,k)\lambda \\ &\quad + \delta u_{12}(k,k)\lambda^2 \end{aligned}$$

Claim 1: $g(M) > 0$ is not possible. For if $g(M) > 0$, then

$$u_{21}(k,k) + [u_{22}(k,k) + \delta u_{11}(k,k)]M + \delta u_{12}(k,k)M^2 > 0$$

We can choose $\varepsilon > 0$ such that

$$\begin{aligned} u_{21}(k,k) &> [(-u_{22}(k,k)) + \delta(-u_{11}(k,k))] \\ &\quad (M+\varepsilon) - \delta u_{12}(k,k)(M^2 - \varepsilon^2) \end{aligned}$$

Given $\varepsilon > 0$, one can find $z^s \to k(z^s \neq k)$ such that

$$[h(k) - h(z^s)]/(k - z^s) \geq (M - \varepsilon) \tag{7.17}$$

One can then find S' and x^s such that $h(x^s) = z^s$ for $s \geq S'$. Then $x^s \to k(x^s \neq k)$ as $s \to \infty$, and one can find $S \geq S'$, such that for $s \geq S$,

$$[h(x^s) - h(k)]/(x^s - k) \leq (M + \varepsilon) \tag{7.18}$$

Using (7.15) with $x = x^s$, (7.17) and (7.18), we have

$$\begin{aligned} \widetilde{u}_{21} &= \left[(-\widetilde{u}_{22}) + \delta(-\overline{u}_{11}) - \delta\overline{u}_{12}\left(\frac{h(z^s)-h(k)}{z^s-k}\right)\right] \\ &\quad \left[\frac{h(x^s)-h(k)}{x^s-x}\right] \\ &\leq [(-\widetilde{u}_{22}) + \delta(-\overline{u}_{11}) - \delta\overline{u}_{12}(M-\varepsilon)]\left[\frac{h(x^s)-h(k)}{x^s-k}\right] \\ &\leq [(-\widetilde{u}_{22}) + \delta(-\overline{u}_{11}) - \delta\overline{u}_{12}(M-\varepsilon)](M+\varepsilon) \end{aligned}$$

Letting $x^s \to k$, we get

$$\begin{aligned} u_{21}(k,k) &\leq [(-u_{22}(k,k)) + \delta(-u_{11}(k,k)) - \delta u_{12}(k,k)(M-\varepsilon)] \\ (M+\varepsilon) &< u_{21}(k,k) \end{aligned}$$

a contradiction, which establishes the claim.

Claim 2: $g(M) < 0$ is not possible. Otherwise,

$$u_{21}(k,k) + [u_{22}(k,k) + \delta u_{11}(k,k)]M + \delta u_{12}(k,k)M^2 < 0$$

We can choose $\varepsilon > 0$ such that

$$\begin{aligned} u_{21}(k,k) &< [(-u_{22}(k,k)) + \delta(-u_{11}(k,k))](M-\varepsilon) - \delta u_{12}(k,k) \\ &\quad (M^2 - \varepsilon^2) \end{aligned}$$

One can find $x^s \to k(x^s \neq k)$ such that

$$[h(k) - h(x^s)]/(k - x^s) \geq (M - \varepsilon) \tag{7.19}$$

Define $z^s = h(x^s)$. Then $z^s \to k(z^s \neq k)$ and we can find S, such that for $s \geq S$,

$$[h(k) - h(z^s)]/(k - z^s) \leq (M + \varepsilon) \tag{7.20}$$

Using (7.15) with $x = x^s$, (7.19) and (7.20), we have

$$\begin{aligned} \widetilde{u}_{21} &= \left[(-\widetilde{u}_{22}) + \delta(-\overline{u}_{11}) - \delta\overline{u}_{12}\left[\frac{h(k) - h(z^s)}{k - z^s}\right]\right] \\ &\quad \left[\frac{h(x^s) - h(k)}{x^s - k}\right] \\ &\geq [(-\widetilde{u}_{22}) + \delta(-\overline{u}_{11}) - \delta\overline{u}_{12}(M+\varepsilon)](M-\varepsilon) \end{aligned}$$

Letting $x^s \to k$, we get

$$\begin{aligned} u_{21}(k,k) &\geq [(-u_{22}(k,k) + \delta(-u_{11}(k,k))](M-\varepsilon) \\ &\quad -\delta u_{12}(k,k)(M^2 - \varepsilon^2) \\ &> u_{21}(k,k), \end{aligned}$$

a contradiction. From Claims 1 and 2, we get $g(M) = 0$, so $M = \lambda_1$ or λ_2.

Step 4: Let $x \in N(k)$, with $x \neq k$. To be precise, let $x < k$, so that $h(x) < k$ and $h^2(x) < k$. [The case $x > k$ can be handled similarly.] Then $V'(x) = u_1(x, h(x))$ and $\delta V'(h(x)) = \delta u_1(h(x), h^2(x)) = -u_2(x, h(x))$. Similarly $V'(k) = u_1(k,k)$ and $\delta V'(k) = \delta u_1(k,k) = -u_2(k,k)$. Thus, using concavity of u, we obtain the following two inequalities:

$$\begin{aligned} u(x,h(x)) + \delta V'(h(x))h(x) - V'(x)x &\geq u(k,k) + \delta V'(h(x))k \\ &\quad -V'(x)k \\ u(k,k) + \delta V'(k)k - V'(k)k &\geq u(x,h(x)) + \delta V'(k)h(x) \\ &\quad -V'(k)x \end{aligned}$$

Adding the inequalities and transposing terms

$$\delta[V'(h(x)) - V'(k)][k - h(x)] \leq [V'(x) - V'(k)][k - x]$$

Iterating on this relationship, we get

$$\delta[V'(h(x)) - V'(k)][k - h^2(x)] \leq [V'(x) - V'(k)][k - x]$$

This yields the inequality

$$\delta^2[k - h^2(x)]/[k - x] \leq [V'(x) - V'(k)]/[V'(h^2(x) - V'(k)] \tag{7.21}$$

We claim now that

$$\delta^2[k - h^2(x)]/[k - x] \leq 1 \tag{7.22}$$

For, if (7.22) were violated, we would have $[k - h^2(x)] > [k - x]/\delta^2 > [k - x]$. Thus, we must have $h^2(x) < x < k$, and $V'(h^2(x)) \geq V'(x) \geq V'(k)$, so that $[V'(x) - V'(k)]/[V'(h^2(x)) - V'(k)] \leq 1$. But then by using (7.21), (7.22) must hold, a contradiction. This establishes (7.22).

Step 5: By the analysis in Step 3, we know that M is either λ_1 or λ_2 and m is either λ_1 or λ_2. We now show that $M = m = \lambda_1$.

By (7.15), we have

$$\widetilde{u}_{21} + [(-\widetilde{u}_{22}) + \delta(-\overline{u}_{11})]\{[h(x) - h(k)]/(k - x)\} + \delta\overline{u}_{12}\{[h^2(k) - h^2(x)]/(k - x)\} = 0$$

Using (7.22) we get

$$[(-\widetilde{u}_{22}) + \delta(-\overline{u}_{11})]\{[h(x) - h(k)]/(x - k)\} \leq \widetilde{u}_{21} + [\overline{u}_{12}/\delta]$$

Thus, we obtain (by letting $x \to k$),

$$|Dh(k)| \leq (1 + \delta)[+u_{12}(k,k)]/\delta[(-u_{22}(k,k)) + \delta(-u_{11}(k,k))]$$

for every Dini-derivative at k. Using C1 we get $|Dh(k)| < (1/\delta) < \lambda_2$. Thus, $M < \lambda_2$ and consequently $M = \lambda_1 < 1$. Since m is either λ_1 or λ_2, and $m \leq M$, $m = \lambda_1$, also.

The above analysis implies that h is differentiable at k, and $h'(k) = \lambda_1$.

We summarize our finding in the following result.

Proposition 7.3: *Under C1, the policy function h is differentiable at k, and*

$$|h'(k)| = |\lambda_1|$$

where λ_1 *is the characteristic root associated with the Ramsey-Euler equation at* (k,k), *with the smaller absolute value.*

8. Uniqueness of the Stationary Optimal Stock

In general, a non-trivial stationary optimal stock need not be unique. However, one can provide a useful sufficient condition on the reduced-form utility function under which a uniqueness result can be obtained, independent of the discount factor. These two topics are discussed in Sections 8.1 and 8.2.

An especially useful theory of uniqueness can be developed when (Ω, u) are given, and the discount factor is restricted to take values close to unity and we examine below the elements of this theory in some detail in Section 8.3.

8.1 An Example of Non-Uniqueness

We examine a situation in the context of Example 3 of Section 2 by specifying (w, f, δ), for which we can verify the existence of two non-trivial stationary optimal stocks.

Let $X = [0, 4]$, and the function $f : X \to X$ be specified by

$$f(x) = 2x^{1/2} \text{ for all } x \in X$$

We show now how two stocks can be shown to be stationary optimal stocks by appropriate choice of (w, δ).

Let us pick $x' = (1/4)$ and $x'' = (1/9)$. Then $f(x') = 2(x')^{1/2} = 1$ and $f(x'') = 2(x'')^{1/2} = 2/3$. Thus, consumption levels associated with stationary programs $(x')_0^\infty$ and $(x'')_0^\infty$ are $c' = (3/4)$ and $c'' = (5/9)$.

In order that these stocks be stationary optimal stocks, it is enough to check that the Ramsey-Euler equations hold for our choice of (w, δ). To make computations easier, we take w to be of the additive separable form:

$$w(c, x) = g(c) + h(x)$$

Then the Ramsey-Euler equations would hold at (x', c') and (x'', c'') if

$$\begin{aligned} g'(c') &= \delta[g'(c')f'(x') + h'(x')] \\ g'(c'') &= \delta[g'(c'')f'(x'') + h'(x'')] \end{aligned}$$

These can be rewritten as:

$$\begin{aligned} g'(c')[1 - \delta f'(x')] &= \delta h'(x') \\ g'(c'')[1 - \delta f'(x'')] &= \delta h'(x'') \end{aligned}$$

Since $f'(x) = [1/x^{1/2}]$, we have $f'(x') = 2$ and $f'(x'') = 3$. Let us choose $\delta = (1/6)$. Then $\delta f'(x') = 1/3$ and $\delta f'(x'') = 1/2$, so that $[1-\delta f'(x')] = (2/3)$ and $[1-\delta f'(x'')] = (1/2)$. Thus, in order for the Ramsey-Euler equations to hold, all we need to ensure is that

$$\begin{aligned} g'(c') &= (3/2)(1/6)h'(x') \\ g'(c'') &= 2(1/6)h'(x'') \end{aligned}$$

We can clearly simplify our calculations by taking h to be linear; so, choose $h(x) = 12x$ for $x \in X$. Then, our restrictions on g are given by

$$\begin{aligned} g'(c') &= 3 \\ g'(c'') &= 4 \end{aligned}$$

So, all one needs is to provide an increasing, concave function g on X with these two-point restrictions. Clearly, a two-parameter function will suffice for our purpose, and we specify g to be the quadratic map:

$$g(c) = ac - bc^2 \text{ for } c \in X$$

Then we need (a, b) to satisfy

$$\begin{aligned} a - 2bc' &= a - 2b(3/4) = 3 \\ a - 2bc'' &= a - 2b(5/9) = 4 \end{aligned}$$

This yields $a = (453/144)$ and $b = (7/72)$. It can be checked that g is strictly increasing and strictly concave on X.

To summarize, with $f(x) = 2x^{1/2}$ for $x \in X$, $w(c, z) = (ac - bc^2) + 12x$ for $(c, x) \in X^2$ (where $a = (453/144)$ and $b = (7/72)$), and $\delta = (1/6)$, both $x' = (1/4)$ and $x'' = (1/9)$ are non-trivial stationary optimal stocks.

8.2 A Sufficient Condition for Uniqueness

Let us assume that we are given a dynamic optimization model (Ω, u, δ) for which u is twice contnuously differentiable on int Ω, with $u_{11}(x, x) < 0$ and $u_{22}(x, x) \leq 0$ for all $(x, x) \in$ int Ω.

If x' and x'' are two distinct stationary optimal stocks, with $x'' > x'$, such that $(x', x') \in \text{int } \Omega$ and $(x'', x'') \in \text{int } \Omega$, then the Ramsey-Euler equations yield:

$$u_2(x', x') + \delta u_1(x', x') = 0 \tag{8.1}$$

$$u_2(x'', x'') + \delta u_1(x'', x'') = 0 \tag{8.2}$$

Using the Mean-Value theorem, we can find z and y in (x', x''), such that

$$u_2(x', x') - u_2(x'', x'') = [u_{21}(z, z) + u_{22}(z, z)](x' - x'') \tag{8.3}$$

$$\delta u_1(x', x') - \delta u_1(x'', x'') = [\delta u_{11}(y, y) + \delta u_{12}(y, y)](x' - x'') \tag{8.4}$$

Then $(y, y) \in \text{int } \Omega$ and $(z, z) \in \text{int } \Omega$, so $u_{11}(y, y) < 0$ and $u_{22}(z, z) < 0$.

Let us assume that the following condition holds:

$$u_{12}(x, x) \leq 0 \text{ for all } (x, x) \in \text{ int } \Omega \tag{8.5}$$

Then adding (8.3) and (8.4), we would get

$$\begin{aligned} 0 = (x'' - x')[(-u_{21}(z, z)) + \\ (-u_{22}(z, z)) + \delta(-u_{11}(y, y)) + \\ \delta(-u_{12}(y, y))] > 0 \end{aligned}$$

a contradiction. Thus, (8.5) is sufficient to ensure uniqueness.

Remark 8.1: (i) In Example 4 of Section 2, we have $u_{12}(x, z) < 0$ for all x, z in $(0, 1) \times (0, 1)$. Thus, (8.5) is satisfied, and a stationary optimal stock must be unique.

(ii) If $u_{12}(x, z) < 0$ for (x, z) in int Ω, then we know that the policy function, h, is decreasing in a neighborhood of x^o, if $(x^o, h(x^o)) \in \text{int } \Omega$. This rules out multiple stationary optimal stocks, x, with (x, x) in the interior of Ω. However, (8.5) is a weaker condition ensuring uniqueness.

(iii) The uniqueness of a non-trivial stationary optimal stock in the one-sector model of optimal growth (Example 1) can be viewed as a special case of the above theory, even though, defining $u(x,z) = w(f(x)-z)$, we have $u_{12}(x,z) = [-w''(f(x)-z)]f'(x) > 0$.

To see this, note that if there are two non-trivial stationary optimal stocks, x', x'', then x' must solve

$$\left.\begin{array}{c}\text{Max } w(f(x)-z)\\ \text{subject to } \delta f(x)-z \geq \delta f(x')-x'\\ (x,z)\in\Omega\end{array}\right\}(P')$$

and x'' must solve

$$\left.\begin{array}{c}\text{Max } w(f(x)-z)\\ \text{subject to } \delta f(x)-z \geq \delta f(x'')-x''\\ (x,z)\in\Omega\end{array}\right\}(P'')$$

But with w strictly increasing (P') and (P'') are the same as:

$$\left.\begin{array}{c}\text{Max } f(x)-z\\ \text{subject to } \delta f(x)-z \geq \delta f(x')-x'\\ (x,z)\in\Omega\end{array}\right\}(P_1')$$

and

$$\left.\begin{array}{c}\text{Max } f(x)-z\\ \text{subject to } \delta f(x)-z \geq \delta f(x'')-x''\\ (x,z)\in\Omega\end{array}\right\}(P_1'')$$

Thus, x' solves (P_1') and x'' solves (P_1''). Then, defining $\mathcal{U}(x,z) = f(x)-z$, x' and x'' are non-trivial stationary optimal stocks for the model $(\Omega, \mathcal{U}, \delta)$. But since $\mathcal{U}_{12} = 0$, $\mathcal{U}$ satisfies (8.5) and with $f'' > 0$ on *int* $X, \mathcal{U}_{11} < 0$ and $\mathcal{U}_{22} = 0$, so we must have uniqueness of the stationary optimal stock; that is, x' must be equal to x''.

8.3 Uniqueness of the Stationary Optimal Stock for High Discount Factors

Let (Ω, u) be given satisfying A1, A2, and A5. Consider the maximization problem:

$$\left.\begin{array}{c}\text{Max } u(x,z)\\ \text{subject to } z-x\geq 0\\ (x,z)\in\Omega\end{array}\right\}(P)$$

Clearly, there is a solution $(\overline{x}, \overline{z})$ to (P), and so $(\overline{x}, \overline{x})$ solves (P) as well. We refer to $\overline{x}$ as a *golden-rule stock.*

Assume, now, that the following condition holds:

C2: (i) $(\overline{x}, \overline{x}) \in \text{int } \Omega$

(ii) u is twice continuously differentiable on int Ω, with the Hessian of u negative definite at $(\overline{x}, \overline{x})$.

Note that there is a neighborhood N (in *int* Ω) containing $(\overline{x}, \overline{x})$, such that the Hessian of u is negative definite on N.

It follows that $(\overline{x}, \overline{x})$ is the only solution to (P). For if (x, z) were any solution to (P), then so is (x, x), and by convexity of Ω and concavity of u, so is $(\lambda x + (1 - \lambda)\overline{x},\ \lambda x + (1 - \lambda)\overline{x})$ for all $0 \leq \lambda \leq 1$. For $0 < \lambda < 1$ and sufficiently close to zero, this is in N, but since u is strictly concave on N, we must have $x = \overline{x}$. Similarly, using the fact that (z, z) solves (P), we can infer that $z = \overline{x}$. Thus $(x, z) = (\overline{x}, \overline{x})$.

We can now find $\varepsilon_1 > 0$ such that $(x, x) \in \text{int } \Omega$ for all $x \in [\overline{x} - \varepsilon_1, \overline{x} + \varepsilon_1] \equiv I_1 \subset N$.

Since the Hessian of u is negative definite on N, we have

$$\alpha = -[u_{21}(\overline{x}, \overline{x}) + u_{22}(\overline{x}, \overline{x}) + u_{11}(\overline{x}, \overline{x}) + u_{12}(\overline{x}, \overline{x})] > 0$$

Thus, there is $0 < \varepsilon_2 < \varepsilon_1$ such that for all x in $I_2 \equiv [\overline{x} - \varepsilon_2, \overline{x} + \varepsilon_2]$, we have

$$-[u_{21}(x, x) + u_{22}(x, x) + u_{11}(x, x) + u_{12}(x, x)] > (\alpha/2)$$

We can now find $0 < \rho_1 < 1$ such that for all $\rho_1 < \rho < 1$ and all $x \in I_2$,

$$-[u_{21}(x, x) + u_{22}(x, x) + \rho u_{11}(x, x) + \rho u_{12}(x, x)] > (\alpha/4)$$

We can find $0 < \beta < [u(\overline{x}, \overline{x}) - u(0,0)]$ such that whenever $(x, x) \in \Omega$ and $|x - \overline{x}| \geq \varepsilon_2$, we have $u(x, x) \leq u(\overline{x}, \overline{x}) - \beta$. Now, pick $0 < \varepsilon_3 < \varepsilon_2$ such that $(\overline{x}, \overline{x} + \varepsilon_3) \in \text{int } \Omega$ and $u(\overline{x}, \overline{x} + \varepsilon_3) > u(\overline{x}, \overline{x}) - \beta$. Next, find $\rho_1 < \rho_2 < 1$ such that for $\rho_2 < \rho < 1$, we have $\rho(\overline{x} + \varepsilon_3) - \overline{x} > 0$.

Now, choose the discount factor, δ, in $(\rho_2, 1)$. Since $(\overline{x}, \overline{x} + \varepsilon_3) \in \Omega$, with $\delta(\overline{x} + \varepsilon_3) > \overline{x}$, we have $(\overline{x} + \varepsilon_3) > \overline{x}/\delta$ and so $(\overline{x}, \overline{x}/\delta) \in \Omega$, and since $u(\overline{x}, \overline{x} + \varepsilon_3) > u(\overline{x}, \overline{x}) - \beta > u(0,0)$, we have $u(\overline{x}, \overline{x}/\delta) > u(0,0)$ and so (Ω, u, δ) is δ-normal. Thus, there exists a discounted golden-rule stock. This is also a non-trivial stationary optimal stock.

We show now that there can be at most one non-trivial stationary stock (SOS), k. We note that if k is a non-trivial SOS, then $(k, k) \in \text{int } \Omega$. To see this note that since $(\overline{x}, \overline{x}) \in \text{int } \Omega$, all $0 < x \leq \overline{x}$ must satisfy $(x, x) \in \text{int } \Omega$ as well. Thus, if k' is a non-trivial SOS such that (k', k') is not in the interior of Ω, then $k' > \overline{x}$. But the program $(k', \overline{x}, \overline{x}, ...)$ from k clearly has a higher discounted sum of utilities than $(k', k', k', ...)$, so k' cannot be an SOS.

We know that any SOS, k, with $(k, k) \in \text{int } \Omega$ is a discounted golden-rule stock, and so (k, k) must solve the problem:

$$\begin{aligned} &\text{Max } u(x, z) \\ &\text{subject to } \delta z - x \geq \delta k - k \\ &(x, z) \in \Omega \end{aligned}$$

Since $\delta(\overline{x} + \varepsilon_3) - \overline{x} > 0 \geq \delta k - k$, we must have $u(k, k) \geq u(\overline{x}, \overline{x} + \varepsilon_3) > u(\overline{x}, \overline{x}) - \beta$. It follows then that $|k - \overline{x}| < \varepsilon_3$. Since $0 < \varepsilon_3 < \varepsilon_2$, $k \in I_2$, and since $1 > \delta > \rho_2$, we have $1 > \delta > \rho_1$, and so

$$-[u_{21}(k, k) + u_{22}(k, k) + \delta u_{11}(k, k) + \delta u_{12}(k, k)] > (\alpha/4)$$

Define $I_3 = [\overline{x} - \varepsilon_3, \overline{x} + \varepsilon_3]$ and $G : I_2 \to \Re$ by

$$G(x) = u_2(x, x) + \delta u_1(x, x)$$

Then $G(k) = 0$ and $G'(k) < 0$.

Let z be any point in I_2 such that $G(z) = 0$. Then $(z, z) \in \text{int } \Omega$ and (z, z) satisfies the Ramsey-Euler equation and is a stationary optimal stock. Thus $z \in I_3$ and $G'(z) < 0$ by the above argument.

It suffices then to show that there can be at most one zero of G in I_2. To this end, define $F : I_2 \to \Re$ by:

$$F(z) = \int_{\overline{x}-\varepsilon_2}^{z} G(x)dx$$

Then F is continuous on I_2 and twice continuously differentiable on int I_2 with $F'(z) = G(z)$ and $F''(z) = G'(z)$ for all $z \in \text{int } I_2$. Further F is quasi-concave on I_2. Otherwise there would exist x, x' in I_2 with $x < x'$, and $0 < \lambda < 1$, such that $F[\lambda x + (1 - \lambda)x'] < \min[F(x), F(x')]$. Then F attains an interior minimum at some z in $[x, x']$, so that $F'(z) = 0$ and $F''(z) \geq 0$. But then $z \in \text{int } I_2$ and $G(z) = 0$ and $G'(z) \geq 0$, a contradiction.

Suppose a and b are distinct zeroes of G in I_2. Then a, $b \in I_3$ and $G(z) = G(b) = 0$ and $G'(a) < 0$ and $G'(b) < 0$. Thus, $F'(a) = F'(b) = 0$, and $F''(a) < 0$ and $F''(b) < 0$. Since F is C^2 on int I_2, we can find a neighborhood $N(a)$ of a, such that for all $x \in N(a)$, $F''(x) < 0$. Then one can find $0 < \lambda < 1$ such that $[\lambda a + (1-\lambda)b] \in N(a)$, and by the strict concavity of F on $N(a)$, we must have $F[\lambda a + (1 - \lambda)b] - F(a) < F'(a)[\lambda a + (1 - \lambda)b - a] = 0$, so that $F[\lambda a + (1 - \lambda)b] < F(a)$. By the quasi-concavity of F, we have $F[\lambda a + (1 - \lambda)b] > \min[F(a), F(b)]$. Thus $F(a) > F(b)$. Exchanging the roles of a and b in the above argument, $F(b) > F(a)$, a contradiction. Thus, there is at most one zero of G in I_2.

9. Global Asymptotic Stability of the Stationary Optimal Stock

Even when there is a unique non-trivial stationary optimal stock, it need not be globally asymptotically stable; that is, optimal programs from other inital stocks need not converge to the *SOS* over time. This topic is discussed in Section 9.1.

Parallel to the theory of uniqueness developed in Section 8.3, when (Ω, u) are given and the discount factor, δ, is restricted to take values close to unity, then a theory of global asymptotic stability of the (unique) non-trivial stationary optimal stock can be obtained. This topic (developed in Section 9.2), which falls under the general subject matter of "turnpike theory", is one of the major developments in dynamic optimization theory. It relies on practically all the material developed in the previous sections, and the argument is subtle. In particular, it is extremely important to obtain the relevant bound on the discount factor in terms of conditions (which can be checked easily) placed solely on (Ω, u).

Our approach to this topic is deliberately one-dimensional, although it uses many ideas which are used in the literature on turnpike theory in the multidimensional setting. We show that for discount factors close to unity, there are no period-two optimal programs, and global asymptotic stability of the non-trivial SOS is a consequence of this fact (in our one-dimensional state variable setting).

This has two advantages. First, it clarifies the relation between turnpike theory and the presence/absence of period-two optimal cycles, an idea which is especially useful as a backdrop in the study of periodic and chaotic optimal programs. Second, the problem of ruling out period-two optimal cycles (when the discount factor is close to unity) is simply an extension of the problem of ruling out multiple stationary optimal stocks. Thus, the idea that is exploited is that a stronger bound on the discount factor than was used to generate uniqueness of the non-trivial stationary optimal stock also ensures global asymptotic stability of this SOS.

9.1 A Counterexample to the Turnpike Property

We examine a situation in the context of Example 2.4, with $\alpha = \beta = 0.5$, the case examined by Weitzman. Thus $X = [0, 1]$, $\Omega = X^2$ and $u : \Omega \to \Re$ is given by

$$u(x, z) = x^{1/2}(1 - z)^{1/2} \tag{9.1}$$

Given any $x \in (0, 1)$, we can show that there is $z \in (0, 1)$ such that $(x, z, x, z, ...)$ is an optimal program from x. To see this simply consider the Ramsey-Euler equation

$$u_2(x,z) + \delta u_1(z,x) = 0 \tag{9.2}$$

and note that (9.1) yields

$$\frac{z}{(1-z)} = \delta^2 \frac{(1-x)}{x} \tag{9.3}$$

Denoting by y the right hand side of (9.3), we note that $y > 0$ and

$$z = y/(1+y) \tag{9.4}$$

leading to a unique choice of z in $(0,1)$, given x in $(0,1)$, to (9.2). Further, with $x \in (0,1)$ and with z defined by (9.4), that is,

$$z = \delta^2(1-x)/[x + \delta^2(1-x)] \tag{9.5}$$

we can also check that

$$u_2(z,x) + \delta u_1(x,z) = 0 \tag{9.6}$$

To see this, use (9.1) in (9.6) to get

$$\frac{z}{(1-z)} = \delta^2 \frac{(1-x)}{x}$$

which is exactly the same equation as (9.3).

Thus, for $x \in (0,1)$, and z defined by (9.5), we have (9.2) and (9.6) holding. Thus $(x, z, x, z, ...)$ is the optimal program from x.

Further, $k = \delta/(1+\delta)$ is the unique stationary optimal stock. And, for $x \in (0,k)$, $z = \delta^2(1-x)/[x+\delta^2(1-x)] > k$, while for $x \in (k,1)$, $z < k$. Thus, for every $x \in (0,1)$, $x \neq k$, there is a period two optimal program. There is, thus, no tendency for optimal programs from $x \neq k$ to converge to the stationary optimal stock, k.

Remark 9.1: The above analysis also demonstrates that the (optimal) policy function, h, for the dynamic optimization model (Ω, u, δ) can be solved explicitly. In fact it is given by (9.5).

9.2 A Turnpike Theorem for High Discount Factors

The example (due to Weitzman and discussed in Samuelson (1973)) presented in the previous subsection has the feature that there are period-two optimal cycles from every initial state (in the interior of the state space) other than the (unique) stationary optimal state, *independent of how high the discount factor is.*

While this destroys the hope of proving a general turnpike theorem for high discount factors, it was observed by Samuelson (1973) that if the utility function has a negative definite Hessian at the golden-rule, then for high discount factors, optimal cycles would be ruled out, and stability (at least of the local kind) of the stationary optimal stock would be ensured.

This observation led to a large literature dealing with the turnpike theorem for high discount factors in a variety of settings. Our objective is to carry through a complete analysis of Samuelson's original idea, to establish an appropriate turnpike theorem. That is, a basic ingredient of our approach is that in the situation described by Samuelson, optimal cycles can be ruled out, and this, in turn, ensures global asymptotic stability of the stationary optimal stock.

We briefly describe the basic steps of our analysis before presenting the material formally. Following Samuelson, we assume that the utility function has a negative definite Hessian at the golden-rule. That is, we maintain condition C2 (used in Section 8.3 to ensure uniqueness of the SOS at high discount factors). As a first step, we choose an appropriately small neightborhoood (N) of the golden-rule, on which the utility function has a negative definite Hessian. Anticipating the proof a bit, this neighborhood is so chosen that when the discount factor (to be chosen later) is high, there are no period-two optimal cycles confined to this neighborhood. The idea here basically mimics the analysis of uniqueness of the SOS for high discount factors, with the single Ramsey-Euler equation (characterizing an interior SOS) replaced by a pair of Ramsey-Euler equations (characterizing an interior period-two cycle).

As a second step, we put bounds on the discount factor, so that the following hold:

(i) there is a unique non-zero SOS, k_δ, in the state space X (which can be ensured, by following the analysis in Section 8.3, to be in the neighborhood N).

(ii) if there is a period-two optimal cycle, it must belong to the neighborhood, N.

(iii) condition C1 of Section 7.2 holds ensuring that the policy function, h_δ, satisfies $|h'_\delta(k_\delta)| < 1$.

(iv) if the initial stock $x \in (0, b]$, $(x, 0, 0, ...)$ is not optimal starting from x.

Findings (i) and (ii) of the second step can be used with the first step to ensure there are no period-two optimal cycles when the discount factor is high. Then, a well-known result from one-dimensional dynamics ensures us that if (x_t) is the optimal program from $x \in X$, then x_t converges to one of the fixed points of h_δ.

Now (iii) and (iv) can be used to ensure that x_t cannot converge to zero. Thus, x_t must converge to the unique non-zero SOS, k_δ, and the turnpike theorem is established.

We now turn to the formal details of turnpike theory. As mentioned above, we maintain condition C2, so that $(\overline{x}, \overline{x})$ is the unique golden-rule (solution to problem (P) in Section 8.3).

Step 1: Using the fact that the Hessian of u is negative definite at $(\overline{x}, \overline{x})$, we can find $\alpha > 0$ such that

$$(-u_{11}(\overline{x}, \overline{x})) + (-u_{22}(\overline{x}, \overline{x})) > \alpha + 2|u_{12}(\overline{x}, \overline{x})| \tag{9.7}$$

Given that $(\overline{x}, \overline{x}) \in \text{int}\ \Omega$, we can find $0 < \varepsilon_1 < \overline{x}$, such that $N_1(\overline{x}) \equiv [\overline{x} - \varepsilon_1, \overline{x} + \varepsilon_1]$ is in the interior of X, and if $(x, z) \in N_1(\overline{x})^2$, then $(x, z) \in \text{int}\ \Omega$, and the Hessian of u is negative definite on $N_1(\overline{x})^2$.

Next, given (9.7), we can find $0 < \varepsilon_2 < \varepsilon_1$, such that $N_2(\overline{x}) \equiv (\overline{x} - \varepsilon_2, \overline{x} + \varepsilon_2) \subset N_1(\overline{x})$, and for all $(x, z) \in N_2(\overline{x})^2$,

$$\det \begin{bmatrix} u_{21}(x, z) + u_{12}(z, x) & u_{22}(x, z) + u_{11}(z, x) \\ u_{22}(x, z) + u_{11}(x, z) & u_{21}(z, x) + u_{12}(x, z) \end{bmatrix} < -(\alpha/2) \tag{9.8}$$

We claim now that we can find $\widehat{\beta} > 0$, such that if $((x, z), (z, x)) \in \Omega^2$ and (x, z) is not in $N_2(\overline{x})^2$, then

$$u(x, z) + u(z, x) < 2u(\overline{x}, \overline{x}) - \widehat{\beta} \tag{9.9}$$

Otherwise, we can find a convergent sequence $(x^s, z^s)_{s=1}^{\infty}$ converging to (x^0, z^0), with $((x^s, z^s), (z^s, x^s))$ in Ω^2 for $s = 1, 2, 3, ...$, and

$$u(x^s, z^s) + u(z^s, x^s) \geq 2u(\overline{x}, \overline{x}) - (1/s) \tag{9.10}$$

Since $(\overline{x}, \overline{x})$ belongs to the interior of $N_2(\overline{x})^2$, we can find $0 < \lambda < 1$ such that $(\widetilde{x}^s, \widetilde{z}^s)$ and $(\widetilde{z}^s, \widetilde{x}^s)$ belongs to $N_3(\overline{x})^2$ for $s = 1, 2, ...$, where $N_3(\overline{x}) = [\overline{x} - 0.5\varepsilon_2, \overline{x} + 0.5\varepsilon_2]$, and

$$(\widetilde{x}^s, \widetilde{z}^s) = \lambda(x^s, z^s) + (1 - \lambda)(\overline{x}, \overline{x}) \tag{9.11}$$

Clearly $(\widetilde{x}^s, \widetilde{z}^s)$ converges to $(\widetilde{x}^0, \widetilde{z}^0) \equiv [\lambda(x^0, z^0) + (1 - \lambda)(\overline{x}, \overline{x})]$, and $(\widetilde{x}^0, \widetilde{z}^0) \in N_3(\overline{x})^2$.

Using (9.10) and (9.11), we have for $s = 1, 2, 3, ...$

$$u(\widetilde{x}^s, \widetilde{z}^s) + u(\widetilde{z}^s, \widetilde{x}^s) \geq 2u(\overline{x}, \overline{x}) - (\lambda/s) \tag{9.12}$$

Letting $s \longrightarrow \infty$, we get

$$u(\widetilde{x}^0, \widetilde{z}^0) + u(\widetilde{z}^0, \widetilde{x}^0) \geq 2(\overline{x}, \overline{x}) \tag{9.13}$$

Note that $(\widetilde{x}^0, \widetilde{z}^0) \neq (\overline{x}, \overline{x})$. Thus, if $\widetilde{x}^0 = \widetilde{z}^0$, (9.13) contradicts the uniqueness of the golden-rule. And if $\widetilde{x}^0 \neq \widetilde{z}^0$, then using the fact that $(0.5\widetilde{x}^0 + 0.5\widetilde{z}^0, 0.5\widetilde{z}^0 + 0.5\widetilde{x}^0)$ is in $N_2(\overline{x})^2$ and u is strictly concave on $N_2(\overline{x})^2$, we get from (9.13)

$$\begin{aligned} u(0.5\widetilde{x}^0 + 0.5\widetilde{z}^0, 0.5\widetilde{z}^0 + 0.5\widetilde{x}^0) &> \\ 0.5u(\widetilde{x}^0, \widetilde{z}^0) + 0.5u(\widetilde{z}^0, \widetilde{x}^0) &\geq u(\overline{x}, \overline{x}) \end{aligned}$$

which clearly contradicts the definition of the golden-rule. This establishes (9.9).

Define $\beta = \min[\widehat{\beta}, u(\overline{x}, \overline{x}) - u(0, 0)]$, and note that $\beta > 0$. Pick $0 < \varepsilon_3 < \varepsilon_2$ such that

$$u(\overline{x}, \overline{x} + \varepsilon_3) > u(\overline{x}, \overline{x}) - (\beta/8). \tag{9.14}$$

Step 2: We now put suitable bounds on the range of the discount factor.

First, using (9.8), we can find $0 < \rho_1 < 1$ such that for all $\rho_1 < \rho < 1$ and all $(x, z) \in N_2(\overline{x})^2$,

$$\det \begin{bmatrix} u_{21}(x,z) + \rho u_{12}(z,x) & u_{22}(x,z) + \rho u_{11}(z,x) \\ u_{22}(z,x) + \rho u_{11}(x,z) & u_{21}(z,x) + \rho u_{12}(x,z) \end{bmatrix} < -(\alpha/4) \tag{9.15}$$

Second, given (9.9) and the definition of β, we can find $\rho_1 < \rho_2 < 1$ such that for all $((x, z), (z, x)) \in \Omega^2$ with (x, z) not in $N_2(\overline{x})^2$, and all $\rho_2 < \rho < 1$, we have

$$u(x,z) + \rho u(z,x) < u(\overline{x},\overline{x}) + \rho u(\overline{x},\overline{x}) - (\beta/2). \tag{9.16}$$

Now, given (9.14), we have for $\rho_2 < \rho < 1$, $(\beta/8) + \rho u(\overline{x}, \overline{x} + \varepsilon_3) > \rho(\beta/8) + \rho u(\overline{x}, \overline{x} + \varepsilon_3) > \rho u(\overline{x}, \overline{x})$. Therefore, for all $((x, z), (z, x)) \in \Omega^2$ with (x, z) not in $N_2(\overline{x})^2$, and all $\rho_2 < \rho < 1$,

$$\begin{aligned} u(x,z) + \rho u(z,x) &< u(\overline{x}, \overline{x} + \varepsilon_3) + (\beta/8) \\ &\quad + \rho u(\overline{x}, \overline{x} + \varepsilon_3) + (\beta/8) - (\beta/2) \end{aligned}$$

so that

$$u(x,z) + \rho(z,x) < u(\overline{x}, \overline{x} + \varepsilon_3) + \rho u(\overline{x}, \overline{x} + \varepsilon_3) - (\beta/4) \tag{9.17}$$

Finally, we pick $\rho_2 < \rho_3 < 1$, such that for all $\rho_3 \leq \rho < 1$, we have

$$\rho \frac{[u(\overline{x},\overline{x}) - u(0,0)}{(1-\rho)} > [u(b,0) - u(\overline{x} - \varepsilon_2, \overline{x})] \tag{9.18}$$

and

$$\rho^2(\overline{x} + \varepsilon_3) > \overline{x} \tag{9.19}$$

Proposition 9.1: *Let (Ω, u) be given satisfying condition C2. If the discount factor δ satisfies $\rho_3 \leq \delta < 1$, an optimal program from $x \in X$ cannot exhibit a period-two cycle.*

Proof: Given (Ω, u), let ρ be a number in $[\rho_3, 1)$. Suppose for (Ω, u, ρ), there is an optimal program from $x' \in X$, such that it exhibits a period-two cycle. Then, there is $z' \in X$, $z' \neq x'$, such that $(x', z', x', z', ...)$ is optimal from x'. Without loss of generality, we can take $x' > z'$. Then $x' > 0$ and $(z', x') \in \Omega$ with $x' > 0$, so we can find $p \in \Re_+$ such that

$$\begin{aligned} & u(x', z') + \rho u(z', x') + p^2 px' - px' \\ \geq \; & u(x'', z'') + \rho u(z'', y'') + \rho^2 py' - px' \end{aligned} \tag{9.20}$$

for all $((x'', z'')(z'', y'')) \in \Omega^2$.

Choosing $(x'', z'') = (\overline{x}, \overline{x})$ and $(z'', y'') = (\overline{x}, \overline{x} + \varepsilon_3)$ in (9.20), we get

$$u(x', z') + \rho u(z', x') \geq u(\overline{x}, \overline{x}) + \rho u(\overline{x}, \overline{x} + \varepsilon_3) \tag{9.21}$$

since $[\rho^2 px' - px'] \leq 0$, and $\rho^2 p((\overline{x} + \varepsilon_3) - p\overline{x} \geq 0$ by (9.19) and the fact that $\rho_3 \leq \rho < 1$. Using (9.21) and (9.17) and $u(\overline{x}, \overline{x}) \geq u(\overline{x}, \overline{x} + \varepsilon_3)$, we see that $(x', z') \in N_2(\overline{x})^2$.

Using (9.14), (9.19) and the definition of β, (Ω, u, ρ) is ρ-normal for $\rho_3 \leq \rho < 1$, so there exists a non-trivial SOS, k_ρ. By the analysis of Section 8.3, k_ρ is the unique non-trivial SOS, and indeed k_ρ is the only non-zero SOS. Further, $k_\rho \in N_2(\overline{x})$.

Let $N(\overline{x}) \equiv [\overline{x} - \varepsilon_2, \overline{x} + \varepsilon_2]$ and define the function, $H : N(\overline{x})^2 \times [0, 1] \to \Re$ by

$$H(x, z; \lambda) = \begin{bmatrix} u_2(x, z) + (\lambda \rho_3 + (1 - \lambda)) u_1(z, x) \\ u_2(z, x) + (\lambda \rho_3 + (1 - \lambda)) u_1(x, z) \end{bmatrix} \tag{9.22}$$

We will show that (i) degree $H(N_2(\overline{x})^2; 0) = -1$, and (ii) degree $H(N_2(\overline{x})^2; \lambda) =$ degree $H(N_2(\overline{x})^2; 0)$ for all λ in $(0, 1]$.

Notice that (9.22) yields:

$$H(x, z; 0) = \begin{bmatrix} u_2(x, z) + u_1(z, x) \\ u_2(z, x) + u_1(x, z) \end{bmatrix} \tag{9.23}$$

Clearly since the golden-rule $(\overline{x}, \overline{x})$ solves problem (P) of Section 8.3, and $(\overline{x}, \overline{x}) \in \text{int } N(\overline{x})^2$, we have $H(\overline{x}, \overline{x}; 0) = 0$. Also, if $(x, z) \in$

$N(\overline{x})^2$ with $(x,z) \neq (\overline{x},\overline{x})$ satisfies $H(x,z;0)=0$, then

$$\begin{aligned} & u(\overline{x},\overline{x}) + u(\overline{x},\overline{x}) - u(x,z) - u(z,x) \\ \leq\ & u_1(x,z)(\overline{x}-x) + u_2(x,z)(\overline{x}-z) + u_1(z,x)(\overline{x}-z) \\ & + u_2(z,x)(\overline{x}-x) \\ =\ & 0 \end{aligned}$$

so that $[0.5u(x,z)+0.5u(z,x)] \geq u(\overline{x},\overline{x})$. If $x=z$ this contradicts the uniqueness of the golden-rule. If $x \neq z$, then since $[0.5x+0.5z, 0.5z+0.5x] \in N(\overline{x})^2$, and u is strictly concave on $N(\overline{x})^2$,

$$\begin{aligned} & u(0.5x+0.5z, 0.5z+0.5x) \\ >\ & 0.5u(x,z) + 0.5u(z,x) \geq u(\overline{x},\overline{x}) \end{aligned}$$

which contradicts the definition of the golden-rule. Thus, $(x,z) = (\overline{x},\overline{x})$ is the only solution to $H(x,z;0)=0$ in $N(\overline{x})^2$. Using (9.8), we have degree of $H(N_2(\overline{x})^2;0) = -1$.

Now for any λ in $(0,1]$, defining $\rho = \lambda\rho_3 + (1-\lambda)$, we have $\rho_2 < \rho_3 \leq \rho < 1$. If (x,z) in $N(\overline{x})^2$ is a solution to $H(x,z;\lambda)=0$ which is on the boundary of $N(\overline{x})^2$, then

$$\begin{aligned} u_2(x,z) + \rho u_1(z,x) &= 0 \\ u_2(z,x) + \rho u_1(x,z) &= 0 \end{aligned}$$

and so either (i) $x=z$, and x is a stationary optimal stock for (Ω,u,ρ); or (ii) $x \neq z$, and (x,z) is a period-two cycle for (Ω,u,ρ). However, as verified above, in case (i), $x = k_\rho \in N_2(\overline{x})$ and in case (ii), $(x,z) \in N_2(x)^2$. Thus, there is no solution (x,z) to $H(x,z;\lambda)=0$ on the boundary of $N(\overline{x})^2$. It follows that the degree of $H(N_2(\overline{x})^2;\lambda) = H(N_2(\overline{x})^2;0) = (-1)$ for all $\lambda \in (0,1]$.

For any solution (x,z) in $N_2(\overline{x})^2$ to the equation $H((x,z);\lambda)=0$ we have an index equal to (-1) by (9.15). Thus, there is exactly one solution to the equation $H((x,z);\lambda)=0$ in $N_2(\overline{x})^2$ for each λ in $[0,1]$.

Now, let (Ω,u,δ) be a dynamic optimization model with $\rho_3 \leq \delta < 1$. Suppose that (Ω,u,δ) has an optimal program exhibiting a period two-cycle, then there is (x',z') with $x' \neq z'$, such that $(x', z', x', z', ...)$ is optimal from x'. Then, as verified above, $(x', z') \in N_2(\overline{x})^2$. But, there is a stationary optimal stock, k_δ, with $(k_\delta, k_\delta) \in N_2(\overline{x})^2$. Defining $\lambda = (1-\delta)/(1-\rho_3)$, we have λ in $(0, 1]$, and multiple solutions to the equation $H((x, z); \lambda) = 0$ in $N_2(\overline{x})^2$ a contradiction, which establishes the result. Q.E.D.

Proposition 9.2: *Let (Ω, u) be given satisfying condition C2. For every discount factor δ satisfying $\rho_3 \leq \delta < 1$, the optimal policy fuction, h_δ, and non-zero SOS, k_δ, satisfy (i) $h_\delta(x) > x$ for $x \in (0, k_\delta)$; (ii) $h_\delta(x) < x$ for $x \in (k_\delta, b]$.*

Proof: We will write h instead of h_δ to simplify notation. We know that there is a unique non-zero SOS, k_δ, which we denote by k, and $k \in N_2(\overline{x})$.

We claim that for $x \in (0, k)$, $h(x) > x$. Otherwise by uniqueness of non-zero SOS, $h(x) < x$ for all $x \in (0, k)$.

Pick $x \in N_2(\overline{x})$, with $x \in (0, k)$ so that $h(x) \in N_2(\overline{x})$ as well. Then, we have

$$\begin{aligned} -u_2(x, h(x)) &= V'(h(x)) \\ -u_2(k, h(k)) &= V'(h(k)) \end{aligned}$$

Thus, using the Mean-Value theorem,

$$(-\widetilde{u}_{21})(k - x) + (-\widetilde{u}_{22})(k - h(x)) = V'(k) - V'(h(x))$$

where $\widetilde{u}_{21}$ and $\widetilde{u}_{22}$ are evaluated at a point in between $(x, h(x))$ and (k, k). Since $h(x) < x < k$, we have $V'(k) \leq V'(h(x))$, and so

$$(-\widetilde{u}_{22})(k - h(x)) \leq \widetilde{u}_{21}(k - x)$$

The Hessian of u is negative definite on $N_2(\overline{x})^2$, and so $(-\widetilde{u}_{22}) > 0$, and $[k - h(x)] > [k - x] > 0$, so we get

$$(-\widetilde{u}_{22})(k - x) \leq \widetilde{u}_{21}(k - x)$$

Thus, $\widetilde{u}_{21} \geq -\widetilde{u}_{22}$, and letting $x \to k$, we obtain $u_{21}(k, k) \geq [-u_{22}(k, k)] > 0$. Using (9.15), we have

$$(1 + \delta)|u_{12}(k, k)| < [-u_{22}(k, k)] + [-u_{11}(k, k)]$$

Also, $u_{11}(k,k) < 0$, $u_{22}(k,k) < 0$, since the Hessian of u is negative definite on $N_2(\overline{x})^2$ and $(k,k) \in N_2(\overline{x})^2$. Thus, condition C1 of Section 7 is satisfied, so that by Proposition 7.3, h is differentiable at k, with $|h'(k)| = |\lambda_1| < 1$. But this contradicts the fact that $h(x) < x$ for all x in $(0,k)$, and establishes our claim and hence (i). The claim (ii) can be established similarly.

Thereom 9.1: *Let (Ω, u) be given satisfying condition C2. For every discount factor δ satisfying $\rho_3 \leq \delta < 1$, the optimal policy function, h_δ, and non-zero SOS, k_δ, satisfy*

$$h_\delta^t(x) \to k_\delta \text{ as } t \to \infty \tag{9.24}$$

for every $x \in X$, with $x > 0$.

Proof: By Proposition 9.1, there are no period-two optimal cycles generated by h_δ. Thus, by Block and Coppel (1992, Proposition 1, p. 121), given any $x \in X$, $h_\delta^t(x)$ converges to one of the fixed points of h_δ.

There are then two cases to consider: (i) $h_\delta(0) \neq 0$, (ii) $h_\delta(0) = 0$.

In case (i), k_δ is the only fixed point of h_δ, and we are done.

In case (ii), we will be done if we can show that $h_\delta^t(x)$ cannot converge to zero for $x \in X$, $x > 0$.

To this end, we establish first that

$$h_\delta(x) > 0 \text{ for } x > 0 \tag{9.25}$$

To prove (9.25), suppose on the contrary there is some $x \in X$ with $x > 0$, such that $h_\delta(x) = 0$. Then, since $h_\delta(0) = 0$, the sequence $(x, 0, 0, 0, ...)$ is the optimal program from x.

By Proposition 9.2, we must have $x > k_\delta$. Since $k_\delta \in N_2(\overline{x})$, $k_\delta > \overline{x} - \varepsilon_2$; and since $(\overline{x} - \varepsilon_2, \overline{x}) \in N_2(\overline{x})^2$, $(k_\delta, \overline{x}) \in \Omega$ and so $(x, \overline{x}) \in \Omega$. Thus $(x, \overline{x}, \overline{x}, \overline{x}, ...)$ is a program from x. Further, the discounted sum of utilities on this program is at least $u(\overline{x} - \varepsilon_2, \overline{x}) + \delta[u(\overline{x}, \overline{x})/(1-\delta)]$. The discounted sum of utilities on the program $(x, 0, 0, ...)$ is $u(x,0) + \delta[u(0,0)/(1-\delta)]$, which is at most $u(b,0) + \delta[u(0,0)/(1-\delta)]$. Thus, we must have

$$u(b,0) + \delta[u(0,0)/(1-\delta)] \geq u(\overline{x} - \varepsilon_2, \overline{x}) + \delta[u(\overline{x}, \overline{x})/(1-\delta)$$

But this contradicts (9.18), since $\rho_3 \leq \delta < 1$, and establishes (9.25).

In view of (9.25), if $h_\delta^t(x)$ converges to zero, we must have some $T \geq 1$, such that for $t \geq T$, $0 < h_\delta^t(x) < k_\delta$. Further, since $h_\delta^t(x)$ converges to zero we can find some $\tau > T$, such that $h_\delta^{\tau+1}(x) < h_\delta^\tau(x)$. But this contradicts (i) of Proposition 9.2. Thus $h_\delta^t(x)$ cannot converge to zero for $x \in X$, $x > 0$. We infer then that in case (ii) as well, $h_\delta^t(x)$ must converge to k_δ as $t \to \infty$, for every $x \in X$, with $x > 0$. This establishes the turnpike theorem. Q.E.D.

10. Bibliographic Notes

Section 1.

The history of our subject has not been systematically surveyed since the masterly reviews by Koopmans. In the first essay of his celebrated "Three Essays on the State of Economic Science" (1957), Koopmans evaluated the accomplishments of general equilibrium theory and intertemporal allocation theory both in terms of the economic content and the newer mathematical techniques used. The literature on pure capital accumulation oriented optimal growth models was reviewed in Koopmans (1964).

For the theory of Ramsey-type optimal growth models, and some open problems in this area of research, see Koopmans (1967a).

Section 2.

For multisectoral versions of the reduced-form model, see Gale (1967) and McKenzie (1968) (for the case in which future utilities are not discounted), and Sutherland (1970), Stokey and Lucas (1989) (for the case in which future utilities are discounted). In contrast to most presentations of the reduced form model, where the utility function is assumed to be continuous, we develop our model with the assumption that the utility function be bounded and upper semicontinuous. This is because the assumption of continuity on the felicity function in the primitive form of the model (where felicity is derived from consumption) does not always ensure the continuity of the reduced form utility function. This issue was raised by Peleg (1973); for a complete discussion, see Dutta and Mitra (1989a).

Example 2.1 is based on the familiar model developed in continuous time by Cass (1965), Koopmans (1965) and Samuelson (1965). The discrete-time version was presented by Koopmans (1967a). A multisector generalization of this model is contained in Peleg and Ryder (1972). Example 2.2 is based on the continuous-time models of Uzawa (1964) and Srinivasan (1964). For a discrete-time version, see Benhabib and Nishimura (1985) and Boldrin and Montrucchio (1986b). Since then, this has become one of the most familiar models giving rise to cyclical and chaotic optimal behavior.

Example 2.3 has been analyzed in detail in Majumdar and Mitra (1994a) for cyclical and chaotic optimal behavior. A similar model (in continuous time) was examined by Kurz (1968) to study optimal growth with wealth effects. In the renewable resources literature, variations of this model in which the production function exhibits an initial phase of increasing returns followed by decreasing returns (an "S-shape") have been analyzed by Clark (1971), and Majumdar and Mitra (1983).

Example 2.4 is a generalization of the Weitzman example, reported in Samuelson (1973). Variations of this example have been analyzed by Scheinkman (1976), McKenzie (1983), Benhabib and Nishimura (1985) and Mitra and Nishimura (1998), among others.

For variations of Example 2.5, see Wan (1993). The point-input flow-output version of the forestry model (an "orchards" model) was analyzed in Mitra, Ray and Roy (1991). A vintage-capital model, with a similar structure, was examined by Benhabib and Rustichini (1993).

Section 3.

The method of dynamic programming, including the optimality principle, was introduced by Bellman (1957). The rigorous foundation of stochastic dynamic programming was developed by Blackwell (1965), Strauch (1966), Denardo (1967) and Maitra (1968). For generalizations of the results presented in this section to non-stochastic models with a multidimensional state space, see for example Dutta and Mitra (1989b), Stokey and Lucas (1989) and Sorger (1992). One can use a version of the maximum theorem of Berge (1963) to establish that the value function is upper semi-continuous ; we provide a self-contained exposition. Continuity of the value function then follows from the "convex structure" of the model.

Section 4.

For a variety of multisector versions of the basic price characterization results, see Peleg (1970), Peleg and Ryder (1972), Weitzman (1973), Cass and Majumdar (1979), and McKenzie (1986), among others. When there is a (modified) golden-rule (see Section 6), it turns out that the transversality condition (which characterizes the optimality of competitive programs) can be replaced by (an infinite number of) period-by-period conditions. For a discussion of this issue, see Brock and Majumdar (1988) for the undiscounted case, and Dasgupta and Mitra (1988) for the discounted case. These papers, together with other papers dealing with the issue of intertemporal decentralization are contained in Majumdar (1992).

Section 5.

The results, on the continuity of the value and policy functions with respect to changes in the parameters of the model, are closely related to the literature on the maximum theorem, mentioned above in the notes on Section 3. For a survey of results on the general problem of "parametric continuity", see Bank et.al. (1983). The use of supermodularity to obtain monotonicity results on the solution to optimization problems (due to Topkis (1978)) has been extended to the ordinal utility setting by Milgrom and Shannon (1994).

Section 6.

The existence of a stationary optimal stock by dynamic programming methods, in the multisector reduced form model, was established by Sutherland (1970). However, Peleg and Ryder (1974) noted that this stationary optimal stock could be trivial. They proved the existence of a non-trivial stationary optimal stock in a multisectoral model, in which felicity is derived from consumption alone, by establishing the existence of a modified golden-rule. Flynn (1980) and McKenzie (1986) obtained the result in the multisectoral reduced-form model by establishing the existence of a discounted golden-rule stock. Khan and Mitra (1986) established the result using the upper semicontinuity (rather than continuity) of the utility function. Then the result in the Peleg-Ryder (1974) framework can be viewed as a special case, as shown in Dasgupta and Mitra (1990).

Section 7.

The differentiability of the value function, in a stochastic version of the one-sector model(discussed in Example 2.1), was established by Mirman and Zilcha (1975). The result of Benveniste and Scheinkman (1979) was developed for a multidimensional state space setting. For an exposition of the result, see Stokey and Lucas (1989). The technique, developed by Benveniste and Scheinkman, was used by Debreu (1976) in his study of least concave utility functions.

The problem of differentiability of the policy function has been investigated by Boldrin and Montrucchio (1989), Araujo (1991), Santos (1991) and Montrucchio (1998). Araujo addressed the problem in the one-dimensional case. Santos provided the definitive solution to the problem in the multisectoral reduced-form model. Our result applies only to the one-dimensional case and only to the stationary optimal stock. However, we do not assume that the Hessian of the utility function is negative definite, an assumption which is crucial to the methods used in all the above cited papers.

Section 8.

In a continuous-time framework of optimal growth with wealth effects, Kurz (1968) provided an example of non-unique steady-states. Sutherland (1970) provided an example of non-unique stationary optimal stocks directly on a reduced-form model, without discussing any primitive form from which it was derived. In the context of the standard forestry model, Mitra and Wan (1985) provided an example of multiple stationary forests. The uniqueness result for high discount factors, discussed in Section 8.3, is based on Benhabib and Nishimura (1979) and McKenzie (1986). The last part of the proof would have to make use of the Poincare-Hopf index theorem, and the notion of degree of a mapping, in the multidimensional case. In our one-dimensional framework, we can provide a self-contained elementary treatment.

Section 9.

Another example, in which there are optimal period-two cycles for all discount factors close to unity, was developed in the context of Example 2.5 by Wan (1989, 1993). In contrast to the example due to Weitzman, the optimal cycles in his example are boundary cycles, and they are robust to perturbations in the utility function. The earlier counterexamples to the turnpike property, provided by Kurz (1968) and Sutherland (1970), were for discount factors, which were not close to unity. The turnpike property for high discount factors presented in Section 9.2, is crucially dependent on the demonstration (in Proposition 9.1) that there cannot be any period-two optimal cycles. This is really a uniqueness argument, since one is establishing the uniqueness of the solution to a pair of Ramsey-Euler equations. But, unlike the uniqueness problem of Section 8, this problem is two-dimensional, prompting us to use index theory. This theory can be found in Milnor (1965), using the method of differential topology, and in Ortega and Rheinboldt (1970), using purely analytic methods.

3

Periodic and Chaotic Programs of Optimal Intertemporal Allocation in an Aggregative Model with Wealth Effects

Mukul Majumdar and Tapan Mitra

1. Introduction

In this chapter, we examine a discrete-time aggregative model of discounted dynamic optimization where the felicity function depends on both consumption and capital stock. The need for studying such a model has been stressed in the theory of optimal growth and also in the economics of natural resources.

In his report on a number of studies in optimal economic growth, Koopmans (1967, p. 2) observed: "In all of the models considered it is assumed that the objective of economic growth depends exclusively on the path of *consumption* as foreseen for the future. That is, the capital stock is not regarded as an end in itself, or as a means to end other than consumption. We have already taken a step away from reality by making this assumption. A large and flexible capital stock has considerable importance for what is usually somewhat inadequately called "defense". The capital stock also helps to meet the cost of retaining all aspects of national sovereignty and power in a highly interdependent world." Interestingly enough, a continuous-time aggregative model in which the objective function is sensitive to both the consumption stream and to the per capita capital stock ("wealth effect") of the society was analyzed in a paper by Kurz (1968). It was noted there that the presence of wealth effects may lead to the appearance of multiple steady states. In fact, two numerical examples definitely established the possibility of such a phenomenon arising in dynamic optimization models. Some further work on a similar model was reported subsequently by Arrow and Kurz (1970).

In environmental economics, incorporating the direct welfare effects of certain types of capital assets [generally referred to as a "stock effect"] is acknowledged to be a significant ingredient in formulating appropriate resource policies. Dasgupta (1982, p. 107) summarizes the importance of such effects as follows: "As a flow DDT is useful in agriculture as an input; as a stock it is hazardous for health. Likewise fisheries and aquifers are useful not only for the harvest they provide: as a stock they are directly useful, since harvesting and extraction costs are low if stocks are large. Likewise, forests are beneficial not only for the flow of timber they can supply: as a stock they prevent soil erosion and maintain a genetic pool". To elaborate somewhat on the standard model of a commercially exploited fishery (which can be viewed as a special case of the aggregative model we study), the harvest is determined by a production function dependingt on the inputs of fishing effort and the stock of fish being exploited ("fish biomass"). This leads to a cost function (where cost is measured in terms of fishing effort) which depends on the harvest itself as well as the fish stock. If benefit from harvest is obtained through a specified revenue function, the net benefit or profit function is seen to be dependent on both the harvest and the stock of fish. [Clark (1976, Chapter 7) and Dasgupta (1982, Chapters 6, 7) develop this model in detail and also examine the nature of optimal harvest policies for some specific cases.]

The aggregative model of optimal intertemporal allocation, in which felicity is derived solely from consumption (which is a special case of the model in this chapter), has of course been studied extensively in the literature under a variety of different technological specifications. It is observed there that optimal programs exhibit monotone behavior over time. In fact, this feature continues to hold even when the technology exhibits increasing returns [see Dechert-Nishimura (1983), Mitra-Ray (1984)] and investment is irreversible [see Majumdar-Nermuth (1983)]. Thus, it is often suggested that a one-sector optimal growth model can only display "simple dynamics". The present exercise indicates that optimal programs in the aggregative model, in the presence of wealth (stock) effects, can exhibit the full spectrum of "complex dynamics" studied in the mathematical theory of chaos.

A few remarks on the relation of this result to those presented in the literature might help to put our chapter in proper perspective. One of the major developments in the last ten years or so, in the area of dynamic economics, has been the identification of relatively simple non-linear models in which "cyclical" or "chaotic" outcomes are consistent with optimizing behavior over time based on complete information. [A number of excellent reviews of this development, from different perspectives, are available in Baumol and Benhabib (1989), Day and Pianigiani (1991), Grandmont (1985, 1986), Brock and Dechert (1991) and Boldrin and Woodford (1990).] Much of this work has been carried out in the so-called "reduced form" model, in the tradition of Gale (1967) and McKenzie (1968). In this version, the principal object of interest is the immediate utility generated by a transition from a stock of goods in one period to another stock in the next period. That is, the focus of attention is the associated concept of the "reduced-form" utility function, which subsumes both the technology set and the welfare function, the "primitives" of the model. [For elaboration, see Section 3, especially 3.2.]

In the context of this (reduced form) model, Benhabib and Nishimura (1985) identified sufficient conditions for the existence of periodic optimal programs. However, commenting on the possibility of chaos in their model, they observed that "it is very difficult to construct examples which generate chaotic dynamics for infinite-horizon models that have concave utility functions." In subsequent papers, Deneckere and Pelikan (1986) and Boldrin and Montrucchio (1986b) provided the kind of examples that Benhabib-Nishimura were referring to. In particular, given any specified C^2 function Boldrin and Montrucchio (1986b) provided a method of constructing a reduced form model, such that the optimal policy function of the model would coincide with the pre-specified C^2 function. [Chaos is exhibited by choosing the C^2 function to be the logistic map, for instance.]

The question naturally arises whether one can specify primitives (technology sets and welfare functions) satisfying standard assumptions, which will give rise to the reduced-form models constructed in the above exercises. It is clear from the above papers that one can specify dynamic optimization models (in terms of primitives) with *two production sectors* which generate the corresponding reduced-form models exhibiting cyclical [see Benhabib-Nishimura (1985)] and chaotic [see Deneckere and Pelikan (1986), Boldrin and Montrucchio (1986b)] optimal trajectories.

The approach taken in this chapter is somewhat different from that of the above literature. We start with a model of dynamic optimization, specified in terms of the primitives: the technology set, the welfare function and the discount factor. The welfare function is defined on consumption and the capital input stock, because the need for capturing such a stock or wealth effect has been repeatedly emphasized in the literature (as noted above) in a number of different intertemporal optimization problems. Our model is kept simple in every other respect, with the standard technological specification of the aggregative model. We are then interested in answering the following question: what kinds of dynamic optimal behavior can this model exhibit? In answering the question, we make considerable use of the results developed for the reduced model; however, our emphasis on the primitives of the particular dynamic optimization model we study is the dominant theme throughout.

Our answer to the above question can be conveniently subdivided into three parts. First, we indicate briefly [by relying on the analysis of Benhabib-Nishimura (1985)] how sufficient conditions on the welfare and production functions can be developed to yield a monotone non-decreasing optimal policy function [see Section 4.1]. We also examine in some detail [in Section 4.2] the restrictiveness of assumptions which have typically been made directly on the reduced-form model to ensure a monontone non-increasing optimal policy function (in the interior of the transition possibility set). The implications of such assumptions on the underlying welfare and production functions need not be at all transparent, as we proceed to show. The methodological point that emerges from this exercise is the need to formalize conditions on the reduced model in terms of justifiable restrictions on the primitive data [see

Section 4 for details].

Second, we show [in Section 5] that our model can exhibit optimal cyclic behavior for *all* discount factors close to one. Clearly, this goes against the spirit of turnpike theorems which propose to show that for discount factors sufficiently close to one, one has global asymptotic stability of optimal programs with respect to the stationary optimal stock. It is, of course, perfectly in harmony with the more general result of the subject, namely the "neighborhood turnpike theorem" of McKenzie (1982). As the discount factor approaches one, the cycles of optimal programs are seen to exhibit smaller amplitude around the stationary optimal stock.

Third, we construct a suitable example [in Section 6.1] to demonstrate that our model can exhibit optimal behavior, which is chaotic in the topological and ergodic senses. Unlike the literature preceding this chapter, we go a step further [in Section 6.2] and enquire whether the (demonstrated) chaotic behavior is robust to small perturbations of the "economy", specified in terms of the primitives (the production and welfare functions and the discount factor). We show that topological chaos is robust to such perturbations by demonstrating that the optimal policy function of each "neighboring" economy satisfies the Li-Yorke condition.

2. Basic Results from the Theory of Dynamical Systems

A *dynamical system* is described by a pair (X, h), where X is a set (called the *state space*) and h a function from X to X describing the *law of motion* of the state variable, $x \in X$. Thus, if $x_1 \in X$ is the state of the system in period t_1 then $x_{t+1} = h(x_t) \in X$ is the state of the system in period $(t+1)$. We will be concerned, in what follows, with dynamical systems for which X is an interval of the real line.

Given a dynamical system (X, h), and any $x \in X$, we define $h^0(x) = x$, and for any integer $k \geq 1$

$$h^k(x) = h(h^{k-1}(x))$$

A point $x \in X$ is called *periodic* if there is $k \geq 2$ such that $h^k(x) = x$. The smallest such k, call it $\mathbf{k}$ is the *period* of x. [In this case, the sequence $\{h^j(x)\}_0^\infty$ is also called periodic with period $\mathbf{k}$]. We denote the set of periodic points in X by $P(X)$. Its complement in X, the set of non-periodic points, is denoted by $N(X)$.

A basic result characterizing the behavior of the dynamical system (X, h) has been given by Li and Yorke (1975), and may be stated as follows.

Theorem 2.1: *Let α, β be in $\Re$, with $\alpha < \beta$. Suppose $X = [\alpha, \beta]$ and $h : X \to X$ is continuous. If there is $x^* \in X$ such that*

$$h^3(x^*) \leq x^* < h(x^*) < h^2(x^*) \qquad \text{(L-Y)}$$

then

(i) for every integer $\mathbf{k} \geq 2$, there is a periodic point $x_k \in X$ with period $\mathbf{k}$;

(ii) there is an uncountable set $W \subset N(X)$ satisfying the following conditions:

(a) If $x, y \in W$ with $x \neq y$, then $\overline{\lim}_{k\to\infty} |h^k(x) - h^k(y)| > 0$; $\underline{\lim}_{k\to\infty} |h^k(x) - h^k(y)| = 0$.

(b) If $x \in W$ and $y \in P(X)$, then $\overline{\lim}_{k\to\infty} |h^k(x) - h^k(y)| > 0$.

Remark 2.1: (1) The result (i) is actually a consequence of Sarkovskii's theorem (see Devaney [1989]).

(2) According to (ii)(a), there are pairs of initial states (in W) such that the sequences of iterates move apart and return close to each other infinitely often. Furthermore, as (ii)(b) states, if the initial state is in the set W, then the system does not converge to any periodic point.

We will say that the dynamical system (X, h) exhibits *topological chaos* if conditions (i) and (ii) of Theorem 2.1 are satisfied. The Li-Yorke condition (L-Y) can be seen as an easily verifiable sufficient condition for topological chaos.

It has been argued that topological chaos may be "unobservable" since the uncountable set W in (ii) of Theorem 2.1 may have Lebesgue measure zero [see, for example Collet and Eckmann (1980), Day and Shafer (1987) for discussions]. This motivates the study of "ergodic chaos" which we turn to next.

Let ξ denote the Lebesgue measure of X (endowed with its Borel σ-algebra). We will say that the dynamical system (X, h) exhibits *ergodic chaos* if there exists a probability measure v satisfying the following conditions:

(i) v is *absolutely continuous* with respect to the Lebesgue measure on X; that is, if B is a Borel set in X, and $\xi(B) = 0$ then $v(B) = 0$;

(ii) v is *invariant* under the action of h; that is, $v(h^{-1}(B)) = v(B)$ for every Borel set $B \subset X$;

(iii) v is *ergodic*; that is, for every v-integrable real valued function, $\phi(on\ X)$, we have

$$\lim_{T\to\infty} (1/T) \sum_{k=1}^{T} \phi(h^k(x)) \to \int \phi \, dv \text{ for } v- \text{ a.e. } x \in X.$$

Condition (iii) above is frequently referred to as: time averages equal space averages. If v is the unique probability measure which satisfies (i)-(iii) above, it is called the *ergodic measure* of h.

If (X, h) has an ergodic measure, v, then for $v-$ a.e. $x \in X$ (and hence, by (i), for x belonging to a subset of X with positive Lebesgue measure) the sequence of iterates $\{h^k(x)\}$ will "fill up" the support of the measure v, and as a consequence, will have extremely complicated trajectories.

In order to provide sufficient conditions for ergodic chaos, we introduce the notion of a Schwarzian derivative [see Devaney (1989) for a more extensive discussion]. Consider a dynamical system (X, h) with $h : X \to X$ of class C^3. The *Schwarzian derivative* $Sh(x)$ is given by

$$Sh(x) = [h'''(x)/h'(x)] - (3/2)[h''(x)/h'(x)]^2$$

for $x \in X$ with $h'(x) \neq 0$.

Theorem 2.2: *Let α, β be in $\Re$ with $\alpha < \beta$. Suppose $X = [\alpha, \beta]$ and $h : X \to X$ satisfying the following conditions:*

(i) h is of class C^3, and there exists $x^* \in (\alpha, \beta)$ such that $h'(x^*) = 0$ and $h''(x^*) < 0$;

$h'(x) > 0$ for all $x \in X$ with $x < x^*$; $h'(x) < 0$ for all $x \in X$ with $x > x^*$.

(ii) $h(x) > x$ *for all* $x \in (\alpha, x^*)$; $h(x^*) \in (x^*, \beta)$; *and* $Sh(x) < 0$ *for all* $x \in X$ *except* $x = x^*$.
(iii) There exists $k \geq 2$ *such that* $y = h^k(x^*)$ *satisfies* $h(y) = y$ *and* $|h'(y)| > 1$.
Then (X, h) *exhibits ergodic chaos.*

A detailed discussion of Theorem 2.2 may be found in Grandmont (1986) and Day and Pianigiani (1991).

Recall from the definition of topological chaos for a dynamical system (X, h), that condition (ii) indicates a certain "sensitive" dependence on initial conditions. Roughly speaking, the extent of this sensitive dependence is measured by the Lyapunov exponent, which we turn to next.

Let (X, h) be a dynamical system, with $h : X \to X$ of class C^1. For any $x_0 \in X$, the *Lyapunov exponent*, $\rho(x_0)$, is defined as

$$\rho(x_0) = \lim_{k \to \infty} (1/k) \ln \left| \frac{dh^k}{dx}(x_0) \right|$$

For sufficiently large k and small $\varepsilon > 0$, the Lyapunov exponent "approximately satisfies"

$$\varepsilon e^{k\rho(x_0)} \approx |h^k(x_0 + \varepsilon) - h(x_0)|$$

The right-hand side indicates how far apart x_0 and $x_0 + \varepsilon$ are under k iterates of h. When $\rho(x_0) > 0$, initially nearby points are stretched (by the successive iterations of h) at a positive exponential rate. [Further discussion of the Lyapunov exponent can be found in Rasband (1990)].

One of the most celebrated examples of a dynamical system (X, h) is one where $X = [0, 1]$ and

$$h(x) = 4x(1 - x) \text{ for all } x \in X$$

It is not difficult to verify that for this example, Theorems 2.1 and 2.2 apply directly. [For Theorem 2.1, choose $x^* = [\sqrt{2}-1]/2\sqrt{2}$ to verify the Li-Yorke condition (L-Y). For Theorem 2.2, choose $x^* = (1/2)$, and note that the Schwarzian derivative $Sh(x) = -6/(1 - 2x)^2$ for all $x \neq (1/2)$, to verify (i) and (ii); further (iii) is verified by choosing $k = 2$]. Thus, this dynamical system exhibits both topological and ergodic chaos.

The density $Q(x)$ of the ergodic measure of this dynamical system has been computed. It is given by

$$Q(x) = [\pi\sqrt{x(1-x)}]^{-1} \text{ for } 0 < x < 1$$

[See Day and Pianigiani (1991) for more on this computation, due to Ulam and von Neumann]. The Lyapunov exponent of this dynamical system is in fact seen to be independent of $x_0 \in X$, and is given by

$$\rho = \rho(x_0) = \ln 2 > 0 \text{ for all } x_0 \in X$$

3. Preliminary Aspects of the Optimization Framework

3.1 The Model

We consider an aggregative model, specified by a *production function*, $f : \Re_+ \to \Re_+$, a *welfare function*, $w : \Re_+^2 \to \Re_+$, and a *discount factor* $\delta \in (0,1)$.

The following assumptions on f are used:

F1: *$f(0) = 0$; f is continuous on $\Re_+$.*
F2: *f is non-decreasing and concave on $\Re_+$.*
F3: *There is some $K > 0$ such that $f(x) > x$ when $0 < x < K$, and $f(x) < x$ when $x > K$.*

We define a set $\Omega \subset \Re_+^2$ as follows:

$$\Omega = \{(x, z) \in \Re_+^2 : z \leq f(x)\}$$

The following assumptions on w are used:

W1: *$w(x,c)$ is continuous on $\Re_+^2$.*
W2: *$w(x,c)$ is non-decreasing in x given c, and non-decreasing in c, given x on $\Re_+^2$.*
Furthermore, if $x > 0$, $w(x,c)$ is strictly increasing in c on Ω.

W3: *$w(x,c)$ is concave on $\Re^2_+$. Furthermore, if $x>0$, $w(x,c)$ is strictly concave in c on Ω.*

A *program* from $\mathbf{x} \geq 0$ is a sequence $\{x_t\}_0^\infty$ satisfying

$$x_0 = \mathbf{x}, 0 \leq x_{t+1} \leq f(x_t) \text{ for all } t \geq 0$$

The *consumption* sequence $\{c_{t+1}\}_0^\infty$ is given by

$$c_{t+1} = f(x_t) - x_{t+1} \text{ for all } t \geq 0$$

It is easy to verify that for any program $\{x_t\}_0^\infty$ from $\mathbf{x} \geq 0$, we have $x_t, c_{t+1} \leq K(\mathbf{x}) \equiv \max(K, \mathbf{x})$ for $t \geq 0$. [In particular, if $\mathbf{x} \in [0, K]$, then $x_t, c_{t+1} \leq K$ for $t \geq 0$.]

A program $\{\hat{x}_t\}_0^\infty$ from $\mathbf{x} \geq 0$ is *optimal* if

$$\sum_{t=0}^{\infty} \delta^t w(\hat{x}_t, \hat{c}_{t+1}) \geq \sum_{t=0}^{\infty} \delta^t w(x_t, c_{t+1})$$

for every program $\{x_t\}_0^\infty$ from $\mathbf{x}$.

A program $\{x_t\}_0^\infty$ from $\mathbf{x}$ is *stationary* if $x_t = \mathbf{x}$ for $t \geq 0$. It is a *stationary optimal program* if it is also an optimal program from $\mathbf{x}$. In this case, $\mathbf{x}$ is called a stationary optimal stock. [Note that 0 is a *stationary optimal stock*]. A stationary optimal stock, x, is *non-trivial* if $x > 0$.

3.2 Conversion to Reduced Form

The model, described in the previous subsection, (which we will call the "primitive form") can be converted to the so-called "reduced form". In reduced form, an intertemporal framework is described by a state space, a transition possibility set, a (reduced form) utility function defined on this set, and a discount factor.

The state space for our purpose will be taken to be $\Re_+$, and we will think of the input (x) as the state variable. The transition possibility set then describes pairs of input levels (x, z), such that if x is the input level today (at t), then it is possible to go to the input level z tomorrow (at $(t+1)$). The utility function then measures the welfare obtained in moving from x today to z tomorrow.

Now, we describe how our framework in primitive form can be viewed as a special case of the reduced form model. Note that Ω, as defined in Section 2, precisely describes the transition possibility set. The utility function $u : \Omega \to \Re_+$ can be defined by

$$u(x, z) = w(x, f(x) - z)$$

Under the assumptions (F.1)-(F.3) on the production function f, the transition possibility set Ω satisfies the following properties:

(Ω.1) $(0, 0) \in \Omega$; $(0, z) \in \Omega$ implies $z = 0$.

(Ω.2) Ω is a closed, convex subset of $\Re^2_+$.

(Ω.3) If$(x, z) \in \Omega$ and $x' \geq x$ and $0 \leq z' \leq z$, then $(x', z') \in \Omega$.

(Ω.4) There is some $K > 0$, such that if $(x, z) \in \Omega$ and $x > K$, then $z < x$.

Furthermore, under the assumption (W.1)-(W.3) on the welfare function w, the (reduced form) utility function u satisfies the following properties

(U.1) $u(x, z)$ is continuous on Ω.

(U.2) If $(x, z) \in \Omega$ and $x' \geq x$ and $0 \leq z' \leq z$, then $u(x', z') \geq u(x, z)$.

(U.3) $u(x, z)$ is concave on Ω; and, given $x > 0, u(x, z)$ is strictly concave in z on Ω.

The concept of a program and that of an optimal program (introduced in Section 3.1) can then be viewed equivalently as follows. a program $\{x_t\}_0^\infty$ from $\mathbf{x} \geq 0$ is seen to be defined by

$$x_0 = \mathbf{x}, (x_t, x_{t+1}) \in \Omega \text{ for all } t \geq 0$$

And an optimal program $\{\hat{x}_t\}_0^\infty$ from $\mathbf{x} \geq 0$ is seen to be a program satisfying

$$\sum_{t=0}^{\infty} \delta^t u(\hat{x}_t, \hat{x}_{t+1}) \geq \sum_{t=0}^{\infty} \delta^t u(x_t, x_{t+1})$$

for every program $\{x_t\}_0^\infty$ from $\mathbf{x}$.

This conversion of our framework in primitive form to the reduced form is particularly helpful because it allows a direct application of the many results developed for the reduced form model [see McKenzie (1986) for a comprehensive survey of the more important results]. We would like to emphasize, however, that our assumptions are exclusively on the primitive form, and these imply that certain properties are satisfied by the corresponding reduced form. If some property is assumed about (Ω, u, δ), this needs to be justified by appropriate assumptions on the primitives (f, w, δ) of the model. We will discuss this point in more detail in Section 4.

3.3 Value and Policy Functions

Given a reduced form model (Ω, u, δ), satisfying $(\Omega.1)$-$(\Omega.4)$, (U.1)-(U.3), standard arguments ensure that there is an optimal program from every $\mathbf{x} \geq 0$. Conditions $(\Omega.2)$ and (U.3) ensure that an optimal program is unique.

We define a *value function* $V : \Re_+ \to \Re$ by

$$V(x) = \sum_{t=0}^{\infty} \delta^t u(\hat{x}_t, \hat{x}_{t+1}) \tag{3.1}$$

and the *optimal policy function* $h : \Re_+ \to \Re_+$ by

$$h(x) = \hat{x}_1 \tag{3.2}$$

where $\{\hat{x}_t\}$ is the optimal program from $x \geq 0$.

It is easy to check that V is a concave function on $\Re_+$. Furthermore, an application of an appropriate version of the maximum theorem yields the continuity of V and h on $\Re_+$ [see, for example, Dutta and Mitra (1986)].

Given any $x \in \Re_+$, the constrained maximization problem

$$\begin{array}{ll} \text{Max} & [u(x,z) + \delta V(z)] \\ \text{Subject to} & (x,z) \in \Omega \end{array}$$

has a unique solution, $H(x)$. It follows from the definitions of V and h that $H(x) = h(x)$ for all $x \in \Re_+$ and further $V(x) = u(x, h(x)) + \delta V(h(x))$. This yields the "optimality equation" for each $x \in \Re_+$:

$$V(x) = \underset{(x,z)\in\Omega}{\text{Max}} [u(x,z) + \delta V(z)]$$

It is not difficult to verify that if ψ is any continuous real valued function on $[0, K]$ satisfying the "functional equation of dynamic programming":

$$\psi(x) = \max_{(x,z)\in\Omega} [u(x,z) + \delta\psi(z)] \text{ for } x \geq 0$$

then $\psi(x) = V(x)$ for all $x \in [0, K]$. That is, V is the *unique* solution (in the class $C[0, K]$) of the functional equation of dynamic programming.

We summarize the above facts in the following theorem, for ready reference.

Theorem 3.1: *(i) The value function, V, defined in (3.1) is the unique continuous real valued function on $[0, K]$, satisfying the functional equation of dynamic programming:*

$$V(x) = \max_{(x,z)\in\Omega} [u(x,z) + \delta V(z)] \tag{3.3}$$

Further, V is concave and non-decreasing on $\Re_+$.

(ii) The policy function h, defined in (3.2) satisfies the following property: for each $x \in \Re_+$, $h(x)$ solves uniquely the constrained maximization problem

$$\begin{array}{ll} \text{Max} & [u(x,z) + \delta V(z)] \\ \text{Subject to} & (x,z) \in \Omega \end{array}$$

Further, h is continuous on $\Re_+$.

3.4 Dual Variables Associated with Optimal Programs

In the course of this chapter, we will have to show that at various points that certain specified programs are optimal. One way of doing this is by finding a value function satisfying the functional equation of dynamic programming. (We use this method to verify a result of Section 6). Another way is by specifying a sequence of dual variables (or "shadow prices") associated with the program, such that the program is "competitive" and satisfies the "transversality condition". (We use this method to verify the result of Section 5). we state this well-known result for ready reference.

Theorem 3.2: *Suppose* $\{\hat{x}_t\}_0^\infty$ *is a program from* $\mathbf{x}$, *and* $\{\hat{p}_t\}_0^\infty$ *is a sequence of non-negative numbers such that*

(i) $\delta^t u(\hat{x}_t, \hat{x}_{t+1}) + \hat{p}_{t+1}\hat{x}_{t+1} - \hat{p}_t\hat{x}_t \geq \delta^t u(x,z) + \hat{p}_{t+1}z - \hat{p}_t x$ *for all* $(x,z) \in \Omega$ *and all* $t \geq 0$

and

(ii) $\lim_{t\to\infty} \hat{p}_t\hat{x}_t = 0$.

Then $\{\hat{x}_t\}_0^\infty$ *is an optimal program from* $\mathbf{x}$.

4. Observations on the Primitive and Reduced Forms

A considerable part of the literature on optimal intertemporal allocation has been developed for the reduced-form model. In this section, we emphasize the need to look at the primitive model, which gives rise to the reduced form, in order to assess the importance and compatibility of some assumptions which are typically made directly on the reduced form.

For this purpose, we consider a reduced form model (Ω, u, δ) as described in Section 3.2, with Ω satisfying $(\Omega.1)$-$(\Omega.4)$, u satisfying (U.1)-(U.3), and $0 < \delta < 1$ and proceed to examine situations for which the optimal policy function will be monotone.

4.1 Monotone Increasing Policy Functions

We first examine the sufficient conditions which ensure that the optimal policy function is monotone non-decreasing. This case has traditionally been of importance because it leads to monotone behavior of optimal input stocks, converging to some stationary optimal input stock.

Denote the interior of Ω by Ω^0. We say that u is supermodular on Ω^0 if whenever (x',z'), (x'',z''), (x',z'') and (x'',z') belong to Ω^0 and $(x'',z'') >> (x',z')$, we have

$$u(x'',z'') + u(x',z') \geq u(x'',z') + u(x',z'') \qquad (\bar{\text{S}})$$

Following Ross (1983) and Benhabib-Nishimura (1985), it can be shown that if u is supermodular on Ω^0, then h is non-decreasing on $\Re_+$. It is worthwhile to state this result formally.

Proposition 4.1: *Suppose* (Ω, u, δ) *satisfies* (Ω.1)-(Ω.4), (U.1)-(U.3), *and* $0 < \delta < 1$. *If* u *is supermodular on* Ω^0, *then the optimal policy function,* h, *is non-decreasing on* $\Re_+$.

There is a convenient way to check the supermodularity of u on Ω^0. If u is C^2 on Ω^0 and $D_{12}u(x,z) \geq 0$ for all $(x,z) \in \Omega^0$, then [following Ross (1983)] u can be shown to be supermodular on Ω^0.

If we turn now to our primitive model, we can try to find conditions on (f, w, δ), which will ensure that u is C^2 on Ω^0 with $D_{12}u(x,z) \geq 0$ for all $(x,z) \in \Omega^0$. If f is C^1 on $\Re_{++}$, and $w(x,c)$ is C^2 on Ω^0, one can verify that for any $(x,z) \in \Omega^0$,

$$D_{12}u(x,z) = [-D_{22}w(x, f(x) - z]\ f'(x) - D_{12}w(x, f(x) - z$$

Thus, if we assume that

$$-D_{22}w(x,c)\ f'(x) \geq D_{12}w(x,c) \text{ for all } (x,c) \in \Omega^0 \tag{4.1}$$

we can ensure $D_{12}u(x,z) \geq 0$ for all $(x,z) \in \Omega^0$. This leads to the following result.

Proposition 4.2: *Suppose* (f, w, δ) *satisfy* (F.1)-(F.3), (W.1)-(W.3), *and* $0 < \delta < 1$. *Assume, further, that* f *is* C^1 *on* $\Re_{++}$, w *is* C^2 *on* Ω^0 *and* (4.1) *is satisfied. Then the optimal policy function,* h, *is non-decreasing on* $\Re_+$.

That all the assumptions of Proposition 4.2 are compatible can be seen by considering the example where $f(x) = 2x^{1/2}$ for $x \in \Re_+$, $w(x,c) = x^{1/2} + c^{1/2}$ for $(x,c) \in \Re_+^2$ and $0 < \delta < 1$.

Under the assumptions of Proposition 4.2, if $\{\hat{x}_t\}_0^\infty$ is an optimal program from $\mathbf{x} > 0$, it is clearly a monotone (non-increasing or non-decreasing) sequence. Thus, the assumptions imply that optimal behavior leads to what can be called "simple dynamics".

4.2 Monotone Decreasing Policy Functions

It might appear that, symmetric to our analysis in Section 4.1, we should be able to get sufficient conditions under which the optimal policy function is monotone non-increasing. However, this turns out to be incorrect except in the extreme case where h is an "extinction" policy (that is, $h(x) = 0$ for all $x \geq 0$). [A strong sufficient condition which ensures that h is *not* an extinction policy is the existence of a non-trivial stationary optimal stock.]

Start, again, with the reduced form model (Ω, u, δ) where Ω satisfies $(\Omega.1)$-$(\Omega.4)$, u satisfies (U.1)-(U.3), and $0 < \delta < 1$.

Now, suppose h where monotone non-increasing for all $x > 0$ and there was some $\tilde{x} > 0$ for which $h(\tilde{x}) > 0$. Then, for all $0 < x < \tilde{x}$, $h(x) \geq h(\tilde{x})$ since h is monotone non-increasing. By continuity of h on $\Re_+$, $h(0) \geq h(\tilde{x}) > 0$. But clearly this contradicts the impossibility of free production $(\Omega.1)$. It is worthwhile to state this formally.

Proposition 4.3: *Suppose* (Ω, u, δ) *satisfies* $(\Omega.1)$-$(\Omega.4)$, (U.1)-(U.3), *and* $0 < \delta < 1$. *If* h *is the optimal policy function, and* $h(\tilde{x}) > 0$ *for some* $\tilde{x} > 0$, h *cannot be non-increasing for all* $x \in (0, \tilde{x}]$.

To relate this result to the sufficient conditions provided in the literature to ensure (locally) non-increasing policy functions, we can proceed as follows. Define u to be *submodular* on Ω^0 if whenever $(x', z'), (x'', z''), (x', z'')$ and (x'', z') belong to Ω^0 and $(x'', z'') >> (x'z,')$, we have

$$u(x'', z'') + u(x', z') \leq u(x'', z') + u(x', z'') \tag{\underline{S}}$$

Following Benhabib-Nishimura (1985), it can be shown that if u is submodular on Ω^0, and $(x^0, h(x^0)) \in \Omega^0$, then h is locally non-increasing at x^0.

Proposition 4.4: *Suppose* (Ω, u, δ) *satisfies* $(\Omega.1)$-$(\Omega.4)$, (U.1)-(U.3), $0 < \delta < 1$ *and* u *is submodular on* Ω^0. *If for some* $\tilde{x}$, *we have* $(\tilde{x}, h(\tilde{x})) \in \Omega^0$, *then there is a neighborhood* $N(\tilde{x})$ *of* $\tilde{x}$, *such that the optimal policy function,* h, *is non-increasing on* $N(\tilde{x})$.

The submodularity of u on Ω^0 can be checked as follows. If u is C^2 on Ω^0 and $D_{12}u(x, z) \leq 0$ for all $(x, z) \in \Omega^0$, then u can be shown to be submodular on Ω^0.

Turning to our primitive model, we can try to find conditions on (f, w, δ) which will ensure that u is C^2 on Ω^0 with $D_{12}u(x, z) \leq 0$ for all $(x, z) \in \Omega^0$. If f is C^1 on $\Re_{++}$, and $w(x, c)$ is C^2 on Ω^0, and for all $(x, c) \in \Omega^0$

$$-D_{22}w(x, c)\ f'(x) \leq D_{12}w(x, c) \tag{4.2}$$

we can ensure $D_{12}u(x, z) \leq 0$ for all $(x, z) \in \Omega^0$.

Now, we can make the following important observation. Condition (4.2) is inconsistent with the following condition:

$$\text{There is some } \hat{x} > 0 \text{ such that } (x, h(x)) \in \Omega^0 \text{ for all } x \in (0, \hat{x}] \tag{4.3}$$

For if (4.3) is satisfied, then by Proposition 4.4, we would have a non-increasing optimal policy function for all $x \in (0, \hat{x}]$. And that is impossible by Proposition 4.3. Thus, *any condition which ensures that (4.3) holds must contradict (4.2).*

A further implication of the fact that we have just noted may be obtained as follows. Suppose we assume condition (4.2), and there is some $\tilde{x} > 0$ such that $h(\tilde{x}) > 0$ [so that the optimal policy is not the "extinction" policy]. Then for some $0 < x \leq \tilde{x}$, we must have $h(x) = f(x)$; that is *for some positive input level, the optimal consumption level must be zero.* We note this formally in the next result.

Proposition 4.5: *Suppose* (f, w, δ) *satisfy* (F.1)-(F.3), (W.1)-(W.3), *and* $0 < \delta < 1$. *Assume, further that* f *is* C^1 *on* $\Re_{++}$, w *is* C^2 *on* Ω^0, *and* (4.2) *is satisfied. If* $h(\tilde{x}) > 0$ *for some* $\tilde{x} > 0$, *then there is some input level* $x' > 0$, *for which* $h(x') = f(x')$.

If we assume the following Inada-type condition:

$$D_2w(x, \lambda c) \to \infty \text{ as } \lambda \to 0 \text{ for all } (x, c) \in \Omega^0 \tag{4.4}$$

then we can ensure that $h(x) < f(x)$ for all $x > 0$. Thus, by Proposition 4.5, Condition (4.2) is inconsistent with Condition (4.4), whenever h is not the extinction policy.

Corollary 4.1: *Under the assumptions of Propositions 4.5, if* $h(\tilde{x}) > 0$ *for some* $\tilde{x} > 0$, *Condition (4.4) must be violated.*

It is worth emphasizing that the relationships between Conditions (4.2) and (4.3) and between Conditions (4.2) and (4.4), might not be readily transparent; they only become so after looking at the *implications* of these conditions for the nature of the optimal policy function.

We make a final remark before concluding this section. In the context of our model, the important contribution of Nishimura and Yano (Chapter 4) can be related to the preceding discussion as follows. If we suppose (in addition to the standard assumptions) that f is C^1 on $\Re_{++}$, w is C^2 on Ω^0, (4.2) is satisfied, and there is a non-trivial stationary optimal stock, then something considerably stronger can be shown than is indicated in Proposition 4.5; viz., there is some input level $x' > 0$ such that $h(x) = f(x)$ for all $0 \leq x \leq x'$, and $h(x) < f(x)$ for all $x > x'$ [and, so, h is non-increasing for $x > x'$ by Proposition 4.4]. This means that the graph of h resembles a "tent-map", and so an interesting route to obtaining cyclical and chaotic optimal programs in this model is indicated, which is quite distinct form the technique pursued by Boldrin and Montrucchio (1986b).

5. Periodic Optimal Programs: an Example

Beyond the "simple dynamics" discussed in Section 4.1, what kinds of optimal dynamic behavior can our model exhibit? In this section, we show by constructing a suitable example, that our model can generate optimal cyclic behavior for *all* discount factors close to one. Examples with a similar flavor have been presented in *multi-commodity* models by Weitzman [reported in Samuelson (1973) and discussed by McKenzie (1983), Benhabib-Nishimura (1985) and others], and Wan (1988). These models can, in fact, be reduced to the form we have discussed in Section 3.2. Indeed, the strategy used in obtaining our example is to start with the reduced-form version of Weitzman's example, and place assumptions on the primitives of our *aggregative* model which will yield that reduced form.

We proceed now to describe the example more precisely. Our purpose is to show that there is a specification of $(f, w, \hat{\delta})$ where

(i) $f : \Re_+ \to \Re_+$ satisfies (F.1)-(F.3);

(ii) $w : \Re^2_+ \to \Re_+$ satisfies (W.1)-(W.3);

(iii) $0 < \hat{\delta} < 1$

such that for *every* δ satisfying $\hat{\delta} < \delta < 1$,

(a) there is a unique non-trivial stationary optimal stock $x(\delta)$

(b) there is an open interval $T(\delta)$, containing $x(\delta)$, such that if $\{x_t\}_0^\infty$ is the optimal program from $\mathbf{x} \in T(\delta)$, with $\mathbf{x} \neq x(\delta)$, then $\{x_t\}_0^\infty$ is periodic with period 2.

Thus, optimal programs continue to be periodic even for discount factors, δ, arbitrarily close to 1, and also for initial input stocks, $\mathbf{x}$, arbitrarily close to the stationary optimal stock, $x(\delta)$.

It is worth mentioning that this example might appear to contradict what is usually called a "turnpike theorem" for discount factors close to one. Actually, it does not, since a correct statement of such a turnpike theorem always includes an assumption of *strict concavity* of the reduced utility function on the transition possibility set. Our example (and Weitzman's) does not have a strictly concave reduced utility function. It should also be observed that the more general theorem of the subject, namely the "neighborhood turnpike theorem" [emphasized notably by McKenzie (1982, 1986)] is completely consistent with our example, since as the discount factor approaches one, optimal programs (even though they exhibit cyclic behavior) are confined in smaller and smaller neighborhoods of the corresponding stationary optimal stock.

Example 5.1: Define $f : \Re_+ \to \Re_+$ by

$$f(x) = \begin{cases} (32/3)x - 32x^2 + (256/3)x^4 & \text{for } x \in [0, 0.25] \\ 1 & \text{for } x \geq 0.25 \end{cases}$$

Define $w : \Re^2_+ \to \Re_+$ by

$$w(x, c) = 2x^{1/2}c^{1/2} \qquad \text{for all } (x, c) \in \Re^2_+$$

Finally, define $\hat{\delta} = (1/3)$.

It can be checked that f satisfies (F.1)-(F.3) with $K = 1$, and w satisfies (W.1)-(W.3).

Now, let δ be any discount factor satisfying $\hat{\delta} < \delta < 1$. We fix this δ in what follows. Define an open interval

$$T(\delta) = \{x : 0.25 < x < 3\delta^2/(1 + 3\delta^2)\}$$

and also a point

$$x(\delta) = \delta/(1 + \delta)$$

It can be shown that

(a) $x(\delta)$ is the unique non-trivial stationary optimal stock [note that $x(\delta) \in T(\delta)$];

(b) for every $\mathbf{x} \in T(\delta)$, with $\mathbf{x} \neq x(\delta)$, the optimal program $\{\hat{x}_t\}_0^\infty$ from $\mathbf{x}$ is periodic with period 2.

6. Chaotic Optimal Programs

6.1 An Example of an Economy Exhibiting Topological and Ergodic Chaos

Optimal cyclic behavior, though "less simple" than monotone behavior, has considerable regularity to it to be viewed as "predictable". In this section, we show by constructing a suitable example that our model can exhibit optimal behavior which is chaotic (in both the "topological" and "ergodic" senses).

Examples similar to ours have been constructed, among others, by Deneckere and Pelikan (1986) and Boldrin and Montrucchio (1986b) in the context of the reduced-form model discussed in Section 3.2. These models in turn have been shown to be reduced-form versions of *two-sector models.* In our example, an *aggregative model* is seen to yield the reduced-form for which the optimal policy function is the logistic function (for an appropriate domain of inputs).

It is, perhaps, worth mentioning that an example of chaotic optimal behavior is obtained in an aggregative model (with wealth effect) by Boldrin and Montrucchio (1986a). While the framework is superficially similar to ours, the production function in their example, is, in fact, "bell-shaped" (initially strictly increasing and thereafter strictly decreasing) and this feature appears to be unattractive in terms of economic modeling, as well as crucial to their construction.

We now describe the nature of our example more formally. Our purpose is to show that there is a specification of $(\mathbf{f}, \mathbf{w}, \bar{\delta})$ where

(i) $\mathbf{f} : \Re_+ \to \Re_+$ satisfies (F.1)-(F.3);

(ii) $\mathbf{w} : \Re_+^2 \to \Re_+$ satisfies (W.1)-(W.3);

(iii) $0 < \bar{\delta} < 1$]

such that the optimal policy function $\mathbf{h}$: $\Re_+ \to \Re_+$ satisfies the logistic equation

$$\mathbf{h}(x) = 4x(1-x) \text{ for } x \in [0,1]$$

Example 6.1: Define $\mathbf{f} : \Re_+ \to \Re_+$ by:

$$\mathbf{f}(x) = \begin{cases} (16/3)x - 8x^2 + (16/3)x^4 & \text{for } x \in [0, 0.5] \\ 1 & \text{for } x \geq 0.5 \end{cases}$$

$$\bar{\delta} = 0.0025$$

The function $\mathbf{w} : \Re_+^2 \to \Re_+$ is defined in a more involved fashion. Define the parameters $L \doteq 98$, $a = 425$. Denote by I the closed interval [0,1], and define the function $\theta : I \to I$ by

$$\theta(x) = 4x(1-x) \text{ for } x \in I$$

Now, define $\mathbf{u}$: $I^2 \to \Re$ by

$$\mathbf{u}(x,z) = ax - 0.5Lx^2 + z\theta(x) - 0.5z^2 - \bar{\delta}[az - 0.5Lz^2 + 0.5\theta(z)^2]$$

Define the set $\mathbf{D} \subset I^2$ by

$$\mathbf{D} = \{(x,c) \in I \times \Re_+ : c \leq \mathbf{f}(x)\}$$

and a function $\mathbf{w}$:$\mathbf{D} \to \Re$ by

$$\mathbf{w}(x,c) = \mathbf{u}(x, \mathbf{f}(x) - c) \text{ for } (x,c) \in \mathbf{D}$$

We extend the definition of $\mathbf{w}$ to the domain Ω. For $(x, c) \in \Omega$ with $x > 1$ (so that $\mathbf{f}(x) = 1$, and $c \leq 1$), define

$$\mathbf{w}(x, c) = \mathbf{w}(1, c)$$

Finally, we extend the definition of $\mathbf{w}$ to the domain $\Re^2_+$. For (x, c) in $\Re^2_+$ with $c > \mathbf{f}(x)$, define

$$\mathbf{w}(x, c) = \mathbf{w}(x, \mathbf{f}(x))$$

It can be checked for this example that $\mathbf{f}$ satisfies (F.1)-(F.3), $\mathbf{w}$ satisfies (W.1)-(W.3).

Furthermore, it can be shown that the optimal policy function $\mathbf{h}$: $\Re_+ \rightarrow \Re_+$ satisfies

$$\mathbf{h}(x) = 4x(1 - x) \text{ for } x \in I$$

Referring to Section 2, we have an example of $(\mathbf{f}, \mathbf{w}, \bar{\delta})$ for which the dynamical system (I, h) exhibits topological and ergodic chaos (and also has a positive Lyapunov exponent).

6.2 Robustness of Topological Chaos

In Section 6.1, we provided a specification of $(\mathbf{f}, \mathbf{w}, \bar{\delta})$ for which topological chaos is seen to occur. A natural question to ask is whether this is fortuitous (so that if the parameters f, w or δ were perturbed ever so little, the property of topological chaos would disappear) or whether this is a "robust" phenomenon (so that small perturbations of f, w or δ would preserve the property of topological chaos).

We proceed now to formalize the above question as follows. Define

$\mathcal{F} = \{f : \Re_+ \rightarrow \Re_+ \text{ satisfying (F.1)-(F.3)}\}$;
$\mathcal{W} = \{w : \Re^2_+ \rightarrow \Re_+ \text{ satisfying (W.1)-(W.3)}\}$;
$\Delta = \{\delta : 0 < \delta < 1\}$.

An *economy*, e, is defined by a triple $(f, w, \bar{\delta}) \in \mathcal{F} \times \mathcal{W} \times \Delta$. The set of economies, $\mathcal{F} \times \mathcal{W} \times \Delta$ is defined by E.

Consider the economy $\mathbf{e} = (\mathbf{f}, \mathbf{w}, \bar{\delta})$ defined in Section 6.1. We would like to demonstrate that all economies $e \in E$ "near" the economy $\mathbf{e}$ will exhibit topological chaos. Thus, the property of topological chaos will be seen to persist for small perturbations of the original economy.

A convenient way to make the above idea precise is to define for $e \in E$, the "distance" between economies e and $\mathbf{e}$ by

$$d(e, \mathbf{e}) = \sup_{x \geq 0} |f(x) - \mathbf{f}(x)| + \sup_{(x,c) \geq 0} |w(x,c) - \mathbf{w}(x,c)| + |\delta - \bar{\delta}|$$

Note that $d(e, \mathbf{e})$may be infinite.

Before we proceed further, we have to clarify two preliminary points. First, if we perturb the original economy $\mathbf{e}$ (that is, choose another economy $e \neq \mathbf{e}$), we will, in general, change the set of programs from any given initial stock, as also the optimal program from any given initial stock. That is, programs and optimal programs [and hence optimal policy functions] are economy specific. Thus, given an economy e, we use expressions like "e-program", "e-optimal program", "e-optimal policy function" with obvious meanings.

Second, recall that for the original economy $\mathbf{e}$, $K = 1$, and so if the initial stock was in I, then for any program from the initial stock, the input stock in every period is confined to I. Furthermore, for any initial stock not in I, the input stock on any program belongs to I from the very next period. In this sense, $(I, \mathbf{h})$ is the "natural" dynamical system for the economy $\mathbf{e}$. When we perturb the economy, we do not wish to restrict the kind of perturbation in any way, and so we would have to allow the new economy's production function to satisfy (F.3) with $K > 1$. This changes the "natural" state space of the dynamical system for the new economy. However, recalling that we are only interested in "small" perturbations, it is surely possible to ensure that $d(e, \mathbf{e}) \leq 1$, so that we can legitimately take the "natural" state space choice to be $J = [0, 2]$.

We can now describe our result formally as follows.

Theorem 6.1: *Let* $\mathbf{e} = (\mathbf{f}, \mathbf{w}, \bar{\delta})$ *be the economy described in Example 6.1. There exists some* $\varepsilon > 0$ *such that for every econ-*

omy $e \in E$ *with* $d(e, \mathbf{e}) < \varepsilon$, *the dynamical system* (J, h) *exhibits topological chaos, where* h *is the e-optimal policy function and* $J = [0, 2]$.

7. Appendix

7.1 Verification of Results of Section 4

(a) Verification of Proposition 4.5

Suppose, contrary to the claim, we have $h(x) < f(x)$ for all $0 < x \leq \tilde{x}$. Then we have two cases to consider. Either (i) $h(x) > 0$ for all $x \in (0, \tilde{x}]$, or (ii) $h(x) = 0$ for some $x \in (0, \tilde{x}]$. In case (i), $(x, h(x)) \in \Omega^0$ for $x \in (0, \tilde{x}]$; that is Condition (4.3) is satisfied. But this is inconsistent with (4.2), so case (i) could not arise. In case (ii), noting that $h(\tilde{x}) > 0$, we actually have $h(x) = 0$ for some $x \in (0, \tilde{x})$. Let $x'' = \sup\{x \in (0, \tilde{x}) : h(x) = 0\}$. Then $h(x'') = 0$, and $0 < x'' < \tilde{x}$ by continuity of h. Also, $f(x) > h(x) > 0$ for all $x'' < x < \tilde{x}$, so that $(x, h(x)) \in \Omega^0$ for $x'' < x < \tilde{x}$. By Proposition 4.4, h is non-increasing in $(x'', \tilde{x})$, and the continuity of h leads to $h(x'') \geq h(\tilde{x}) > 0$, a contradiction. Thus case (ii) could not arise. Since these are the only two cases possible, the claim is established.

(b) Verification of Corollary 4.1

We claim that Condition (4.4) implies $h(x) < f(x)$ for all $x > 0$. For if the claim is false, there is some $x > 0$ for which $h(x) = f(x)$. This means that if $\{\bar{x}_t\}_0^\infty$ is the optimal program from x, then $\bar{x}_1 = f(x)$. Consider the sequence $\{x'_t\}_0^\infty$ defined by $x'_0 = x$, and $x'_t = (1 - \lambda)x_t$ for $t \geq 0$, where $0 < \lambda < 0.5$. Clearly $\{x'_t\}_0^\infty$ is a program. Now, we have

$$\sum_0^\infty \delta^t w(x'_t, c'_{t+1}) \geq w(x, \lambda f(x)) + (1 - \lambda(\delta V(f(x))$$

$$= [w(x, \lambda f(x)) - w(x, 0)] + [w(x, 0) + \delta V(f(x))] + [(1 - \lambda)\delta V(f(x)) - \delta V(f(x))]$$

$$\geq D_2 w(x, \lambda f(x))\lambda f(x) + V(x) - \lambda\delta V(f(x))$$

Thus $\sum_0^\infty \delta^t w(x'_t, c'_{t+1}) - V(x) \geq \lambda[D_2 w(x, \lambda f(x)) - \delta V(f(x))]$.

Now $(x, f(x)/2) \in \Omega^0$ and so by Condition (4.4),
$D_2 w(x, 2\lambda(f(x)/2)) \to \infty$ as $2\lambda \to 0$.
Thus for λ small enough and positive, we can ensure
$\lambda[D_2 w(x, \lambda f(x)) - \delta V(f(x))] > 0$. This implies that

$$\sum_0^\infty \delta^t w(x'_t, c'_{t+1}) > V(x)$$

which is a contradiction to the definition of V. This establishes our claim.

Corollary 4.1 now follows directly from Proposition 4.5.

7.2 Verification of Example 5.1

(a) Verification of Assumptions (F.1)-(F.3), (W.1)-(W.3)

Note that $f(0) = 0$ and $f(x) \to 1$ as $x \to 0.25$ from the left, so f is clearly continuous on $\Re_+$ verifying (F.1).

Next, observe that for $0 \leq x < 0.25$, f is C^2 and

$$f'(x) = (32/3) - 64x + (1024/3)x^3$$
$$f''(x) = -64 + 1024x^2$$

As $x \to 0.25$ from the left, $f'(x) \to 0$ and $f''(x) \to 0$, so f is in fact C^2 on $\Re_+$. It is clear that $f''(x) < 0$ for $0 \leq x < 0.25$ and so $f'(x) > 0$ for $0 \leq x < 0.25$ [since $f'(0.25) = 0$]. Thus, f is non-decreasing and concave on $\Re_+$ verifying (F.2).

Note that (F.3) is satisfied by choosing $K = 1$. For if $x > 1$, then $[f(x)/x] = (1/x) < 1$. If $0.25 \leq x < 1$, $[f(x)/x] = (1/x) > 1$. Also, since $f(0) = 0$ and f is concave, $[f(x)/x]$ is non-increasing in x. Thus, for $0 < x < 0.25$, $[f(x)/x] \geq [f(0.25)/0.25] = 4 > 1$.

The verification of assumptions (W.1)-(W.3) for the given welfare function is too obvious to be spelt out in detail.

(b) Verification of the unique non-trivial stationary optimal stock $x(\delta)$

Since $0.25 < x(\delta) < 1$, we have $f(x(\delta)) = 1 > x(\delta)$, and so $c(\delta) \equiv f(x(\delta)) - x(\delta) = [1/(1+\delta)] > 0$. Define $\{\bar{x}_t\}_0^\infty$ by $\bar{x}_t = x(\delta)$ for $t \geq 0$. Then, $\{\bar{x}_t\}_0^\infty$ is a stationary program from $x(\delta)$, and $\bar{c}_{t+1} = f(\bar{x}_t) - \bar{x}_{t+1} = c(\delta) > 0$ for $t \geq 0$. Next, define a sequence $\{p_t\}_0^\infty$ by $p_t = (\delta^t/\delta^{1/2})$ for $t \geq 0$.

Recalling our definition of Ω and u (see Section 3.2), we note that u is C^1 in the interior of Ω (that is, Ω^0), and continuous and concave on Ω. Thus, for any $(x, z) \in \Omega$, and for any $t \geq 0$,

$$\begin{aligned} u(x,z) - u(\bar{x}_t, \bar{x}_{t+1}) \leq {} & D_1 u(\bar{x}_t, \bar{x}_{t+1})(x - \bar{x}_t) + \\ & D_2 u(\bar{x}_t, \bar{x}_{t+1})(z - \bar{x}_{t+1}) \end{aligned}$$

Now, for any $(x, z) \in \Omega^0$, $x > 0$ and $f(x) - z > 0$. So

$$\left.\begin{aligned} D_2 u(x,z) &= -D_2 w(x, f(x) - z) \\ D_1 u(x,z) &= D_1 w(x, f(x) - z) + f'(x) D_2 w(x, f(x) - z) \end{aligned}\right] \tag{7.1}$$

Since $0.25 < x(\delta) < 1$, $f'(x(\delta)) = 0$, we have
$[D_2 u(x(\delta), x(\delta)) = -D_2 w(x(\delta), c(\delta))]$ and
$[D_1 u(x(\delta), x(\delta)) = D_1 w(x(\delta), c(\delta))]$.
So

$$\begin{aligned} u(x,z) - u(\bar{x}_t, \bar{x}_{t+1}) &= D_1 w(x(\delta), c(\delta))(x - \bar{x}_t) - \\ & \quad D_2 w(x(\delta), c(\delta))(z - \bar{x}_{t+1}) \\ &= (1/\delta^{1/2})(x - \bar{x}_t) - \delta^{1/2}(z - \bar{x}_{t+1}) \end{aligned}$$

Thus, multiplying through by δ^t, transposing terms, and substituting $p_t = (\delta^t/\delta^{1/2})$, we get

$$\delta^t u(x,z) + p_{t+1} z - p_t x \leq \delta^t u(\bar{x}_t, \bar{x}_{t+1}) + p_{t+1}\bar{x}_{t+1} - p_t \bar{x}_t$$

Also, $p_t \bar{x}_t = p_t x(\delta) \to 0$ as $t \to \infty$. Thus, by Theorem 3.2, $\{\bar{x}_t\}_0^\infty$ is an optimal program from $x(\delta)$, and $x(\delta)$ is a non-trivial stationary optimal stock.

To check that $x(\delta)$ is the only non-trivial stationary optimal stock, we proceed as follows. Suppose $\bar{x}$ is any non-trivial stationary optimal stock. Then, by stationarity, $\bar{x} \in [0,1]$. Since $\bar{x}$ is non-trivial, $\bar{x} \in (0,1]$. By optimality, $\bar{x} \in (0,1)$. Since $f(x) > x$ for all $x \in (0,1)$, $(\bar{x},\bar{x})$ is in Ω^0. Denote $f(\bar{x}) - \bar{x}$ by $\bar{c}$; then $\bar{c} > 0$.

Now, for each $t \geq 0$, $\bar{x}$ must solve the following problem

$$\begin{array}{ll} \text{Max} & u(\bar{x},x) + \delta u(x,\bar{x}) \\ \text{Subject to} & (\bar{x},x) \in \Omega \text{ and } (x,\bar{x}) \in \Omega \end{array}$$

Since $(\bar{x},\bar{x})$ is in the interior of Ω, we have the first-order condition

$$D_2 u(\bar{x},\bar{x}) + \delta D_1 u(\bar{x},\bar{x}) = 0$$

Thus, using (7.1), we get

$$(\bar{x}/\bar{c})^{1/2} = \delta(\bar{x}/\bar{c})^{1/2} f'(\bar{x}) + \delta(\bar{c}/\bar{x})^{1/2}$$

Since $f'(\bar{x}) \geq 0$, we then have $\bar{x} \geq \delta\bar{c} = \delta[f(\bar{x}) - \bar{x}]$. This yields

$$\delta\{[f(\bar{x})/\bar{x}] - 1\} \leq 1$$

Since f is concave and $f(0) = 0$, $[f(x)/x]$ is non-increasing in x. Thus, for $x \leq 0.25$, $[f(x)/x)] \geq [f(0.25)/0.25] = 4$, and $\delta[f(x)/x] - 1 \geq 3\delta > 1$. Consequently, $\bar{x} > 0.25$.

Since $\bar{x} > 0.25$, we can return to the above first-order condition, and observe that $f'(\bar{x}) = 0$; so, we obtain $\bar{x} = \delta\bar{c} = \delta[f(\bar{x}) - \bar{x}] = \delta(1 - \bar{x})$. This means $\bar{x} = [\delta/(1+\delta)] = x(\delta)$.

(c) Verification of the Periodic Optimal Program

Given any $\tilde{x} \in T(\delta)$, $\tilde{x} \neq x(\delta)$, define

$$\tilde{y} = \delta^2(1-\tilde{x})/[\delta^2(1-\tilde{x}) + \tilde{x}]$$

Given the definition of $\tilde{y}$, it is easy to check that $\tilde{y} \neq \tilde{x}$, and

$$[\delta^2(1-\tilde{y}) + \tilde{y}][\delta^2(1-\tilde{x}) + \tilde{x}] = \delta^2 \tag{7.2}$$

a relation we will find useful in what follows.

We note that $\tilde{y} < \delta^2(1-\tilde{x})/[\delta^2(1-\tilde{x})+0.25] < \delta^2(1-0.25)/[\delta^2(1-0.25)+0.25] = 3\delta^2/[1+3\delta^2]$. Also, $\delta^2(1-\tilde{x})/\tilde{x} = \delta^2[(1/\tilde{x})-1] > \delta^2[\{1+3\delta^2)/3\delta^2\}-1] = (1/3)$. So, $\tilde{y} = \delta^2(1-\tilde{x})/[\delta^2(1-\tilde{x})+\tilde{x}] = 1-\{\tilde{x}/[\delta^2(1-\tilde{x})+\tilde{x}]\} = 1-(1/[\{\delta^2(1-\tilde{x})/\tilde{x}\}+1]) > 1-(1/[(1/3)+1] = 0.25$. Thus, $\tilde{y} \in A(\delta)$.

Define a sequence $\{\tilde{x}_{zt}\}_0^\infty$ as $\tilde{x}_t = \tilde{x}$ for $t = 0, 2, 4, ...$; $\tilde{x}_t = \tilde{y}$ for $t = 1, 3, 5, ...$ We can verify that $\{\tilde{x}_t\}_0^\infty$ is a program from $\tilde{x}$. To see this, note that for $t = 0, 2, 4, ..., \tilde{x}_{t+1} = \tilde{y} < 1$, and $f(\tilde{x}_t) = f(\tilde{x}) = 1$, so that for $t = 0, 2, 4$,

$$\tilde{c}_{t+1} \equiv f(\tilde{x}_t) - \tilde{x}_{t+1} = \tilde{x}/[\delta^2(1-\tilde{x})+\tilde{x}]$$

Similarly, for $t = 1, 3, 5, ..., \tilde{x}_{t+1} = \tilde{x} < 1$, and $f(\tilde{x}_t) = f(\tilde{y}) = 1$, so that for $t = 1, 3, 5...$

$$\tilde{c}_{t+1} = f(\tilde{x}_t) - \tilde{x}_{t+1} = (1-\tilde{x})$$

It remains to verify that $\{\tilde{x}_t\}_0^\infty$ is an optimal program from $\tilde{x}$. To this end, define a sequence $\{\tilde{p}_t\}_0^\infty$ as follows. For $t = 0, 2, 4, ...,$

$$\tilde{p}_{t+1} = \delta^t[\delta^2(1-\tilde{x})+\tilde{x}]^{1/2} \tag{7.3}$$

For $t = 1, 3, 5, ...,$

$$\tilde{p}_{t+1} = \delta^{t+1}/[\delta^2(1-\tilde{x})+\tilde{x}]^{1/2} \tag{7.4}$$

Further, define

$$\tilde{p}_0 = 1/[\delta^2(1-\tilde{x})+\tilde{x}]^{1/2} \tag{7.5}$$

Since $(\tilde{x}_t, \tilde{x}_{t+1}) \in \Omega^0$ for $t \geq 0$, we obtain, for any $(x, z) \in \Omega$, and $t \geq 0$, [using (7.1)]

$$\begin{aligned} u(x,z) - u(\tilde{x}_t, \tilde{x}_{t+1}) &\leq D_1u(\tilde{x}_t, \tilde{x}_{t+1})(x-\tilde{x}_t) + \\ &\quad D_2u(\tilde{x}_t, \tilde{x}_{t+1})(z-\tilde{x}_{t+1}) \\ &= D_1w(\tilde{x}_t, \tilde{c}_{t+1})(x-\tilde{x}_t) - \\ &\quad D_2w(\tilde{x}_t, \tilde{c}_{t+1})(z-\tilde{x}_{t+1}) \end{aligned}$$

For $t = 0, 2, 4, \ldots$, we then have

$$u(x, z) - u(\tilde{x}_t, \tilde{x}_{t+1}) \leq \{1[\delta^2(1-\tilde{x}) + \tilde{x}]\}^{1/2}(x - \tilde{x}) - [\delta^2(1-\tilde{x}) + \tilde{x}]^{1/2}(z - \tilde{y}).$$

So, multiplying through by δ^t, and using (7.3)-(7.5),

$$\delta^t[u(x, z) - u(\tilde{x}_t, \tilde{x}_{t+1})] \leq \tilde{p}_t(x - \tilde{x}) - \tilde{p}_{t+1}(z - \tilde{y}).$$

Transposing terms, for any $(x, z) \in \Omega$, and $t = 0, 2, 4, \ldots$,

$$\delta^t u(\tilde{x}_t, \tilde{x}_{t+1}) + \tilde{p}_{t+1}\tilde{x}_{t+1} - \tilde{p}_t\tilde{x}_t \geq \delta^t u(x, z) + \tilde{p}_{t+1}z - \tilde{p}_t x \quad (7.6)$$

For $t = 1, 3, 5, \ldots$, we have

$$\begin{aligned} &u(x, z) - u(\tilde{x}_t, \tilde{x}_{t+1}) \\ &\leq \{1/[\delta^2(1-\tilde{y}) + \tilde{y}]\}^{1/2}(x - \tilde{y}) - [\delta^2(1-\tilde{y}) + \tilde{y}]^{1/2}(z - \tilde{x}) \\ &= (1/\delta)[\delta^2(1-\tilde{x}) + \tilde{x}]^{1/2}(x - \tilde{y}) - \\ &\{\delta/[\delta^2(1-\tilde{x}) + \tilde{x}]\}^{1/2}(z - \tilde{x}) \end{aligned}$$

using the relation (7.2). Thus, multiplying through by δ^t, we have

$$\delta^t[u(x, z) - u(\tilde{x}_t, \tilde{x}_{t+1})] \leq \tilde{p}_t(x - \tilde{y}) - \tilde{p}_{t+1}(z - \tilde{x})$$

using (7.3) and (7.4). Transposing terms, we obtain, for any $(x, z) \in \Omega$, and $t = 1, 3, 5, \ldots$

$$\delta^t u(\tilde{x}_t, \tilde{x}_{t+1}) + \tilde{p}_{t+1}\tilde{x}_{t+1} - \tilde{p}_t\tilde{x}_t \geq \delta^t u(x, z) + \tilde{p}_{t+1}z - \tilde{p}_t x \quad (7.7)$$

Note finally that $\tilde{p}_t \to 0$ as $t \to \infty$ and $\tilde{x}_t$ is either $\tilde{x}$ or $\tilde{y}$ for each $t \geq 0$. Hence, we have

$$\lim_{t\to\infty} \tilde{p}_t\tilde{x}_t = 0 \quad (7.8)$$

Using (7.6)-(7.8) and Theorem 3.2, $\{\tilde{x}_t\}_0^\infty$ is an optimal program from $\tilde{x}$.

7.3 Verification of Example 6.1

(a) Verification of Assumptions (F.1)-(F.3), (W.1)-(W.3)

Note that $\mathbf{f}(0) = 0$, and as $x \to 0.5$ from the left, $\mathbf{f}(x) \to 1$. So $\mathbf{f}$ is clearly continuous on $\Re_+$, verifying (F.1).

Next, note that for $0 \leq x < 0.5$, $\mathbf{f}$ is C^2, and

$$\mathbf{f}'(x) = (16/3) - 16x + (64/3)x^2$$
$$\mathbf{f}''(x) = -16 + 64x^2$$

As $x \to 0.5$ from the left, $\mathbf{f}'(x) \to 0$ and $\mathbf{f}''(x) \to 0$, so $\mathbf{f}$ is in fact C^2 on $\Re_+$. It is clear that $\mathbf{f}''(x) < 0$ for $0 \leq x < 0.5$, and so $\mathbf{f}'(x) > 0$ for $0 \leq x < 0.5$ [since $\mathbf{f}'(0.5) = 0$]. Thus, $\mathbf{f}$ is non-decreasing and concave on $\Re_+$, verifying (F.2).

Note that (F.3) is satisfied by choosing $K = 1$. For if $x > 1$, then $[\mathbf{f}(x)/x] = (1/x) < 1$. If $0.5 \leq x < 1$, $[\mathbf{f}(x)/x] = (1/x) > 1$. Also, since $\mathbf{f}(0) = 0$, and $\mathbf{f}$ is concave, $[\mathbf{f}(x)/x]$ is non-increasing in x. Thus, for $0 < x < 0.5$, $[\mathbf{f}(x)/x] \geq [\mathbf{f}(0.5)/0.5] = 2 > 1$.

Before proceeding to verify the assumptions (W.1)-(W.3), we note an important property of $\mathbf{f}$, namely

$$\mathbf{f}(x) \geq \theta(x) \text{ for } x \in I$$

To see this, note that $\mathbf{f}(x) = 1$ for $0.5 \leq x \leq 1$, and $\theta(x) \in I$ for all $x \in I$ so $\mathbf{f}(x) \geq \theta(x)$ for $0.5 \leq x \leq 1$. Also, $\mathbf{f}(0) = \theta(0) = 0$. For $0 < x < 0.5$, we define $\beta(x) \equiv [\mathbf{f}(x) - \theta(x)]/x$. Then $\beta(x) = (4/3) - 4x + (16/3)x^3$, and $\beta'(x) = -4 + 16x^2$. Thus $\beta'(x) < 0$ for $0 < x < 0.5$, and β is a decreasing function. As $x \to 0.5$ from the left $\beta(x) \to 0$. Thus, $\beta(x) > 0$ for $0 < x < 0.5$, and so $\mathbf{f}(x) > \theta(x)$ for $0 < x < 0.5$.

We now verify (W.1)-(W.3). Recalling the definition of $\mathbf{u} : I^2 \to$

$\Re$, we can compute the following derivatives:

$$\begin{aligned}
D_1\mathbf{u}(x,z) &= a - Lx + 4z(1-2x) \\
D_2\mathbf{u}(x,z) &= 4x(1-x) - z - \bar{\delta}a + \bar{\delta}Lz - 16\bar{\delta}z + \\
&\quad 48\bar{\delta}z^2 - 32\bar{\delta}z^3 \\
D_{11}\mathbf{u}(x,z) &= -L - 8z \\
D_{12}\mathbf{u}(x,z) &= 4(1-2x) = D_{21}u(x,z) \\
D_{22}\mathbf{u}(x,z) &= -1 + \bar{\delta}L - 16\bar{\delta} + 96\bar{\delta}z - 96\bar{\delta}z^2
\end{aligned}$$

We first check (W.1)-(W.3) on the set $\mathbf{D}$. Clearly $\mathbf{w}$ is C^2 on $\mathbf{D}$, and we can compute the first partials of w as follows:

$$\begin{aligned}
D_2\mathbf{w}(x,c) &= -D_2\mathbf{u}(x,\mathbf{f}(x)-c) \\
D_1\mathbf{w}(x,c) &= D_1\mathbf{u}(x,\mathbf{f}(x)-c) + D_2\mathbf{u}(x,\mathbf{f}(x)-c)\mathbf{f}'(x)
\end{aligned}$$

Now,

$$\begin{aligned}
&D_2\mathbf{u}(x,\mathbf{f}(x)-c \\
&\leq 1 - \bar{\delta}a + \bar{\delta}z(L-16) - z + 48\bar{\delta} \\
&\leq 1 - \bar{\delta}a + 48\bar{\delta} \text{ [since } \bar{\delta}(L-16) \leq 1] \\
&\leq 1 - \bar{\delta}(a-48) \\
&< 0 \text{ [since } \bar{\delta}(a-48) > 1]
\end{aligned}$$

Thus $D_2\mathbf{w}(x,c) > 0$ for $(x,c) \in \mathbf{D}$. Also, $D_1\mathbf{u}(x,z) \geq a-L-4 \geq 323$, $\mathbf{f}'(x) \leq \mathbf{f}'(0) = (16/3)$, and

$$\begin{aligned}
&D_2\mathbf{u}(x,z) \\
&\geq -\bar{\delta}a - z + \bar{\delta}z(L-16) + 16\bar{\delta}z^2(3-2z) \\
&\geq -\bar{\delta}a - z \text{ [since } L > 16] \\
&\geq -2.0625
\end{aligned}$$

Thus, $D_2\mathbf{u}(x,z)\mathbf{f}'(x) \geq -16$, and $D_1w(x,c) \geq 323 - 16 > 0$. Thus w is strictly increasing in c (given x) and in x (given c) on $\mathbf{D}$.

The second partials of $\mathbf{w}$ are:

$$\begin{aligned}
D_{22}\mathbf{w}(x,c) &= D_{22}\mathbf{u}(x,\mathbf{f}(x)-c) \\
D_{21}\mathbf{w}(x,c) &= -D_{21}\mathbf{u}(x,\mathbf{f}(x)-c) - D_{22}\mathbf{u}(x,\mathbf{f}(x)-c)\mathbf{f}'(x) \\
&= D_{12}\mathbf{w}(x,c) \\
D_{11}\mathbf{w}(x,c) &= D_{11}\mathbf{u}(x,\mathbf{f}(x)-c) + D_{12}\mathbf{u}(x,\mathbf{f}(x)-c)\mathbf{f}'(x) \\
&\quad + [D_{21}\mathbf{u}(x,\mathbf{f}(x)-c) \\
&\quad + D_{22}\mathbf{u}(x,\mathbf{f}(x)-c)\mathbf{f}'(x)]\mathbf{f}'(x) + \\
&\quad D_2\mathbf{u}(x,\mathbf{f}(x)-c)\mathbf{f}''(x)
\end{aligned}$$

Now, $D_{22}\mathbf{u}(x,z) \le -1+\bar{\delta}L+96\bar{\delta}z \le -1+\bar{\delta}(L+96) < -0.5 < 0$; and $D_{22}\mathbf{u}(x,z) \ge -1+\bar{\delta}(L-16) > -1$. Also, $D_{11}\mathbf{u}(x,z) \le -L < 0$; and $|D_{12}\mathbf{u}(x,z)| \le 4$. These estimates imply that

$$(D_{11}\mathbf{u})(D_{22}\mathbf{u}) - (D_{12}\mathbf{u})^2 > (L/2) - 16 = 33 > 0$$

Also, $D_2\mathbf{u}(x,z) > -2.0625$ and $D_{22}\mathbf{u}(x,z) > -1$ as checked above; while $\mathbf{f}''(x) = -16 + 64x^2$ for $0 \le x \le 0.5$, so $0 \ge \mathbf{f}''(x) \ge -16$. Thus, $(D_{22}\mathbf{u})(D_2\mathbf{u})\mathbf{f}''(x) > -33$.

Now, $(D_{11}\mathbf{w})(D_{22}\mathbf{w}) - (D_{12}\mathbf{w})^2 = (D_{22}\mathbf{u})(D_{11}\mathbf{u})+ 2(D_{12}\mathbf{u})(D_{22}\mathbf{u})\mathbf{f}'(x) + (D_{22}\mathbf{u})^2\mathbf{f}'(x)^2 + (D_{22}\mathbf{u})(D_2\mathbf{u})\mathbf{f}''(x) - (D_{12}\mathbf{u})^2 - (D_{22}\mathbf{u})^2\mathbf{f}'(x)^2 - 2(D_{12}\mathbf{u})(D_{22}\mathbf{u})\mathbf{f}'(x) = (D_{22}\mathbf{u})(D_{11}\mathbf{u}) - (D_{12}\mathbf{u})^2 + (D_{22}\mathbf{u})(D_2\mathbf{u})\mathbf{f}''(x)$. Thus, using the above estimates, we have

$$D_{11}\mathbf{w}(x,c)D_{22}\mathbf{w}(x,c) - (D_{12}\mathbf{w}(x,c))^2 > 33 - 33 = 0$$

Also, $D_{22}\mathbf{w}(x,c) = D_{22}\mathbf{u}(x,\mathbf{f}(x)-c) < 0$, so that $\mathbf{w}(x,c)$ is strictly concave on the set $\mathbf{D}$.

Next, we check that w satisfies (W.1)-(W.3) on Ω. By definition of w on Ω, it satisfies (W.1) and (W.2). To check (W.3), let (x,c) and $(\bar{x},\bar{c})$ belong to Ω and let $0 < \lambda < 1$. Then $\mathbf{w}(\lambda(x,c) + (1-\lambda)(\bar{x},\bar{c})) = \mathbf{w}(\lambda x + (1-\lambda)\bar{x}, \lambda c + (1-\lambda)\bar{c})$. Now $[\lambda x + (1-\lambda)\bar{x}, \lambda c + (1-\lambda)\bar{c}]$ is in $\mathbf{D}$ if $[\lambda x + (1-\lambda)\bar{x}] \le 1$. If x and $\bar{x}$ are both ≤ 1, then concavity of $\mathbf{w}$ follows from concavity of $\mathbf{w}$ on $\mathbf{D}$. So, consider without loss of generality that $x > 1$ while $\bar{x} < 1$. Then $\mathbf{w}(\lambda x + (1-\lambda)\bar{x}, \lambda c + (1-\lambda)\bar{c}) \ge \mathbf{w}(\lambda + (1-\lambda)\bar{x}, \lambda c + (1-\lambda)\bar{c})$

[using (W.2)] $\geq \lambda \mathbf{w}(1,c) + (1-\lambda)\mathbf{w}(\bar{x},\bar{c})$ [since $(1,c)$ and $(\bar{x},\bar{c})$ belong to $\mathbf{D}$] $= \lambda \mathbf{w}(x,c) + (1-\lambda)\mathbf{w}(\bar{x},\bar{c})$. If $[\lambda x + (1-\lambda)\bar{x}] > 1$, then $\mathbf{w}(\lambda x + (1-\lambda)\bar{x}, \lambda c + (1-\lambda)\bar{c}) = \mathbf{w}(1, \lambda c + (1-\lambda)\bar{c}) \geq \lambda \mathbf{w}(1,c) + (1-\lambda)\mathbf{w}(1,\bar{c})$ [since $(1,c)$ and $(1,\bar{c})$ belong to $\mathbf{D}$] $\geq \lambda \mathbf{w}(x,c) + (1-\lambda)\mathbf{w}(\bar{x},\bar{c})$.

If $0 \leq x \leq 1$ is fixed, then $\mathbf{w}(x,c)$ is strictly concave in c [since $D_{22}w(x,c) < 0$ on $\mathbf{D}$]. If $x > 1$ is fixed, $\mathbf{w}(x,c) = \mathbf{w}(1,c)$ is strictly concave in c [since $D_{22}w(x,c) < 0$ on $\mathbf{D}$].

Finally, we check that $\mathbf{w}$ satisfies (W.1)-(W.3) on $\Re^2_+$. Again, $\mathbf{w}$ clearly satisfies (W.1) and (W.2). To check (W.3), let (x,c) and $(\bar{x},\bar{c})$ belong to $\Re^2_+$ and let $0 < \lambda < 1$. Denote $\min[c, \mathbf{f}(x)]$ by $G(x,c)$; $\min[\bar{c}, \mathbf{f}(\bar{x})]$ by $G(\bar{x},\bar{c})$. Then, we have $\mathbf{w}(\lambda x + (1-\lambda)\bar{x}, \lambda c + (1-\lambda)\bar{c}) \geq \mathbf{w}(\lambda x + (1-\lambda)\bar{x}, \lambda G(x,c) + (1-\lambda)G(\bar{x},\bar{c}))$. Also, $\lambda G(x,c) + (1-\lambda)G(\bar{x},\bar{c}) \leq \lambda \mathbf{f}(x) + (1-\lambda)\mathbf{f}(\bar{x}) \leq \mathbf{f}(\lambda x + (1-\lambda)\bar{x})$. Further $G(x,c) \leq \mathbf{f}(x)$ and $G(\bar{x},\bar{c}) \leq \mathbf{f}(\bar{x})$. Thus, $(x, G(x,c)), (\bar{x}, G(\bar{x},\bar{c}))$ and $[\lambda x + (1-\lambda)\bar{x}, \lambda G(x,c) + (1-\lambda)G(\bar{x},\bar{c})]$ all belong to Ω. Thus, $\mathbf{w}(\lambda x + (1-\lambda)\bar{x}, \lambda c + (1-\lambda)\bar{c}) \geq \lambda \mathbf{w}(x, G(x,c)) + (1-\lambda)\mathbf{w}(\bar{x}, G(\bar{x},\bar{c}))$. If $\min[c, \mathbf{f}(x)] = c$ then $\mathbf{w}(x, G(x,c)) = \mathbf{w}(x,c)$; if $\min[c, \mathbf{f}(x)] \neq c$, then $c > \mathbf{f}(x)$, and $\mathbf{w}(x, G(x,c)) = \mathbf{w}(x, \mathbf{f}(x)) = \mathbf{w}(x,c)$; thus in either case, $\mathbf{w}(x, G(x,c)) = \mathbf{w}(x,c)$. Similarly, $\mathbf{w}(\bar{x}, G(\bar{x},\bar{c})) = \mathbf{w}(\bar{x},\bar{c})$. Hence, $\lambda \mathbf{w}(x, G(x,c)) + (1-\lambda)\mathbf{w}(\bar{x}, G(\bar{x},\bar{c})) = \lambda \mathbf{w}(x,c) + (1-\lambda)\mathbf{w}(\bar{x},\bar{c})$, completing our demonstration of the concavity of $\mathbf{w}$ on $\Re^2_+$.

It is worth noting that by (W.1), $\mathbf{w}(x,c) \geq \mathbf{w}(0,0) = \mathbf{u}(0, f(0)-0) = 0$ for all $(x,c) \in \Re^2_+$. Thus $\mathbf{w}$ maps $\Re^2_+$ to $\Re_+$. Also, for all $(x,c) \in \Re^2_+$, $\mathbf{w}(x,c) \leq \mathbf{w}(1,c) \leq \mathbf{w}(1, f(1)) = \mathbf{w}(1,1) = \mathbf{u}(1,0) = 376$.

(b) Verification of the Optimal Policy Function

This part of our verification relies heavily on the technique of Boldrin-Montrucchio (1986b).

Define $\phi : I^2 \to \Re$ by

$$\phi(x,z) = ax - 0.5Lx^2 + z\theta(x) - 0.5z^2$$

Clearly ϕ is C^2 on I^2, and we can compute the following derivatives:

$$\begin{aligned}
D_1\phi(x,z) &= a - Lx + z\theta'(x) \\
D_2\phi(x,z) &= \theta(x) - z \\
D_{11}\phi(x,z) &= -L + z\theta''(x) \\
D_{12}\phi(x,z) &= \theta'(x) = D_{21}\phi(x,z) \\
D_{22}\phi(x,z) &= -1
\end{aligned}$$

Now $\theta'(x) = 4 - 8x$ and $\theta''(x) = -8$. Thus $|\theta'(x)| \leq 4$ for $x \in I$. Also, looking at the Hessian matrix of ϕ, we note that $D_{11}\phi(x,z) < 0$ since $L > 0$ and $\theta''(x) < 0$. And $(D_{11}\phi)(D_{22}\phi) - (D_{12}\phi)^2 = L - z\theta''(x) - [\theta'(x)]^2 \geq L - 16 > 0$. Thus ϕ is strictly concave on I^2.

Clearly $\theta(x) \in I$ for all $x \in I$. And, for any $z \in I$, $z \neq \theta(x)$, we have by the strict concavity of ϕ,

$$\phi(x,z) - \phi(x,\theta(x)) < D_2\phi(x,\theta(x))(z - \theta(x)) = 0$$

Hence, for every $x \in I$, $\theta(x)$ uniquely solves the following constrained maximization problem:

$$\left.\begin{array}{ll} \text{Max} & \phi(x,z) \\ \text{Subject to} & z \in I \end{array}\right\} \tag{P}$$

We define $\psi : I \to I$ by

$$\psi(x) = \phi(x,\theta(x)) \text{ for all } x \in I$$

Then by definitions of ϕ and θ, we have

$$\psi(x) = ax - 0.5Lx^2 + \theta(x)^2 - 0.5\theta(x)^2 = ax - 0.5Lx^2 + 0.5\theta(x)^2$$

Clearly ψ is a continuous function on I.

Now, note that for $(x,z) \in I^2$, the definitions of ϕ, ψ and $\mathbf{u}$ yields the equation

$$\mathbf{u}(x,z) = \phi(x,z) - \bar{\delta}\psi(z)$$

Thus for every $x \in I$, $\theta(x)$ uniquely solves the problem:

$$\left.\begin{array}{ll} \text{Max} & \mathbf{u}(x,z)+\bar{\delta}\psi(z) \\ \text{Subject to} & z \in I \end{array}\right\} \qquad \text{(Q)}$$

But since $\theta(x) \leq \mathbf{f}(x)$ for $x \in I$, we can clearly also conclude that $\theta(x)$ uniquely solves the problem

$$\left.\begin{array}{ll} \text{Max} & \mathbf{u}(x,z)+\bar{\delta}\psi(z) \\ \text{Subject to} & (x,z) \in \Omega \end{array}\right\} \qquad \text{(Q}'\text{)}$$

This means that for all $x \in I$,

$$\underset{(x,z)\in\Omega}{\text{Max}}\,[\mathbf{u}(x,z)+\bar{\delta}\psi(z)] = \mathbf{u}(x,\theta(x)) + \bar{\delta}\psi(\theta(x)) =$$
$$\phi(x,\theta(x)) = \psi(x)$$

Applying Theorem 3.1 (i), we can conclude that $\psi(x) = \mathbf{V}(x)$ for all $x \in I$. Thus $\theta(x)$ uniquely solves the problem:

$$\left.\begin{array}{ll} \text{Max} & \mathbf{u}(x,z)+\bar{\delta}\mathbf{V}(z) \\ \text{Subject to} & (x,z) \in \Omega \end{array}\right\} \qquad \text{(Q}''\text{)}$$

Applying Theorem 3.1(ii), $\theta(x) = \mathbf{h}(x)$ for all $x \in I$.

7.4 Verification of Theorem 6.1

We describe our strategy of proof briefly before providing the formal details. First, we consider an **e**-optimal program from a *specific* initial input [we actually choose $\mathbf{x}_0 = (1/8)$] for which

$$\mathbf{h}^3(\mathbf{x}_0) < \mathbf{x}_0 < \mathbf{h}(\mathbf{x}_0) < \mathbf{h}^2(\mathbf{x}_0)$$

This enables us to find a number $\eta > 0$ [we actually have $\eta = (1/32)$] such that if $\{x'_t\}_0^\infty$ is any sequence with $x'_0 = \mathbf{x}$ and $|x'_t - \mathbf{x}_t| \leq \eta$ for $t = 1, 2, 3$, then

$$x'_3 + \eta < x'_0 < x'_1 - \eta < x'_2 - 2\eta$$

Thus, it is enough to show that for an economy e near $\mathbf{e}$, the e-optimal program from $\mathbf{x}$ stays within η of the $\mathbf{e}$-optimal program from $\mathbf{x}$ *in the first three periods* to invoke the Li-Yorke theorem (Theorem 2.1) and conclude robustness of topological chaos. This appears to be intuitively plausible, but it does involve checking the details that we provide below.

We proceed formally as follows. Let $\{\mathbf{x}_t\}_0^\infty$ be the $\mathbf{e}$-optimal program from $\mathbf{x} = (1/8)$. Then $\mathbf{x}_0 = (1/8)$, $\mathbf{x}_1 = \mathbf{h}(\mathbf{x}_0) = (7/16)$, $\mathbf{x}_2 = \mathbf{h}^2(x_0) = (63/64)$ and $\mathbf{x}_3 = \mathbf{h}^3(\mathbf{x}_0) = (63/1024)$. Clearly, we have

$$\mathbf{x}_3 < \mathbf{x}_0 < \mathbf{x}_1 < \mathbf{x}_2$$

We now break up our proof into several steps:
Step1: Let $\{x_t'\}_0^\infty$ *be any sequence with* $x_0' = \mathbf{x} = (1/8)$.
If $\max_{t=1,2,3} |x_t' - \mathbf{x}_t| \leq (1/32)$, *then*

$$x_3' + (1/32) < x_0' < x_1' - (1/32) < x_2' - (1/16)$$

To see Step 1, note first that $x_1' - x_0' = x_1' - \mathbf{x} = (x_1' - \mathbf{x}_1) + (\mathbf{x}_1 - \mathbf{x}) \geq (\mathbf{x}_1 - \mathbf{x}) - (1/32) = (5/16) - (1/32) = (9/32) > (1/32)$. Second, $x_2' - x_1' = (x_2' - \mathbf{x}_2) + (\mathbf{x}_2 - \mathbf{x}_1) + (\mathbf{x}_1 - x_1') \geq (\mathbf{x}_2 - \mathbf{x}_1) - (1/32) - (1/32) = (35/64) - (4/64) = (31/64) > (1/32)$. Finally, $x_0' - x_3' = \mathbf{x} - x_3' = (\mathbf{x} - \mathbf{x}_3) + (\mathbf{x}_3 - x_3') \geq (\mathbf{x} - \mathbf{x}_3) - (1/32) > (1/8) - (1/16) - (1/32) = (1/32)$.

Define a set of programs, Y as follows:

$$Y = \{\{x_t\}_0^\infty : \{x_t\}_0^\infty \text{ is an } \mathbf{e}\text{-program from } \mathbf{x} = (1/8), \text{ such that}$$

$$\max_{t=1,2,3} |x_t - \mathbf{x}_t| \geq (1/32)\}$$

Step 2: There is $\beta > 0$ *such that if* $\{x_t'\}_0^\infty$ *is any* $\mathbf{e}$*-program belonging to* Y, *then*

$$\sum_0^\infty \bar{\delta}^t \mathbf{w}(x_t', c_{t+1}') \leq \mathbf{V}(\mathbf{x}) - \beta \tag{7.9}$$

To verify this claim, suppose on the contrary for each integer $n \geq 1$, there is an **e**-program $\{x_t^n\}_0^\infty$ in Y satisfying

$$\sum_0^\infty \bar{\delta}^t \mathbf{w}(x_t^n, c_{t+1}^n) \geq \mathbf{V}(\mathbf{x}) - (1/n)$$

We can then pick a subsequence (retain notation) such that for each $t \geq 1$

$$x_t^n \to \bar{x}_t \text{ as } n \to \infty$$

Clearly $\{\bar{x}_t\}_0^\infty$ is an **e**-program which belongs to Y. And, since $\{\mathbf{x}_t\}_0^\infty$ is the unique **e**-optimal program from $\mathbf{x}$, there is $\alpha > 0$ such that

$$\sum_0^\infty \bar{\delta}^t \mathbf{w}(\bar{x}_t, \bar{c}_{t+1}) \leq \mathbf{V}(\mathbf{x}) - \alpha$$

Given α, we can pick a positive integer T large enough so that $\bar{\delta}^T M_1(1-\bar{\delta}) < (\alpha/4)$, where $M_1 \equiv \mathbf{w}(1,1) > 0$. Then we can pick N large enough so that $n \geq N$ implies that for all $t \in [0, 1, ..., T]$,

$$|\mathbf{w}(x_t^n, c_{t+1}^n) - \mathbf{w}(\bar{x}_t, \bar{c}_{t+1})| < [\alpha(1-\bar{\delta})/4]$$

Then, for $n \geq N$,

$$\begin{aligned}
\sum_0^\infty \bar{\delta}^t \mathbf{w}(\bar{x}_t, \bar{c}_{t+1}) &\geq \sum_0^T \bar{\delta}^t \mathbf{w}(\bar{x}, \bar{c}_{t+1}) \\
&\geq \sum_0^T \bar{\delta}^t \mathbf{w}(x_t^n, c_{t+1}^n) - (\alpha/4) \\
&\geq \sum_0^\infty \bar{\delta}^t \mathbf{w}(x_t^n, c_{t+1}^n) - (\alpha/2) \\
&\geq \mathbf{V}(\mathbf{x}) - (1/n) - (\alpha/2)
\end{aligned} \tag{7.10}$$

Combining (7.9) and (7.10),

$$\mathbf{V}(\mathbf{x}) - \alpha \geq \mathbf{V}(\mathbf{x}) - (\alpha/2) - (1/n)$$

Thus $(1/n) \geq (\alpha/2)$ for all $n \geq N$, a contradiction which establishes the claim.

We now introduce some notation for the next (and crucial) step in the proof. Define $\hat{\alpha} = f(\mathbf{x}) - \mathbf{x}$. Clearly $\hat{\alpha} > 0$. Note that $2\bar{\delta} < 1$, and define $M_2 = [1/(1-2\bar{\delta})]$. Now choose $0 < \lambda < 1$ with λ sufficiently close to 1 so that

$$2(1-\lambda)M_1M_2 \leq (\beta/4) \tag{7.11}$$

where β is given in Step 2. Next, define $M_3 = \sum_0^{\infty}(t+1)(2\bar{\delta})^t$. Note that $\mathbf{w}$ is C^1 on D, and define $M_4 = \max_{(x,c)\in D} D_2\mathbf{w}(x,c)$. Finally define

$$\varepsilon = \min[(1-\lambda)8, 0.00125, M_1, \{\beta/4[2M_1M_3 + 2M_2 + M_2M_4]\}] \tag{7.12}$$

Step 3: Consider any $e \in E$ with $d(e, \mathbf{e}) < \varepsilon$, where ε is defined by (7.12). Let $\{x'_t\}_0^\infty$ be the e-optimal program from $\mathbf{x} = (1/8)$. Consider the sequence $\{x''_t\}_0^\infty$ defined by $x''_t = \lambda x'_t + (1-\lambda)\mathbf{x}$ for $t \geq 0$, where $0 < \lambda < 1$ satisfies (7.11). Then $\{x''_t\}_0^\infty$ is an e-program from x which does not belong to Y.

Note that for $e \in E$ with $d(e, \mathbf{e}) < \varepsilon$, $f(x) \leq \mathbf{f}(x) + \varepsilon$ for all $x \geq 0$, so that $f(x) \leq 1 + \varepsilon$ for $x \geq 0$. This means that $K \leq (1+\varepsilon) < 2$. Thus, defining $J = [0, 2]$ we certainly have $x_t \in J$ for all $t \geq 0$, for every e-program $\{x_t\}_0^\infty$ from every initial stock in J.

Using the definition of $\{x''_t\}_0^\infty$ it is routine to check that $\{x''_t\}_0^\infty$ is an e-program as well as an **e**-program from **x**.

Define another sequence $\{\bar{x}_t\}_0^\infty$ by $\bar{x}_t = \lambda\mathbf{x}_t + (1-\lambda)\mathbf{x}$ for $t \geq 0$. Then again it is straightforward to check that $\{\bar{x}_t\}_0^\infty$ is an e-program as well as an **e**-program from **x**.

We have to show that $\{x''_t\}_0^\infty$ does not belong to Y. Suppose it did. Then by Step 2 we would have

$$\sum_0^{\infty} \bar{\delta}^t \mathbf{w}(x''_t, \mathbf{f}(x''_t) - x''_{t+1}) \leq \mathbf{V}(\mathbf{x}) - \beta \tag{7.13}$$

Now using $d(e, \mathbf{e}) < \varepsilon$, and the fact that $\{\bar{x}_t\}_0^\infty$ is close to $\{\mathbf{x}_t\}_0^\infty$, one can obtain the following estimate

$$\sum_0^\infty \delta^t w(\bar{x}_t, f(\bar{x}_t) - \bar{x}_{t+1}) \geq \mathbf{V}(\mathbf{x}) - \boldsymbol{\varepsilon} M_1 M_3 - \boldsymbol{\varepsilon} M_2 - (1-\lambda) M_1 M_2 \tag{7.14}$$

Comparing $\{x_t''\}_0^\infty$ as an e-program and as an $\mathbf{e}$-program, and using $d(e, \mathbf{e}) < \boldsymbol{\varepsilon}$, we can also obtain the estimate

$$\sum_0^\infty \delta^t w(x_t'', f(x_t'') - x_{t+1}'') \leq \sum_0^\infty \bar{\delta}^t \mathbf{w}(x_t'', \mathbf{f}(x_t'') - x_{t+1}'') + \boldsymbol{\varepsilon}[M_1 M_3 + M_2 + M_2 M_4] \tag{7.15}$$

Combining (7.13)-(7.15)

$$\sum_0^\infty \delta^t w(\bar{x}_t, f(\bar{x}_t) - \bar{x}_{t-1}) \geq \sum_0^\infty \delta^t w(x_t', f(x_t') - (x_{t+1}') + (\beta/4)$$

which contradicts the fact that $\{x_t'\}_0^\infty$ is e-optimal from $\mathbf{x}$.

Step 4: Consider any $e \in E$, with $d(e, \mathbf{e}) < \boldsymbol{\varepsilon}$ where $\boldsymbol{\varepsilon}$ is defined by (7.12). Let h be the e-optimal policy function. Then

$$h^3(\mathbf{x}) < \mathbf{x} < h(\mathbf{x}) < h^2(\mathbf{x}) \tag{7.16}$$

To see Step 4, consider the sequences $\{x_t'\}_0^\infty$ and $\{x_t''\}_0^\infty$ defined in Step 3. Since we concluded that $\{x_t''\}_0^\infty$ is an $\mathbf{e}$-program which does *not* belong to Y, we must have $\max_{t=1,2,3} |x_t'' - \mathbf{x}_t| < (1/32)$. Thus, by Step 1, we have

$$x_3'' + (1/32) < x_0'' < x_1'' - (1/32) < x_2'' - (1/16) \tag{7.17}$$

Now for any $s, t \geq 0$, $(x_s'', x_t'') = [\lambda x_s' + (1-\lambda)\mathbf{x}] - [\lambda x_t' + (1-\lambda)\mathbf{x}] = \lambda(x_s' - x_t')$. Thus, using, in turn, $s = 1, t = 0$; $s = 2, t = 1$; and $s = 0, t = 3$; and (7.17), we get

$$x_3' < x_0' < x_1' < x_2' \tag{7.18}$$

Since h is the e-optimal policy function, and $\{x'_t\}_0^\infty$ the e-optimal program from $\mathbf{x}$, (7.18) yields (7.16).

We can now apply the Li-Yorke theorem (Theorem 2.1) to the dynamical system (J, h) where $J = [0, 2]$ and h is the e-optimal policy function for $e \in E$ satisfying $d(e, \mathbf{e}) < \varepsilon$ [and ε is given by (7.12)]. Given (7.16), (J, h) exhibits topological chaos, which establishes Theorem 6.1.

4

Optimal Chaos, Nonlinearity and Feasibility Conditions

Kazuo Nishimura and Makoto Yano

1. Introduction

This chapter presents a new characterization of chaotic optimal capital accumulation by which a chaotic optimal path can be constructed in a simple systematic manner. In the existing literature, it has been demonstrated that optimal capital accumulation may be chaotic in the sense of Li and Yorke (1975); see Boldrin and Montrucchio (1986b) and Deneckere and Pelikan (1986).[1] As Scheinkman's survey (1990) discusses, this finding indicates that the deterministic equilibrium model of a dynamic economy may explain various complex dynamic behaviors of economic variables. And, in fact, the search for such explanations has already begun.[2] In the existing literature, however, not much as been revealed with respect to the circumstances under which optimal accumulation exhibits complex nonlinear dynamics.[3]

In this chapter, we construct a class of dynamic models in which optimal paths are generated by nonlinear transition functions similar to a tent map. In that class, the optimal transition function, which relates a beginning-of-a-period stock level to the corresponding optimal choice of an end-of-a-period stock level, is monotone decreasing in the interior of the set of feasible activities.[4]

---

[1]Prior to these results, Benhabib and Day (1982) and Grandmont (1985) demonstrated the possibility of chaotic accumulation in overlapping generation models.

[2]See Brock (1986) and Scheinkman and LeBaron (1989).

[3]In the existing literature, two types of models are known to have chaotic optimal paths. The examples of Denerckere and Pelikan (1986), Boldrin and Montrucchio (1986b) and Boldrin and Deneckere (1990) are based on two-sector models while that of Majumdar and Mitra (Chapter 3) focuses on the shape of utility functions.

[4]This structure follows Benhabib and Nishimura (1985), which provides a condition under which the optimal transition function is monotone in the interior of the feasible set.

Under this circumstance, the optimal transition function must lie on the boundary of the feasible set, which is increasing, for any beginning-of-a-period stock below a certain level; above that level, the optimal choice appears in the interior of the feasible set and is monotone decreasing.[5] In short, the optimal transition function must look like a tent map. This fact demonstrates that optimal accumulation may follow complex nonlinear dynamics for a rather wide range of models.

More importantly, that fact indicates an intuitive and widely applicable way to construct a chaotic optimal transition function, which produces a cyclical path of period 3 (according to Li and Yorke's theorem, as is well-known a transition function that produces a cyclical path of period 3 is chaotic). Our idea is to focus cyclical optimal paths of period 3 that alternate to be on the boundary of the feasible set for two periods and to be in its interior for one period. The characterization of such an optimal path is not obvious and requires a rigorous mathematical argument, to which we devote a large part of this study. Once, however, this characterization is made, it leads to an example of a chaotic optimal path almost immediately, as we demonstrate below.

In what follows, Section 2 introduces the model. Section 3 explains Li and Yorke's result and the chaotic dynamics created by a tent map. Section 4 demonstrates the monotonicity of an optimal transition function for interior activities. Section 5 establishes that the transition function is similar to a tent map. Section 6 gives sufficient conditions under which it is actually a chaotic map. Section 7 presents an example of chaotic optimal accumulation. Section 8 makes concluding remarks.

[5] Boldrin and Deneckere (1990) discuss that a non-linear optimal transition function may arise from the existence of the boundary imposed by the limit of capital depreciation. However, that study does not consider a condition under which the existence of the boundary leads to chaotic optimal dynamics.

2. Model and Some Definitions

We consider a one-sector version of the standard optimal growth model (McKenzie, 1986). Think of $\tilde{k} \geq 0$ as the minimum level of a capital stock that can sustain an economy. Define $K = \{k \in R | k \geq \tilde{k}\}$. A function $s : K \to R_+$ relates each level of a capital stock existing at the beginning of a period, say k, to the maximum level of a capital stock that can be achieved at the end of that period, $h = s(k)$. In this sense, we call function $s : K \to R_+$ a feasibility function. Feasibility function s defines the set of feasible activities in a period as

$$D = \{(k, h) \in K \times R_+ | 0 \leq h \leq s(k)\}. \qquad (2.1)$$

Denote by k_t the stock level at time t. An accumulation path $\{k_t\}_{t=0}^{\infty}$ is *feasible* from initial stock $\bar{k}$ if $k_0 = \bar{k}$ and $(k_{t-1}, k_t) \in D$ for $t = 1, 2, \ldots$. If, in particular, it satisfies $(k_{t-1}, k_t) \in D^{\circ}$ for $t = 1, 2, \ldots$, it is called an *interior path.* A function that relates a feasible activity, $(k, h) \in D$, to the maximum utility that the economy can achieve in one period if it has stock k at the beginning of the period and plans to leave stock h at the end of the period.

The optimal growth model stipulates that a feasible path be chosen so as to maximize the discounted utility sum;

$$\max_{k_0, k_1, \ldots,} \sum_{t=1}^{\infty} \rho^{t-1} v(k_{t-1}, k_t) \text{s.t. } k_0 = \bar{k}. \qquad (2.2)$$

A path $\{k_t\}$ that solves (2.2) is called an *optimal path.* If an optimal path $\{k_t\}$ satisfies $k_t = k$ for all k, it is called a *stationary optimal path*, and the stock k is called a *modified golden rule state.* In what follows, we describe an optimal growth model by a pair of an objective function and a feasibility condition $M_\rho = \{\sum_{t=1}^{\infty} \rho^{t-1} v(k_{t-1}, k_t), s(k)\}$. We assume the following:

A1: *Feasibility function $s : K \to R_+$ is continuous and concave. Moreover, it satisfies the following:*

(i) $k < k'$ implies $s(k) \leq s(k')$;

(ii) there are $k_H > k_L \geq \tilde{k}$ such that $k_L = s(k_L)$ and $k_H = s(k_H)$;

(iii) $s(k)$ *is continuously differentiable on* $k > \tilde{k}$.

A2: *The reduced-form utility function* $v : D \to R$ *is continuous, concave on* D *and continuously twice differentiable on* D°.

For $v = v(k, h)$, denote $v_1 = \partial v/\partial k$, $v_2 = \partial v/\partial h$, $v_{11} = \partial^2 v/\partial k^2$, $v_{22} = \partial v^2/\partial h^2$ and $v_{12} = \partial^2 v/\partial k \partial h$. Assume the following:

A3: *Let* $(k, h) \in D^\circ$.
(i) $v_1(k, h) > 0$ *and* $v_2(k, h) < 0$;
(ii) $v_{11}(k, h) < 0$ *and* $v_{22}(k, h) < 0$.

The above assumptions are more or less standard in the literature on optimal growth theory (McKenzie, 1986).

3. Li and Yorke's Theorem

In order to describe the complex nonlinear dynamics of optimal accumulation, we use the well-known theorem of Li and Yorke, which demonstrates that "period-3 cycles imply chaos". Take a function defined on an interval J into itself, $f : J \to J$. We denote by $f^{(m)} = f \circ ... \circ f$ the composite map consisting of f's as many as m. Call a point $x \in J$ a *period-m periodic point* of function f if $f^{(m)}(x) = x$ and $f^{(n)}(x) \neq x$ for $n = 1, ..., m-1$. Call a path $\{x_t\}$ a *period-m period path* if it starts from a period-m periodic point i.e., if $x_t = f^{(t)}(x)$ and if x is a periodic point with period m. Call, moreover, a function $f : J \to J$ a *chaotic map* in the sense of Li and Yorke if the following two conditions are satisfied:

Condition I: There is an uncountable set $C \subset J$ (containing no periodic points) that satisfies the following:

A. For every p, $q \in C$ with $p \neq q$.

$$\lim\ \sup_{n\to\infty} |f^{(n)}(p) - f^{(n)}(q)| > 0$$

and

$$\lim\ \inf_{n\to\infty} |f^{(n)}(p) - f^{(n)}(q)| = 0.$$

B. For every $p \in C$ and periodic point $q \in J$,

$$\lim\ \sup_{n\to\infty} |f^{(n)}(p) - f^{(n)}(q)| > 0.$$

Condition II: For every $k = 1, 2, ...$, there is a periodic point with J with period k.

The property stated in Condition I.A implies that the paths from any two-distinct points in C are not periodic, do not converge to each other, but come arbitrarily close to each other from time to time. That in Condition I.B implies that the non-periodic path from a point in C does not converge to a periodic path. In this sense, we may call a point in set C a *chaotic point* and a path starting from a chaotic point a *chaotic path.* Li and Yorke's theorem states the following:

Lemma 3.1: *(Li and Yorke, 1975): Let J be an interval and $f : J \to J$ be continuous. Assume that there is a point $a \in J$ that satisfies*

$$f^{(3)}(a) \leq a < f(a) < f^{(2)}(a). \tag{3.1}$$

Then, function f is chaotic map in the sense of Li and Yorke.

A well-known example of a function that generates a chaotic path is a tent map. An example of a tent map that generates chaos is shown in Figure 1. $f(x) = 2x$ on $I[0, 1/2]$ and $f(x) = 2 - 2x$ on $I[1/2, 1]$ as is shown in Figure 1. Because $f(2/9) = 4/9$ and $f \circ f(2/9) = 8/9$ and $f \circ f \circ f(2/9) = 2/9$, $a = 2/9$ satisfy condition (3.1) in Lemma 3.1, f is a chaotic map in the sense of Li and Yorke.

4. Monotonicity in the Interior of the Feasible Set

This section proves the monotonicity of an optimal accumulation function, h_ρ, in the interior of the feasible set by extending Benhabib and Nishimura (1985). As is discussed in the Introduction, this study deals with the case in which an optimal path is not an interior path. Therefore, it is important to distinguish the case in which an optimal path is to hit the boundary of the feasible set from that in which it is to stay in the interior of the feasible set. In this respect, Theorem 4.1 below clarifies Benhabib and Nishimura (1985), which does not explicitly deal with the boundary case.

We say that function $f : J \to J$, $J \subset R$, is *weakly monotone* if $x < x'$ implies $f(x) \leq f(x')$ and *strictly monotone* if $x < x'$ implies

$f(x) < f(x')$. Moreover, we adopt the following convention in describing intervals (closed, half-closed and open):

$$I[\alpha,\beta] = \{x \in R | \alpha \leq x \leq \beta\}; \tag{4.1}$$

$$I[\alpha,\beta) = \{x \in R | \alpha \leq x < \beta\}; \tag{4.2}$$

$$I(\alpha,\beta] = \{x \in R | \alpha < x \leq \beta\}; \tag{4.3}$$

$$I(\alpha,\beta) = \{x \in R | \alpha < x < \beta\}; \tag{4.4}$$

We first demonstrate that the value function, given below, is well defined and continuous on $I[k_L, \infty)$.

$$V_\rho(h) = \max_{k_0,k_1,\ldots} \sum_{t=1}^{\infty} \rho^{t-1} v(k_{t-1}, k_t) \text{ s.t. } k_0 = h. \tag{4.5}$$

It is well-known that this value function is well defined and concave (for a proof, for example, see McKenzie, 1983). Moreover, the next lemma holds; a proof is routine and omitted.

Lemma 4.1: *Given ρ, $0 < \rho < 1$, function V_ρ is well-defined and continuous on $I[k_L, \infty)$.*

The next lemma implies that the optimal accumulation function can be defined by

$$h_\rho(k) = \operatorname{argmax}_{h \in I[k_L, s(k)]} [v(k,h) + \rho V_\rho(h)] \tag{4.6}$$

(a proof is routine and omitted). Note that if path k_t, $t = 0, 1, \ldots$, is optimal,

$$k_t = h_\rho(k_{t-1}) \tag{4.7}$$

for $t = 0, 1, \ldots$ (the optimality principle).

Lemma 4.2: *The optimal accumulation function $h_\rho(k)$ is well defined and continuous on $I[k_L, \infty)$. Moreover, it maps $I[k_L, k_H]$ into itself.*

The following theorem demonstrates the monotonicity of optimal accumulation function h_ρ. We denote the left- and right-derivatives of a function, f, as follows.

$$f'_{-}(x) = \lim_{\Delta x \to 0-}[f(x+\Delta x) - f(x)]/\Delta x; \tag{4.8}$$

$$f'_{+}(x) = \lim_{\Delta x \to 0+}[f(x+\Delta x) - f(x)]/\Delta x. \tag{4.9}$$

Because, by Lemma 4.1, value function $V_\rho(k)$ is concave on K, both $V'_{\rho+} > 0$ and $V'_{\rho-} > 0$ exist on interior K, and $V'_{\rho+} > 0$ exists on boundary K.

Theorem 4.1: *Suppose $v_{12} < 0$ in D°. If $(k, h_\rho(k)) \in D^\circ$, $h_\rho(k)$ is non-increasing in a neighborhood of k. If, in addition, $(h_\rho(k), h_\rho^{(2)}(k)) \in D^\circ$, $h_\rho(k)$ is strictly decreasing in a neighborhood of k.*

Proof: Let $(k, h_\rho, (k)) \in D^\circ$. Define $h = h_\rho(k)$ and $h + \Delta h = h_\rho(k + \Delta k)$. Using the continuity of h_ρ (Lemma 4.2), choose Δk small so that $(k + \Delta k,\ h + \Delta h) \in D^\circ$. Take the case of $\Delta k > 0$. Suppose that h_p is increasing at k. Then, $\Delta h > 0$. Thus, $V'_{\rho+}(h) > V'_{\rho-}(h+\Delta h)$. This implies that by the differentiability of v at (k, h) and $(k + \Delta k,\ h + \Delta k)$, (k, h) and $(k + \Delta k,\ h + \Delta h)$ must satisfy the first order condition

$$\begin{aligned} -v_2(k,h) &\geq \rho V'_{\rho+}(h) > \rho V'_{\rho-}(h+\Delta h) \\ &\geq -v_2(k+\Delta k,\ h+\Delta h) \geq \rho V'_{\rho+}(h+\Delta h). \end{aligned} \tag{4.10}$$

However, by $\Delta k > 0$ and $\Delta h > 0$, this contradicts $-v_2(k,h) < -v_2(k+\Delta k, h) < -v_2(k+\Delta k,\ h+\Delta h)$, where the first inequality follows from $v_{21} < 0$ and the second from $v_{22} < 0$ (Assumption 3.ii). In a similar way, we may derive a contradiction in the case of $\Delta k < 0$; h_ρ is non-increasing at k.

Let, in addition, $(h_\rho(k),\ h_\rho^{(2)}(k)) \in D^\circ$. Then, v is differentiable at $(k, h_\rho(k))$ and $(h_\rho(k), h_\rho^{(2)}(k))$. This implies that $(k_{t-1}, k_t, k_{t+1}) =$

$(k, h_\rho(k), h_\rho^{(2)}(k))$ must satisfy the Euler equation

$$v_2(k_{t-1}, k_t) + \rho v_1(k_t, k_{t+1}) = 0. \tag{4.11}$$

If $h_\rho(k)$ is not strictly decreasing, there is $\Delta k \neq 0$ such that, by the continuity of h_ρ, $(k+\Delta k,\, h_\rho(k+\Delta k)) \in D^\circ$ and $h_\rho(k+\Delta k) = h_\rho(k)$. Since this implies $h_\rho^{(2)}(k+\Delta k) = h_\rho^{(2)}(k)$, $(k_{t-1}, k_t, k_{t+1}) = (k+\Delta k,\, h_\rho(k),\, h_\rho^{(2)}(k))$ must also satisfy the Euler equation, (4.11). This is impossible since $v_{21} < 0$ and $\Delta k \neq 0$. Q.E.D.

As a corollary of Theorem 3.1, a similar result can be established for the case in which $v_{12} > 0$.

Corollary 4.1: *Suppose $v_{12} > 0$ on D°. If $(k, h_\rho(k)) \in D^\circ$, $h_\rho(k)$ is non-decreasing in a neighborhood of k. If, in addition, $(h_\rho(k), h_\rho^{(2)}(k)) \in D^\circ$, $h_\rho(k)$ is strictly increasing in a neighborhood of k.*

5. Nonlinearity in the Optimal Accumulation Function

In what follows, we demonstrate that if v_{12} is always negative on D°, the optimal accumulation function must have a shape similar to a tent map. In order to make an explicit discussion, we introduce the following terminology. Take a closed interval $I[x_L, x_H]$. A function $f : I[x_L, x_H] \to I[x_L, x_H]$ is a *semi-tent map* if there is $\hat{x} \in I(x_L, x_H)$ such that (i) $f(x)$ is non-decreasing and non-constant over $I[x_L, \hat{x}]$, (ii) $f(x)$ is non-increasing and non-constant over $I[\hat{x}, x_H]$, and (iii) $f(\hat{x}) > \hat{x}$.

Define $r(k) = -v_2(k,k)/v_1(k,k)$; i.e., $r(k)$ is the inverse of the slope of an indifference curve, $v(k,h) =$ constant, on the 45° line. The next theorem implies that h_ρ is a semi-tent map.

Theorem 5.1: *Suppose $v_{12} < 0$ on D°, and that there are k' and k'' in $I[k_L, k_H]$ such that*

$$r(k') < \rho < r(k''). \tag{5.1}$$

The optimal accumulation function $h_\rho(k)$ of model M is a semi-tent map. In particular, there are $\hat{k}$ and $k_\rho \in I(k_L, k_H)$ such that the following holds:

A. If $k_L < k \leq \hat{k}$, $h_\rho(k)$ coincides with $s(k)$ and is strictly increasing.

B. If $\hat{k} < k \leq k_H$, $h_\rho(k)$ is non-increasing.

C. $h_\rho(\hat{k}) = s(\hat{k}) > h_\rho(k_\rho) = k_\rho > h_\rho(k_H)$.

Proof: By (5.1) and the continuity of v, there is $k_\rho \in I(k_L, k_H)$ such that $k_{t-1} = k_t = k_{t+1} = k_\rho$ satisfies the Euler equation (4.11). This implies $k_\rho = h_\rho(k_\rho)$. Since $k_L < k_\rho < k_H$ implies $k_\rho < s(k_\rho)$ by Assumption 1, $(k_\rho, h_\rho(k_\rho)) = (k_\rho, h_\rho^{(2)}(k_\rho)) = (k_\rho, k_\rho) \in D^\circ$. Thus, by Theorem 3.1, h_ρ is strictly decreasing at k_ρ. If k increases from k_ρ, by Theorem 3.1 and the continuity of h_ρ, h_ρ must first decrease and then never increase as k is increased to k_H. If, instead, k decreases from k_ρ, h_ρ must first increase and then never decrease. Thus, point $(k, h_\rho(k))$ must reach the boundary of D before k reaches k_L. Once $(k, h_\rho(k))$ reaches the boundary, by Theorem 3.1 and the continuity of h_ρ, it must stay on the boundary as k decreases further from $\hat{k}$ to k_L. In summary, there is $k' \in I(k_L, k_H)$ such that $h_\rho(k) = s(k)$ for any $k \in I[k_L, k']$ and $h_\rho(k) < s(k)$ for any $k \in I(k', k_H]$. Moreover, by Assumption 1, there is $k'' \in I(k_L, k_H)$ such that $s(k)$ is strictly increasing on $I[k_L, k'')$ and constant on $I[k'', k_H]$. With $\hat{k} = \min\{k', k''\}$, A, B and C hold. Q.E.D.

Theorem 5.1 implies that if $v_{12} < 0$ globally, the optimal accumulation function involves a tent-shaped nonlinearity, as illustrated in Figure 2. This demonstrates that there is a wide range of economic models in which optimal capital accumulation may follow complex nonlinear dynamics. In the literature on optimal growth, it is well-known that for each given value of discount factor ρ, a stationary state can be either locally stable or locally unstable and that, regardless of the sign of v_{12}, there is a wide range of reduced form utility functions with the stationary state is locally unstable (see McKenzie, 1986, for an explanation). Because globally negative v_{12} implies a unique stationary state, given a reduced form utility function that belongs to that range and satisfies $v_{12} < 0$ globally, optimal accumulation follows the complex nonlinear dynamics resulting from the tent-shaped optimal accumulation function with a (unique) unstable stationary state.

Remark 5.1: Theorem 5.1 implies the non-differentiability of optimal accumulation function h_ρ at $\hat{k}$, provided that the derivative of feasibility function $s(k)$ is assumed to be always positive. This fact may be of interest in relation to the recent work of Araujo (1988), Boldrin and Montrucchio (1989) and Santos (1990), which considers the differentiability of optimal accumulation functions.

6. Optimal Chaos: Characterization

Theorem 4.2 indicates that $v_{12} < 0$ may lead to chaotic optimal accumulation rather than excludes it. If, as in the existing studies, we focus on interior optimal paths, it is obvious from the result of Benhabib and Nishimura that $v_{12} < 0$ excludes the possibility of chaotic optimal accumulation. Theorem 5.1, however, implies that if $v_{12} < 0$ globally, in general, optimal paths are not confined to the interior of the feasible set; the optimal accumulation function is generally tent-shaped and may result in chaotic optimal accumulation.

By using this fact, we may construct an easy-to-use characterization of a chaotic optimal path. Suppose that a period-3 cycle path k, $s(k)$, $s^{(2)}(k)$, $k, \ldots$ is optimal. Then, it satisfies

$$h_\rho^{(3)}(k) = k < s(k) = h_\rho(k) < s^{(2)}(k) = h_\rho^{(2)}(k). \tag{6.1}$$

This implies, by Li and Yorke's result (see Lemma 3.1), that the optimal accumulation function, h_ρ, is chaotic. With these considerations, we construct a set of first order conditions under which path k, $s(k)$, $s^{(2)}(k)$, $k, \ldots$ is optimal.

In order to derive the set of first order conditions, it is necessary to take special care of boundary conditions. Optimal growth theory, traditionally, has not paid much attention to non-interior optimal paths *per se*, although various studies have been concerned with general characterizations covering both interior and non-interior optimal paths. The existing studies on chaotic optimal accumulation, too, are concerned with interior optimal paths.

As a result, a chaotic optimal path can be characterized by the Euler equation (4.11). Deneckere and Pelikan (1986) first construct such an example by looking for a reduced-form utility function, v, with the following property: There are k, k' and k'', $k < k' < k''$, such that $(k_{t-1}, k_t, k_{t+1}) = (k, k', k'')$, (k', k'', k) and (k'', k, k') all satisfy the Euler equation, (4.11). Since, in this case, the period-3 periodic path $k, k', k'', k, ...$ is optimal, it holds that $k' = h_\rho(k)$, $k'' = h_\rho^{(2)}(k)$ and $k = h_\rho^{(3)}(k))$, which implies, by Li and Yorke's theorem (Lemma 3.1), that h_ρ is a chaotic map.[6]

In this study, dealing with non-interior optimal paths, we replace the Euler equation with a new set of conditions involving inequality conditions. Because such a characterization is not obvious, it may be useful to explain first the basic idea behind the characterization. Because we look for a period-e cyclical optimal path, it is natural to work with the three-period reduced-form utility function is follows:

$$w(x, y) = \max_{(k_1,k_2)}[v(x, k_1) + \rho v(k_1, k_2) + \rho^2 v(k_2, y)]. \tag{6.2}$$

With this function, consider the following three-period optimal growth model:

$$\max_{x_0,x_1,...} \sum_{\tau=1}^{\infty} (\rho^3)^\tau w(x_{t-1}, x_t) \text{ subject to } x_0 = \bar{k}. \tag{6.3}$$

If path k, $s(k)$, $s^{(2)}(k)$, k, ..., is optimal in the original optimization model (2.2), a (period-3) stationary path $x_\tau = k$, $\tau = 0, 1, ...$, solves (6.3) for $\bar{k} = k$; i.e., it is a stationary optimal path in the three-period optimal growth model (6.3). Conversely, a path k, $s(k)$, $s^{(2)}(k)$, k, ... is optimal in the original model, (2.2), if the following two conditions are satisfied (see Theorem 6.1 for a proof):

Condition 1: Vector $(k_1, k_2) = (s(k), s^{(2)}(k))$ solves the maximization problem (6.2) for $(x, y) = (k, k)$.

Condition 2: Path $x_t = k$, $t = 0, 1, ...$, is a stationary optimal path of the three-period optimal growth model (6.3).

[6]The examples of Boldrin and Montrucchio (1986b), Boldrin and Deneckere (1990) and Majumdar and Mitra (Chapter 3) are also based on this idea.

The above discussion demonstrates that it suffices to characterize Conditions 1 and 2 in order to prove the optimality of period-3 cyclical path k, $s(k)$, $s^{(2)}(k)$, $k, \ldots$ In the following part of this study we deal with the case in which the one-sided partial derivatives of $v(k,h)$ exist on the boundary of D. Denote these one-sided partial derivatives by $v_1(k,h)$ and $v_2(k,h)$ as well. The next lemma characterizes Condition 1 by focusing the first order conditions of optimization problem (6.2).

Lemma 6.1: *Let $x > k_L$ and $(k_1, k_2) = (s(x), s^{(2)}(x))$. Suppose $(k_2, y) \in D^\circ$. Then, (k_1, k_2) solves the maximization problem in (6.2) if the following conditions are satisfied:*

$$v_2(k_1,k_2) + \rho v_1(k_2,y) \geq 0; \tag{6.4}$$

$$v_2(x,k_1) + \rho v_1(k_1,k_2) + \rho[v_2(k_1,k_2) + \rho v_1(k_2,y)]s'(k_1) \geq 0. \tag{6.5}$$

Proof: Since v is continuous and concave, the following three expressions hold:

$$v(x,k_1) - v_2(x,k_1)k_1 \geq v(x,\kappa_1) - v_2(x,k_1)\kappa_1 \tag{6.6}$$

for any $(x,\kappa_1) \in D$;

$$\begin{aligned} &v(k_1,k_2) - v_2(k_1,k_2)k_2 - v_1(k_1,k_2)k_1 \\ &\geq v(\kappa_1,\kappa_2) - v_2(k_1,k_2)\kappa_2 - v_1(k_1,k_2)\kappa_1 \end{aligned} \tag{6.7}$$

for any $(\kappa_1,\kappa_2) \in D$;

$$v(k_2,y) - v_1(k_2,y)k_2 \geq v(\kappa_2,y) - v_1(k_2,y)\kappa_2 \tag{6.8}$$

for any $(\kappa_2, y) \in D$. By adding these expressions with appropriate weights,

$$\begin{aligned} &v(x,k_1) + \rho v(k_1,k_2) + \rho^2 v(k_2,y) - [v(x,\kappa_1) + \\ &\rho v(\kappa_1,\kappa_2) + \rho^2 v(\kappa_2,y)] \end{aligned}$$

$$\geq [v_2(x,k_1)+\rho v_1(k_1,k_2)](k_1-\kappa_1)+\rho[v_2(k_1,k_2)+ \rho v_1(k_2,y)](k_2-\kappa_2) \tag{6.9}$$

for any $(x,\kappa_1) \in D$, $(\kappa_1,\kappa_2) \in D$ and $(\kappa_2,y) \in D$.

Since $x > k_L$, $k_1 = s(k) > k_L$. Thus, $s'(k)$ exists by Assumption 1. Since $k_2 = s(k_1)$ and s is concave, $s(k_1) - s'(k_1)k_1 \geq s(\kappa_1) - s'(k_1)\kappa_1$ for any $\kappa_1 \in K$. Since $(\kappa_1,\kappa_2) \in D$ implies $\kappa_2 \leq s(\kappa_1)$, this implies

$$k_2 - \kappa_2 \geq s(k_1) - s(\kappa_1) \geq s'(k_1)(k_1-\kappa_1) \tag{6.10}$$

for any $\kappa_1 \in K$.

By (6.4), (6.9) and (6.10) imply

$$v(x,k_1)+\rho v(k_1,k_2)+\rho^2 v(k_2,y) - [v(x,\kappa_1)+\rho v(\kappa_1,\kappa_2)+\rho^2 v(\kappa_2,y)]$$

$$\geq [v_2(x,k_1)+\rho v_1(k_1,k_2)](k_1-\kappa_1)+\rho[v_2(k_1,k_2)+ \rho v_1(k_2,y)]s'(k_1)(k_1-\kappa_1)$$

$$\geq \{v_2(x,k_1)+\rho v_1(k_1,k_2)+\rho[v_2(k_1,k_2)+ \rho v_1(k_2,y)]s'(k_1)\}(k_1-\kappa_1) \tag{6.11}$$

for any $(x,\kappa_1) \in D$, $(\kappa_1,\kappa_2) \in D$ and $(\kappa_2,y) \in D$. Since $k_1 = s(x)$ and $(x,\kappa_1) \in D$, $k_1 = s(x) > \kappa_1$. Thus, by (6.5), the last expression in the above inequalities is non-negative. This implies

$$v(x,k_1)+\rho v(k_1,k_2)+\rho^2 v(k_2,y) - [v(x,\kappa_1)+\rho v(\kappa_1,\kappa_2)+\rho^2 v(\kappa_2,y)] \geq 0 \tag{6.12}$$

for any $(x,\kappa_1) \in D$, $(\kappa_1,\kappa_2) \in D$ and $(\kappa_2,y) \in D$. Q.E.D.

In order to characterize Condition 2 by first order (Kuhn-Tucker) conditions, we need to work with support prices of $w(x,y)$ instead of its partial derivative (McKenzie, 1986); in general, w may not be differentiable. Note that a vector (p, q) that has the following property is called a support price vector:

$$w(x,y) + qy - px \geq w(\zeta,\xi) + q\xi - p\zeta \tag{6.13}$$

for any $(\zeta, \xi) \in E$, where E is the domain of w given by

$$E = \{(x,y) \in R^2 | (x,k_1) \in D, (k_1,k_2) \in D, (k_2,y) \in D\}. \tag{6.14}$$

The next lemma characterizes the support price vector for the case in which $(k_1, k_2) = (s(x), s^{(2)}(x))$ solves maximization problem (6.2).

Lemma 6.2: *Let $x > k_L$ and $(k_1, k_2) = (s(x), s^{(2)}(x))$. Suppose $(k_2, y) \in D^\circ$ and that (k_1, k_2) satisfies (6.4) and (6.5). Then, a support price vector of $w(x, y)$ is given by $(p, q) = (p(x, k_1, k_2, y), q(k_2, y))$, defined as follows:*

$$p(x,k_1,k_2,y) = v_1(x,k_1) + [v_2(x,k_1) + \rho v_1(k_1,k_2)]s'(x)$$

$$+\rho[v_2(k_1,k_2) + \rho v_1(k_2,y)]s'(x)s'(k_1); \tag{6.15}$$

$$q(k_2,y) = -\rho^2 v_2(k_2,y). \tag{6.16}$$

Proof: The proof of this lemma proceeds in a similar way as that of Lemma 6.1. By an argument similar to that in the proof of Lemma 6.1, the following holds:

$$\begin{aligned} &v(x,k_1) - v_2(x,k_1)k_1 - v_1(x,k_1)x \\ &\geq v(\zeta,\kappa_1) - v_2(x,k_1)\kappa_1 - v_1(x,k_1)\zeta \end{aligned} \tag{6.17}$$

for any $(\zeta, \kappa_1) \in D$;

$$\begin{aligned} &v(k_2,y) - v_2(k_2,y)y - v_1(k_2,y)k_2 \\ &\geq v(\kappa_2,\xi) - v_2(k_2 - y)\xi - v_1(k_2,y)\kappa_2 \end{aligned} \tag{6.18}$$

for any $(\kappa_2, \xi) \in D$. By adding these expressions to (6.7) with appropriate weights, we have

$$v(x,k_1) + \rho v(k_1,k_2) + \rho^2 v(k_2,y) - \\ [v(\zeta,\kappa_1) + \rho v(\kappa_1,\kappa_2) + \rho^2 v(\kappa_2,\xi)]$$

$$\geq v_1(x,k_1)(x-\zeta) + [v_2(x,k_1) + \rho v_1(k_1,k_2)](k_1-\kappa_1)$$

$$+\rho[v_2(k_1,k_2) + \rho v_1(k_2,y)](k_2-\kappa_2) + \rho^2 v_2(k_2,y)(y-\xi) \quad (6.19)$$

for any $(\zeta, \kappa_1) \in D$, $(\kappa_1, \kappa_2) \in D$ and $(\kappa_2, \xi) \in D$.

Since (6.10) holds for any $\kappa_1 \in K$, (6.19) implies

$$v(x,k_1) + \rho v(k_1,k_2) + \rho^2 v(k_2,y) - \\ [v(\zeta,\kappa_1) + \rho v(\kappa_1,\kappa_2) + \rho^2 v(\kappa_2,\xi)]$$

$$\geq v_1(x,k_1)(x-\zeta) +$$

$$\{v_2(x,k_1) + \rho v_1(k_1,k_2) + \rho[v_2(k_1,k_2) + \rho v_1(k_2,y)]s'(k_1)\} \\ (k_1-\kappa_1) + \rho^2 v_2(k_2,y)(y-\xi) \quad (6.20)$$

for any $(\zeta, \kappa_1) \in D$, $(\kappa_1, \kappa_2) \in D$ and $(\kappa_2, \xi) \in D$. In a similar way as that in which (6.10) is derived, we may derive

$$k_1 - \kappa_1 \geq s(x) - s(\zeta) \geq s'(x)(x-\zeta) \quad (6.21)$$

for any $\zeta \in K$. Thus, by (6.21), (6.20) implies

$$v(x,k_1) + \rho v(k_1,k_2) + \rho^2 v(k_2,y) - \\ [v(\zeta,\kappa_1) + \rho v(\kappa_1,\kappa_2) + \rho^2 v(\kappa_2,\xi)]$$

$$\geq \{v_1(x,k_1) + [v_2(x,k_1) + \rho v_1(k_1,k_2)]s'(x)$$

$$+\rho[v_2(k_1,k_2)+\rho v_1(k_2,y)]s'(k_1)s'(x)\}(x-\zeta)+\rho^2 v_2(k_2,y)(y-\zeta)$$

for any $(\zeta,\kappa_1)\in D$, $(\kappa_1,\kappa_2)\in D$ and $(\kappa_2,\xi)\in D$. This readily implies the lemma. Q.E.D.

The next lemma implies that if $w(k,k)$ is supported by a quasi-stationary price vector $(p,q)=(p,\rho^3,p)$, then path $x_\tau=k$, $\tau=0,1,...$, is stationary optimal path of model (6.3) (Condition 2).

Lemma 6.3: *Let $k>k_L$ and $(k_1,k_2)=(s(k),s^{(2)}(k))$. Suppose that (k_1,k_2) satisfies (6.4), (6.5) and*

$$q(k_2,k)=\rho^3 p(k,k_1,k_2,k). \tag{6.22}$$

Then, path $x_\tau=k$, $\tau=0,1,...$, is a stationary optimal path of model (6.3).

Proof: By $k>k_L$, $s^{(2)}(k)\in I(k_L,k_H]$. Thus, $k<s^{(2)}(k)\leq s(s^{(2)}(k))$. This implies $(s^{(2)}(k),k)\in D^\circ$. Thus, by Lemma 6.2 and (6.22), (6.13) can be written as

$$w(k,k)+\rho^3 pk-pk\geq w(\zeta,\xi)+\rho^3 p\xi-p\zeta \tag{6.23}$$

for any $(\zeta,\xi)\in E$, where $p=p(k,k_1,k_2,k)$. Thus,
$\sum_{\tau=1}^{\infty}(\rho^3)^\tau w(k,k)\geq\sum_{\tau=1}^{\infty}(\rho^3)^\tau w(\zeta_{\tau-1},\zeta_\tau)$ for any path ζ_τ, $\tau=0,1,...$, such that $\zeta_0=k$ and $(\zeta_{\tau-1},\zeta_\tau)\in E$, $\tau=1,2,...$ Q.E.D.

In order to summarize the above discussion, we introduce the following notation.

$$\Gamma(k_{t-1},k_t,k_{t+1})=v_2(k_{t-1},k_t)+\rho v_1(k_t,k_{t+1}); \tag{6.24}$$

$$\Gamma_{(1)}(k)=\Gamma(s(k),s^{(2)}(k),k); \tag{6.25}$$

$$\Gamma_{(2)}(k)=\Gamma(k,s(k),s^{(2)}(k))+\rho\Gamma_{(1)}(k)s'(s(k)); \tag{6.26}$$

$$\Gamma_{(3)}(k) = \Gamma(s^{(2)}(k), k, s(k)) + \rho\Gamma_{(2)}(k)s'(k). \tag{6.27}$$

Recall that, in the above expressions, $v_i(k, s(k))$ and $v_i(s(k), s^{(2)}(k))$, $i = 1, 2$, are the one-sided partial derivatives introduced above.

Theorem 6.1: *Suppose that there is $k \in I(k_L, k_H)$ that satisfies the following:*

$$\Gamma_{(1)}(k) \geq 0; \tag{6.28}$$

$$\Gamma_{(2)}(k) \geq 0; \tag{6.29}$$

$$\Gamma_{(3)}(k) = 0. \tag{6.30}$$

If $k < s(k) < s^2(k)$, the optimal accumulation function of model M is chaotic in the sense of Li and Yorke.

Proof: By definition, conditions (6.28), (6.29) and (6.30) are equivalent to (6.23), (6.24) and (6.22), respectively. This implies, by Lemma 6, that $x_\tau = k$, $\tau = 0, 1, \ldots$, solves (6.3), given $\bar{k} = k$. Moreover, by Lemma 4, $(k_1, k_2) = (s(k), s^{(2)}(k))$ solves (6.2), given $(x, y) = (k, k)$. Thus, a period-3 cyclical path $k, s(k), s^{(2)}(k), k, \ldots$ solves (2.2) for $\bar{k} = k$. Since $k < s(k) < s^{(2)}(k)$ by hypothesis, (6.1) holds; therefore, as is discussed above, h_ρ is a chaotic map. Q.E.D.

In order to find a parametric example in which chaotic optimal paths exist, it suffices to find a model in which there is k that satisfies the conditions (6.28) through (6.30). In a practical sense, however, this is not easy because it requires to find a solution to equation (6.30); in a numerical example of a model, only an approximated value of such a solution may be obtained. For this reason, we prove the following theorem, which makes it possible to overcome this difficulty.

Theorem 6.2: *Suppose that there are $k \in I(k_L, k_H)$ and $\kappa \in I(k_L, k_H)$ that satisfy the following:*

$$\Gamma_{(1)}(k) \geq 0; \tag{6.31}$$

$$\Gamma_{(2)}(k) \geq 0; \tag{6.32}$$

$$\Gamma_{(3)}(k) \leq 0; \tag{6.33}$$

$$\Gamma_{(3)}(\kappa) > 0. \tag{6.34}$$

If $k < s(k) < s^{(2)}(k)$, the optimal accumulation function of model M is chaotic in the sense of Li and Yorke.

Proof: Note $d\Gamma_{(1)}/dk < 0$. Since $d\Gamma(k, s(k), s^{(2)}(k))/dk < 0$, and since s is concave and since $\Gamma_{(1)}(k) \geq 0$, $\Gamma_{(2)}$ is strictly decreasing in k. Since $d\Gamma(s^{(2)}(k), k, s(k))/dk < 0$ and since $\Gamma_{(2)}(k) \geq 0$, $\Gamma_{(3)}$ is also strictly decreasing in k. Since $\Gamma_{(3)}(\kappa) > 0$ by (6.35), these facts together with (6.31), (6.32) and (6.33) imply that there is k', $\kappa < k' < k$, such that $\Gamma_{(1)}(k') > \Gamma_{(1)}(k) \geq 0$, $\Gamma_{(2)}(k') \geq \Gamma_{(2)}(k) \geq 0$ and $\Gamma_{(3)}(k') = 0$. Since s is concave and non-decreasing, and since $k < s(k) < s^{(2)}(k)$ and $k' < k$, $k' < s(k') < s^{(2)}(k')$. By Theorem 3, therefore, h_ρ is a chaotic map. Q.E.D.

7. An Example

This section constructs a parametric model of optimal growth in which optimal capital accumulation follows complex dynamics for a wide range of parameter values. Moreover, it demonstrates that for specific values of parameters, the optimal accumulation function of the model is a chaotic map in the sense of Li and Yorke.

Define θ, ε and ε' in such a way that $\varepsilon + \varepsilon' = 2/\theta$ and $0 < \varepsilon < \varepsilon' < 1$. Choose a feasibility function, $s_\theta(x)$, as follows.

$$s_\theta(x) = \begin{cases} \theta x & \text{if } 0 \leq x \leq \varepsilon \\ g(x) & \text{if } \varepsilon \leq x \leq \varepsilon' \\ 1 & \text{if } \varepsilon' \leq x \leq 1, \end{cases} \tag{7.1}$$

where

$$g(x) = -\frac{\theta}{2(\varepsilon' - \varepsilon)}(x - \varepsilon')^2 + 1.$$

In other words, the set of feasible activities is

$$D_\theta = \{(x, y) \in R_+^2 | y \leq s_\theta(x), 0 \leq x \leq 1\}. \tag{7.2}$$

On the feasible set, D_θ, define a reduced form utility function $v : D_\theta \to R$,

$$v(x, y) = x^\alpha(1 - y)^\beta. \tag{7.3}$$

This reduced form utility function is taken over from Weitzman's example reported in Samuelson (1973). In the choice of a feasible set, D_θ, the present example differs from that of Weitzman, in which the feasible set is assumed to be $D_w = \{(x, y) \in R_+^2 | 0 \leq y \leq 1\}$. In Weitzman's setting, therefore, it is possible to produce a capital stock at the end of a period even if the economy has no capital stock at the beginning of the period. In contrast, we assume the necessity of a capital stock for production.[7]

Denote the optimal growth model as

$$M(\alpha, \beta, \rho) = \left\{ \sum_{t=1}^{\infty} \rho^{t-1}(k_{t-1})^\alpha(1 - k_t)^\beta, s_\theta \right\}. \tag{7.4}$$

Assume the following relationship for α, β and θ :

$$1/2 < \alpha < 1; \tag{7.5}$$

$$0 < \beta < 1/2; \tag{7.6}$$

$$\alpha + \beta < 1; \tag{7.7}$$

$$\theta > 1. \tag{7.8}$$

Lemma 7.1: *$M(\alpha, \beta, \alpha, \rho)$ satisfies Assumptions 1 through 3. Moreover, it satisfies $v_{12} < 0$ on D_θ° and condition (5.1).*

[7]See Boldrin and Deneckere (1990) and Nishimura and Yano (Chapter 8) for the possibility of chaotic optimal accumulation in a two-sector growth model.

Proof: Note $k_L = 0$, $k_H = 1$, $s_\theta(0) = 0$ and $s_\theta(1) = 1$. Moreover, $s_\theta(x) > x$ if $0 < x < 1$, and $s_\theta(x)$ is non-decreasing and continuously differentiable. Thus, Assumption 1 is satisfied. Since function v is continuous in k_{t-1} and k_t on D_θ, concave and continuously twice differentiable, Assumption 2 is satisfied. In order to complete the proof, note the following.

$$v_1(x, y) = \alpha x^{\alpha-1}(1-y)^\beta; \tag{7.9}$$

$$v_2(x, y) = -\beta x^\alpha (1-y)^{\beta-1}; \tag{7.10}$$

$$v_{11}(x, y) = \alpha(\alpha-1)x^{\alpha-2}(1-y)^\beta; \tag{7.11}$$

$$v_{22}(x, y) = -\beta(\beta-1)x^\alpha(1-y)^{\beta-2}; \tag{7.12}$$

$$v_{12}(x, y) = -\alpha\beta x^{\alpha-1}(1-y)^{\beta-1}. \tag{7.13}$$

Since $v_1 > 0$ and $v_2 < 0$, Assumption 3 is satisfied. Moreover, by (7.13), $v_{12} < 0$ and D_θ°. Since $r(x) = \beta x/[\alpha(1-x)]$, $\lim_{x \to 0} r(x) = \infty$. Thus, condition (5.1) is satisfied for any ρ such that

$$0 < \rho < 1. \tag{7.14}$$

Thus, the lemma is proved (for any values of parameters ρ, θ, α and β satisfying (7.4), (7.5), (7.6), (7.7) and (7.8)). Q.E.D.

Lemma 7.1 demonstrates that the above results are applicable to model $M(\alpha, \beta, \theta, \rho)$. In model $M(\alpha, \beta, \theta, \rho)$, by Theorem 5.1, the optimal accumulation function of model $M(\alpha, \beta, \theta, \rho)$ is tent-shaped. In the model, therefore, there is a unique stationary state. In particular, the stationary state is locally unstable if

$$\rho < \frac{\beta(2\alpha-1)}{\alpha(1-2\beta)} \tag{7.15}$$

(see Benhabib and Nishimura, (1985)). In model $M(\alpha, \beta, \theta, \rho)$, therefore, there is a wide range of parameter values with which optimal accumulation follows the complex nonlinear dynamics resulting from the semi-tent shape map with a locally unstable stationary state.

We use Theorem 6.2 in order to find specific values of the parameters under which the optimal accumulation function h_ρ is a chaotic map. To this end, note that in model $M(\alpha, \beta, \theta, \rho)$,

$$\Gamma(k_{t-1}, k_t, k_{t+1}) = -\beta k_{t-1}^{\alpha}(1 - k_t)^{\beta-1} + \rho\alpha k_t^{\alpha-1}(1 - k_{t+1})^{\beta}. \tag{7.16}$$

Theorem 7.1: *Let* α, β, θ, ρ, ε, *and* ε' *satisfy*

$$\begin{aligned} &\alpha = 0.75, \beta = 0.125, \theta = 1.1, \rho = 0.01, \\ &\varepsilon = 99/110 \text{ and } \varepsilon' = 101/110. \end{aligned} \tag{7.17}$$

The optimal accumulation function of model $M(\alpha, \beta, \theta, \rho)$ is a chaotic map in the sense of Li and Yorke.

Proof: Let $k^* = 0.04946 \in [0, 1]$. Then, $k_1^* = \theta k^* = 0.054406$ and $k_2^* = \theta k_1^* = 0.0598466$. First, we demonstrate that $k = k^*$ satisfies conditions (6.31)-(6.33) in Theorem 6.2. Since $k_1^* < k_2^* = \varepsilon < 1/\theta = 1/1.1$, $(k_1^*, k_2^*) = (s_\theta(k^*), s_\theta^{(2)}(k^*))$. Moreover, $v_i(k^*, k_1^*), v_i(k_1^*, k_2^*), s'_\theta(k^*)$ and $s'_\theta(k_1^*)$ are all well defined. Thus, the functions $\Gamma_{(1)}$, $\Gamma_{(2)}$ and $\Gamma_{(3)}$ defined by (6.25)-(6.27) are all well defined. By a simple calculation, the following can be obtained:

$$\Gamma_{(1)}(k^*) = 0.00020507596 > 0; \tag{7.18}$$

$$\Gamma_{(2)}(k^*) = 0.00164451767 > 0; \tag{7.19}$$

$$\Gamma_{(3)}(k^*) = 0.00000027421 < 0. \tag{7.20}$$

These facts imply that conditions (6.31)-(6.33) are all satisfied.

Let $\kappa = 0.04945$. Then $\Gamma_{(3)}(\kappa) = 0.00000312402 > 0$, which implies (6.35) is satisfied. Q.E.D.

Note that, in the above example, $k = 0.04946$ approximates a periodic point of period 3.

8. Concluding Remarks

We have demonstrated that if the reduced form utility function v of an optimal growth model always has a negative partial derivative, $v_{12} < 0$, on the interior of its domain, the optimal accumulation function must have a shape similar to a tent map. The reason can be explained by the following facts: (i) the set of expansible stock levels can be partitioned into two intervals; (ii) if a stock level at the beginning of a period falls in the lower interval, the optimal choice of a stock level at the end of the period must hit the boundary of the feasible set, which is generally increasing; (iii) if a stock level at the beginning of a period falls in the upper interval, the optimal choice must lie in the interior of the feasible set and is (weakly) decreasing in the beginning-of-a-period stock level.

The result demonstrates that if $v_{12} < 0$, optimal capital accumulation may follow complex nonlinear dynamics for a wide range of economic models. In consideration of the result, we have constructed a characterization of a chaotic optimal accumulation function, which provides an easy way to construct chaotic optimal paths. Using that characterization, we have produced an example of chaotic optimal accumulation.

A possible extension is to apply our method to the analysis of the interlinkage of international business cycles, which Nishimura and Yano (1993a,c) initiate.

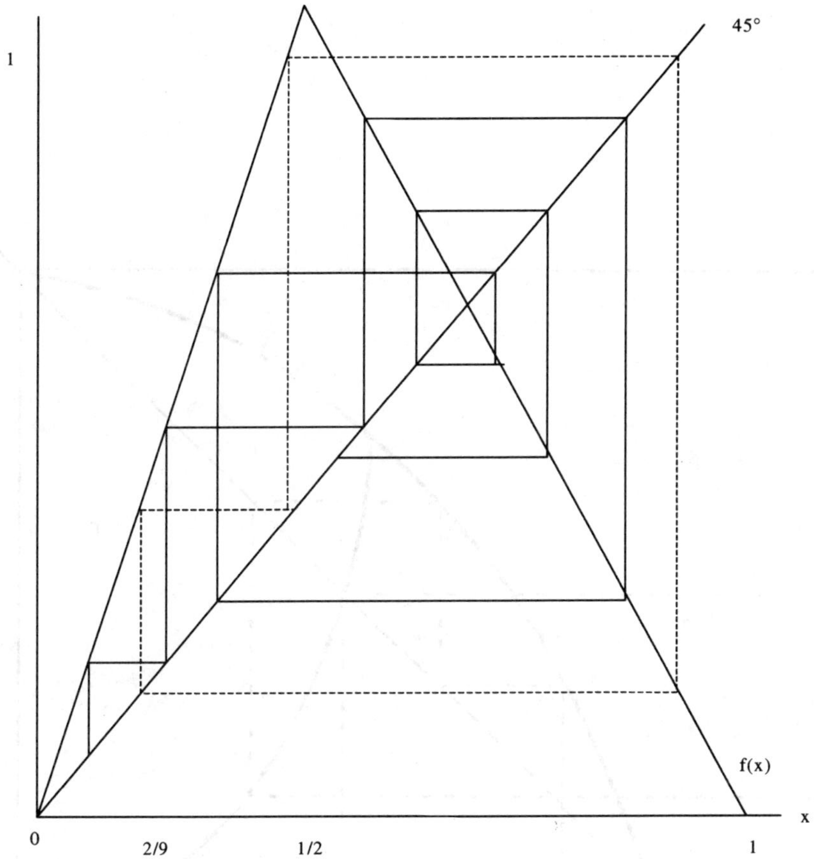

Figure 1

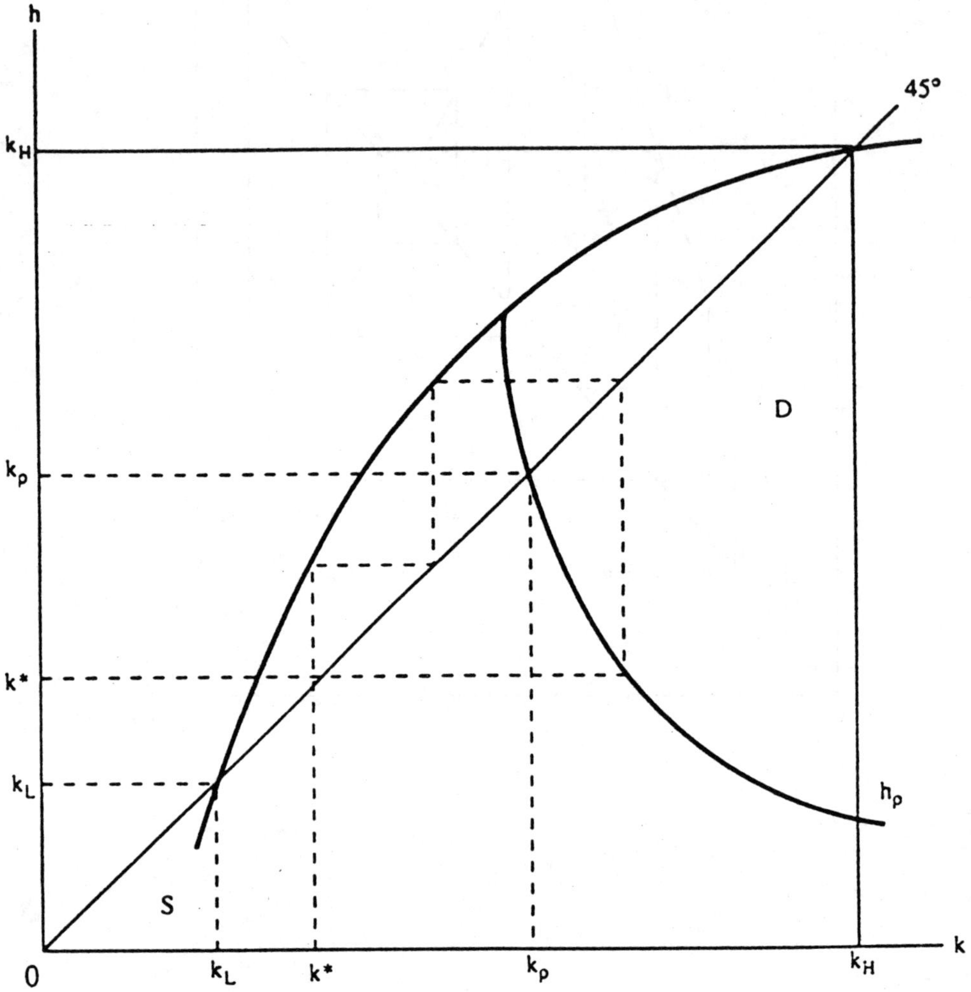

Figure 2

5

The Local Bifurcation of Ramsey Equilibrium

Robert A. Becker and Ciprian Foias

1. Introduction

The hypothesis that aggregate fluctuations may represent an endogenous feature of dynamic competitive economies with incomplete markets has been advanced in several papers.[1] The role of incomplete markets seems essential for the appearance of cycles in one sector models.[2] Becker and Foias (1987) and Woodford (1988a) have pointed out that the elasticity of substitution of the production function plays a fundamental role in the existence of cyclic equilibrium paths. In those papers cycles are generated if the substitutability between capital and labor is not too great. Heterogeneity of households is also a crucial component of their fluctuation theories.

Central to our paper is the exploration of the role of the elasticity of substitution in the generation of cyclic equilibrium trajectories. We study the properties of a two dimensional dynamical system derived from the perfect foresight equations governing equilibrium in an economy with a single capital good and many infinitely lived households. A local bifurcation of the steady state to a stable two-period orbit is found when the elasticity of substitution is sufficiently smaller than one. The model originates in the work of Ramsey (1928) and is further developed in Becker (1980), Becker

[1]See Bewley (1986), Becker and Foias (1987), and Woodford (1988a). Boldrin and Woodford (1990) and Guesnerie and Woodford (1992) survey those equilibrium models.

[2]Dechert (1984) shows that the one sector Ramsey model exhibits a monotonic aggregate capital stock along an optimal program for a wide variety of technologies. Benhabib, Jafarey, and Nishimura (1988) show that standard turnpike properties may be recovered for a class of recursive preferences in Pareto optimal growth with many consumers.

and Foias (1987, 1990), and Becker, Boyd and Foias (1991).[3] The model differs from the related one introduced by Woodford (1988a) in several respects. His entrepreneur does not have labor income and his workers may exhibit a labor-leisure choice. The objective function of the entrepreneur is special in the specification of a logarithmic one-period return function. His agents, like ours, may have different discount factors although differences between the workers and entrepreneurial discount factor play no role in his analysis. By contrast, differences between agents discount factors lies at the heart of our model.[4] Moreover, our sole agent with capital also has a wage income; capital markets exist unlike the Woodford model so workers in our framework always have the option of acquiring capital even though they will not choose to do so in the class of equilibria investigated in this chapter.

The underlying economic model is a standard neoclassical model of capital accumulation with discrete time. There are several infinitely lived agents each maximizing lifetime utility subject to a sequence of budget constraints, one for each time. Each household's preference ranking over alternative consumption profiles is specified by a time additive separable objective with a fixed utility discount factor; there is a least and most impatient household.

Given a sequence of wages and rental rates for capital, an agent chooses a pattern of consumption and capital holdings at each time such that consumption plus next period's capital equals the wage and total capital income. The sequential budget constraint combined with the nonnegativity of capital implies there is a borrowing constraint - future income cannot be discounted to the present. A *Ramsey equilibrium* occurs at a path of rental and wage rates such that the demand for consumption and the supply of next period's capital goods equals total output at each time.[5]

Equilibrium programs are known to converge to the stationary

---

[3]See Boyd (1990) and Hernández (1991) for additional research on this model. Bewley (1986) seems the first to draw attention to market incompleteness as a potential source of equilibrium fluctuations.

[4]See Woodford (1988b) for a critique of the closely related cash-in-advance models where laborers choose not to borrow due to high discount rates.

[5]Ramsey equilibrium programs were shown to exist with general utility functions by Becker, Boyd and Foias (1991).

Ramsey equilibrium found in Becker (1980) when the total capital income is an increasing function of the capital stock.[6] A periodic equilibrium may otherwise exist.[7] Cyclic equilibria are constructed using a local bifurcation analysis in this chapter. Hence, we are interested in the behavior of the orbits of a dynamical system representing the evolution of equilibrium capital stocks when capital income is no longer an increasing function of the total capital stock. The formulation of the bifurcation problem turns on the specification of an appropriate "parameterization" of the production function. The change in the parameter should bring about a change in the elasticity of substitution between capital and labor. If capital is not too easily substituted for labor, then a decrease in the elasticity of substitution is associated with oscillations of total capital income as total capital changes. This may be sufficient to induce fluctuations in the income of the capitalist agent (the sole agent with capital) while all other agents merely consumer their wages.[8]

A dynamical system may be defined for the case where the most patient agent holds all the capital. We perturb the production function in the neighborhood of the steady state equilibrium stock to change only the curvature of the production function and thereby induce a change in the elasticity of substitution. A center manifold reduction leads to a one parameter family of one dimensional maps. For a fixed value of the parameter, the iterates of the corresponding map represent the evolution of the equilibrium capital stock. The construction of the periodic equilibrium path is then based on a local bifurcation argument.

In order to apply standard bifurcation theory, we show how to construct one parameter perturbations of the production function in order to reduce our problem to the conventional bifurcation framework. Our approach to the bifurcation problem emphasizes the variation of the technology as a source of cyclic equilibria.

---

[6]See Becker and Foia (1987). Henández (1991) recovers this result in the case where all households have a common discount factor.

[7]Becker and Foias (1987) provide an *ad hoc* construction of an equilibrium 2-cycle.

[8]This is also the intuition for the 2-cycle example built by Becker and Foias (1987). The intertemporal tolerance for consumption variation and discounting also enter the analysis.

This emphasis on one of the economic primitives is illustrative of a more general concern for explaining the origin of endogenous cycles in terms of the underlying preferences and technology. The local bifurcation analysis conducted here represents one way to capture a change in an underlying production or utility function lying in an infinite dimensional function space. Hence, the method of reduction of the otherwise infinite dimensional parameter dynamical system to a scalar parameter model may be useful in other dynamic economic models.[9]

The particular type of bifurcation constructed is a *flip bifurcation.* The linearization of the unperturbed dynamical system around the steady state equilibrium produces two eigenvalues: both eigenvalues are distinct with the larger eigenvalue always greater than one. If the income monotonicity hypothesis fails, we give a condition for the other eigenvalue to equal -1; this is the first step in verifying there is a flip bifurcation. Our main result is the demonstration that generically speaking, for any production function meeting this eigenvalue test at the rest point there exist nearby production functions for which the model has nonstationary stable periodic Ramsey equilibria. In fact, the original steady state equilibrium loses stability and a two period cycle is born.

The outline of the chapter is as follows: Section 2 reviews the basic Ramsey equilibrium model and defines the dynamical system under study. The local analysis of a steady state equilibrium including the condition for one eigenvalue to have the value -1 is presented in Section 3. We also present the center manifold reduction of the dynamical system when one eigenvalue is -1. This yields the form of the dynamical system that is studied in order to produce a flip bifurcation. The center manifold reduction implies that our analysis of the bifurcation problem is local. Two subsets of the space of all production functions are introduced in Section 4. Examples of economies (production functions and capitalist's temporal utility function) which give rise to elements of

[9]Majumdar and Mitra (Chapter 3) examine changes in the utility and production functions in a one-sector Ramsey model with wealth effects. They demonstrate the possibility of topological chaos for an open set of production and felicity functions that are the economic primitives of the model.

those sets are also given in that section. The perturbations of the original production function are constructed as elements of these crucial spaces. The main theorem is presented in Section 5. We report there the calculations supporting the conclusion that the conditions for a stable flip bifurcation hold. Concluding comments follow in Section 6. Proofs of some technical lemmas stated in Section 4 are collected in the Appendix.

2. The Ramsey Equilibrium Model

The Ramsey equilibrium model is briefly summarized together with the reduction to the dynamical system governing the equilibrium profiles explored in this chapter.

2.1 The Basic Model

There are $H \geq 1$ households indexed by $h = 1, ..., H$. There is a single commodity available for consumption or investment at each time. At time zero, households are endowed with capital stocks $k^h \geq 0$. Put $K = \sum_h k^h$ and assume $K > 0$. Time is taken in discrete intervals, $t = 1, 2, ...$. Let c_t^h, x_t^h denote the consumption and capital stock of household h at time t. Household h has temporal utility function u_h; c_t^h is the argument of u_h. Household h discounts future utilities by the factor δ_h with $0 < \delta_h < 1$. We assume (put $\Re_+ = [0, \infty)$ and $\Re_{++} = (0, \infty)$):

A1: *For each* h, $u_h : \Re_+ \to \Re$ *is* $C^{(2)}$ *on* $\Re_{++}$ *with* $u_h' > 0$, $u_h'' < 0$, $\lim_{c \to 0} u'(c) = \infty$, *and* $\lim_{c \to \infty} u'(c) = 0$.
A2: $1 > \delta_1 > \delta_2 > ... > \delta_H > 0$.

Assumption (A1) is standard; assumption (A2) defines the first household as the most patient. Every household supplies one unit of labor at each time and labor services are assumed homogeneous.

The production sector is characterized by the one sector neoclassical production function f; inputs precede outputs by one period. Capital is the only variable factor. The technology is described by the standard assumption:

A3: $f : \Re_+ \to \Re_+$, $f(0) = 0$, f *is* $C^{(2)}$ *on* $\Re_{++}$, $f' > 0$,

$\lim_{x \to 0} f'(x) = \infty$, $\lim_{x \to \infty} f'(x) = 0$, *and* $f'' < 0$.

This assumption implies there is a maximum sustainable stock, K^M, with $0 < f(K^M) = K^M < \infty$. We regard f as derived from a neoclassical constant returns to scale production function with labor input fixed at H.

Let $\{1 + r_t, w_t\}$ be a sequence of one period rental factors and wage rates, respectively. The sequences $\{1 + r_t, w_t\}$ are always taken to be nonnegative and nonzero. Households are competitive agents and perfectly anticipate the profile of factor returns $\{1 + r_t, w_t\}$. Given $\{1 + r_t, w_t\}$, h solves

$$P(h) : \sup \sum_{t=1}^{\infty} \delta^{t-1} u_h(c_t^h)$$

by choice of nonnegative numbers $\{c_t^h, x_t^h\}$ satisfying $x_0^h = k^h$ and

$$c_t^h + x_t^h = w_t + (1 + r_t)x_{t-1}^h, \ (t = 1, 2, ...). \tag{2.1}$$

The *no arbitrage* necessary conditions for $\{c_t^h, x_t^h\}$ to solve $P(h)$ are $c_t^h > 0$ and

$$\delta_h (1 + r_{t+1}) u_h'(c_{t+1}^h) \leq u_h'(c_t^h) \tag{2.2}$$

with equality whenever $x_t^h > 0$.

The production sector is modeled by the myopic profit maximization problem $P(F)$ at each t, where

$$P(F) : \sup[f(x_{t-1}) - (1 + r_t)x_{t-1}]$$

by choice of $x_{t-1} \geq 0$. The assumption of a point-input point-output production function that a competitive rental market implies the maximization of discounted profits is equivalent to $P(F)$.

Sequences $\{1 + r_t, w_t, K_{t-1}, c_t^h, x_{t-1}^h\}$ constitute a *Ramsey Equilibrium* provided:

(A) *For each* h, $\{c_t^h, x_{t-1}^h\}$ *solves* $P(h)$ *given* $\{1 + r_t, w_t\}$.
(B) *For each* t, K_{t-1} *solves* $P(F)$ *given* $1 + r_t$.
(C) $Hw_t = f(K_{t-1}) - (1 + r_t)K_{t-1} (t = 1, 2, ...)$.
(D) $\sum_h x_{t-1}^h = K_{t-1} (t = 1, 2, ...)$, $0 < K = K_0 \leq K^M$.

Thus, consumers maximize utility (A) and producers maximize profits (B). The labor market clearing condition is expressed in (C). The capital market clearing condition is (D). The output market balance follows by combining (A)-(D). Hence

$$\sum_h (c_t^h + x_t^h) = f(K_{t-1}). \tag{2.3}$$

Note that equilibrium consumption and capital sequences are bounded from above by the maximum sustainable stock.

If $0 < 1 + r_t < \infty$, then (A3) implies there is a unique positive stock K_{t-1} which solves $P(F)$ at each t; clearly

$$f'(K_{t-1}) = 1 + r_t; \tag{2.4}$$

furthermore, the corresponding $\{w_t\}$ defined by

$$Hw_t = f(K_{t-1}) - (1 + r_t)K_{t-1} \tag{2.5}$$

is positive.

2.2 Equilibrium Programs

An important example of a Ramsey equilibrium is the *stationary Ramsey equilibrium* defined as a time invariant Ramsey equilibrium. Becker (1980) showed there is a unique stationary Ramsey equilibrium. Moreover, in that equilibrium configuration only the most patient household had a positive capital stock; all other agents consumed only wages. This steady state equilibrium level of the first agent's capital stock (which is also the aggregate capital stock) is denoted by k and satisfies the equation $\delta_1 f'(k) = 1$. Becker and Foias (1987) found equilibria converged to this steady state when *capital income*, $f'(K)K$, is a strictly increasing function of K. This restricts the elasticity of substitution of capital for labor. It is satisfied by the Cobb-Douglas function $f(K) = AK^\alpha$ where $A > 0$ and $0 < \alpha < 1$. We note the elasticity of substitution of the production function can be written as $\sigma_f(x) = -f'(x)[f(x) - f'(x)x]/xf(x)f''(x)$ under our neoclassical assumptions. In particular, $\sigma_f(x) = 1$ for all x in the Cobb-Douglas case.

An equilibrium program exhibits the *turnpike property* if all households but the most patient eventually achieve and maintain a zero capital position. We will consider the cases where the equilibrium profile with the turnpike property exist and explore the dynamics resulting from that specification of the model. To this end we first state the necessary and sufficient conditions for the existence of such equilibria.[10] A Ramsey equilibrium for an economy evolves in the case where the turnpike property holds according to the equations

$$x_{t+1} + c^1_{t+1} = g(x_t), \;\; c^h_{t+1} = w_{t+1}(h \geq 2); \tag{2.6}$$

$$u'(c^1_t) = \delta_1(1 + r_{t+1})u'_1(c^1_{t+1}) \tag{2.7}$$

where

$$g(x_t) = (1 + r_{t+1})x_t + w_{t+1} = \left(1 - \frac{1}{H}\right) f'(x_t)x_t + \frac{1}{H} f(x_t).$$

We call g the first household's *income function.* Equation (2.6) expresses the budget constraint for each agent; equation (2.7) is the first agent's no arbitrage condition. The equilibrium condition (C) may be used to track agents $h \geq 2$ consumption. The no arbitrage condition for those agents is expressed by the inequality given below.

Turnpike Property Characterization Theorem: *Suppose* $\{1 + r_t, w_t, K_{t-1}, c^h_t, x^h_{t-1}\}$ *is a Ramsey equilibrium with the properties* $\liminf_{t\to\infty} x_t > 0$ *and* $\liminf_{t\to\infty} w_t > 0$ *where* $\sum_{h=1}^{H} x^h_t = x_t$. *Suppose further that* $\limsup c^h_t \leq K^M$ *and* $\limsup_{t\to\infty} x_t \leq K^M$. *Then a necessary and sufficient condition*

[10] Becker and Foias (1987) give a special version of these condtions for the case of a cycle of period 2.

for $x_{t-1}^h = 0$ *and* $c_t^h = w_t$ *for* $h \geq 2$ *and all* t *is that*

$$c_{t+1}^1 + x_{t+1} = g(x_t),$$
$$\delta_1 f'(x_t) u_1'(c_{t+1}^1) = u_1'(c_t^1),$$
$$\delta_h f'(x_t) u_h'(w_{t+1}) \leq u_h'(w_t), \ h \geq 2,$$

where $x_t = x_t^1$.

The proof of this proposition is omitted since it is trivial: the transversality and no arbitrage conditions are sufficient for each agent's optimization problem. We observe the equilibrium level of aggregate capital depends only upon the magnitude of δ_1 but not on the values of the other agents' discount factors. Whenever we have a sequence of aggregate capital stocks $\{K_t\}$ with $K_t = x_t$ satisfying the first two conditions of the Turnpike Property Characterization Theorem, then just taking δ_h small enough for $h \geq 2$ will inusre the third condition holds and that the sequence is a Ramsey equilibrium.

2.3 The Dynamical System

The Turnpike Property Characterization Theorem allows us to represent those equilibria in terms of a 2-dimensional dynamical system. Using the conditions for a Ramsey equilibrium (and ignoring agents $h \geq 2$) we rewrite (2.6) and (2.7) with the notation $c_t^1 = y_t$ and $K^M = a$ as the system

$$x_{t+1} + y_{t+1} = g(x_t) \tag{2.8}$$

$$u'(y_t) = \delta f'(x_t) u'(y_{t+1}) \tag{2.9}$$

where $x_t > 0$, $y_t > 0$ for $t = 0, 1, \ldots$. The monotonicity properties of u and f imply the existence of a continuous function $F : \Re_{++} \times \Re_{++} \to \Re_{++}$ such that

$$u'(y) = \delta f'(x) u'(F(x, y)). \tag{2.10}$$

Moreover, F is strictly increasing in y for each x and strictly decreasing in x for each y. The properties of F are independent of assumptions governing g. We may write a Ramsey equilibrium, after the turnpike property takes hold, as the system

$$\begin{aligned} y_{t+1} &= F(x_t, y_t) \\ x_{t+1} + y_{t+1} &= g(x_t) \end{aligned} \tag{2.11}$$

where (x_0, y_0) are given and $0 < x_t,\ y_t < a$ for all t. Notice that a stationary Ramsey equilibrium is determined by the relation $\delta f'(\bar{x}) = 1$ for $\bar{x} = k$; clearly $F(\bar{x}, y) \equiv y$.[11]

We want to obtain a dynamical system from (2.11) defined on $[0, a] \times [0, a]$. We will extend F to $([0, a] \times [0, a]) \backslash \{(0, 0)\}$ by the following definitions:

$$\begin{aligned} F(x, 0) &= \lim_{y \downarrow 0} F(x, y),\ x > 0; \\ F(0, y) &= \lim_{x \downarrow 0} F(x, y),\ y > 0. \end{aligned}$$

Notice that $F(0, y) = +\infty$ if $y > 0$. We observe the fact that $a \geq x > \bar{x}$ implies $F(x, y) \leq F(\bar{x}, y) = y$. Hence $F(\bar{x}, 0) = 0$. Indeed, $F(x, 0) = 0$ for all $x > 0$.

The dynamical system on $[0, a] \times [0, a]$ is

$$y_v = \min\{F(x, y), g(x)\},\ x_v + y_v = g(x) \tag{2.12}$$

for $(x, y) \in (0, a) \times (0, a)$. If $x = 0$, $0 < y \leq a$, then $y_v = 0$ and $x_v = 0$. We define $F(0, 0) \equiv \alpha$ where α is arbitrarily chosen in $\Re_+$. If $y = 0$, $0 \leq x \leq a$, then $y_v = 0$ and $x_v = g(x)$. The initial data is specified by the pair (x_0, y_0), where x_0 is given and y_0 has to be determined optimally. For example, in the saddle point case y_0 has to be chosen such that (x_0, y_0) lies on the stable manifold.

The sign of g' is crucial for the analysis of the orbits of (2.12). Assumptions (A1)-(A3) along do not imply $g' > 0$. The capital income monotonicity property implies $g' > 0$. We have already noted that there are periodic equilibria if g' is not positive. In Becker and Foias (1990) we considered the properties of solutions to (2.12) given in the following assumption:

---

[11]We have dropped the subscript on δ since the meaning is clear.

Income Monotonicity Assumption: *The income function is strictly increasing.*

We then showed that a Ramsey equilibrium is equivalent to an orbit of (2.12) lying entirely in $(0, a) \times (0, a)$; given the Income Monotonicity Assumption, we found monotonic convergence of $\{x_t\}$ to $\bar{x}$. We also developed a local analysis based on a linear approximation system for (2.12) in a neighborhood of the steady state and found that $x_t \to \bar{x}$ monotonically provided $|x_0 - \bar{x}|$ is small enough. Let Φ denote near $(\bar{x}, \bar{y})$ with $\bar{y} = g(\bar{x}) - \bar{x}$; $(\bar{x}, \bar{y})$ is a fixed point of Φ.[12] Given the Income Monotonicity assumption, the linear analysis in the neighborhood of the fixed point leads to the conclusion that $(\bar{x}, \bar{y})$ is a *saddle point* in the two dimensional phases space. The Jacobian of Φ at $(\bar{x}, \bar{y})$ has the distinct eigenvalues λ_1, λ_2 satisfying $0 < \lambda_1 < 1 < \lambda_2$ and the conclusion follows.[13] In this last case, the Ramsey equilibria must lie entirely on the stable manifold of $(\bar{x}, \bar{y})$. We notice that a sufficient condition for Income Monotonicity is $\sigma_f(x) \geq 1$ for all x. Thus, Income Monotonicity holds in the Cobb-Douglas case.

3. Local Dynamics and the Bifurcation Condition

3.1 The Linear Analysis without Income Monotonicity

The linear analysis of (2.12) is developed in this section when the Income Monotonicity hypothesis no longer holds. In fact, we will assume without loss of generality that $g'(\bar{x}) < 0$. In particular, this implies $\sigma_f(\bar{x}) < 1$. For (x, y) near $(\bar{x}, \bar{y})$ we have a map Φ defined by

$$\Phi : \begin{pmatrix} x \\ y \end{pmatrix} \mapsto \begin{pmatrix} x_v \\ y_v \end{pmatrix} = \begin{pmatrix} g(x) - F(x, y) \\ F(x, y) \end{pmatrix}. \tag{3.1}$$

[12]See equation (3.1) below for a precise definition of Φ.

[13]See Becker and Foias (1990), Section IV.

The differential of this map evaluated at $(\bar{x}, \bar{y})$ is:

$$J = \begin{pmatrix} g'(\bar{x}) - F_x(\bar{x}, \bar{y}) - F_y(\bar{x}, \bar{y}) \\ F_x(\bar{x}, \bar{y}) & F_y(\bar{x}, \bar{y}) \end{pmatrix}$$

where $F_x(\bar{x}, \bar{y}) = -\delta f''(\bar{x})u'(\bar{y})/u''(\bar{y}) = -\eta < 0$, $F_y(\bar{x}, \bar{y}) = 1$, and $g'(\bar{x}) = (1 - 1/H) \times f''(\bar{x})\bar{x} + f'(\bar{x}) \equiv \gamma < 0$.

The characteristic polynomial for J is

$$\lambda^2 - (\gamma + \eta + 1)\lambda + \gamma = 0. \tag{3.2}$$

Obviously both roots λ_1, λ_2 of (3.2) are real, $\lambda_1\lambda_2 < 0$ (hence by a suitable notation $\lambda_1 < 0 < \lambda_2$), and $\lambda_1 + \lambda_2 = \gamma + \eta + 1$. Therefore, $\eta > 0$ implies $\lambda_1\lambda_2 = \gamma < \gamma + \eta = \lambda_1 + \lambda_2 - 1$; it follows that $1 - \lambda_2 < \lambda_1(1 - \lambda_2)$ and neither root equals 1. If $0 < \lambda_2 < 1$, then $1 - \lambda_2 > 0$ implies $\lambda_1(1 - \lambda_2) > 0$ and consequently $\lambda_1 > 0$, which is impossible. So we can make the following:

Remark 3.1: $\lambda_2 > 1$.

As long as $-1 < x_1 < 0$, $(\bar{x}, \bar{y})$ is still a saddle point and locally the dynamics behaves as in the Income Monotonicity case. In particular, $(\bar{x}, \bar{y})$ has a stable manifold and the trajectory near $(\bar{x}, \bar{y})$ is a Ramsey equilibrium if and only if it lies on the stable manifold of $(\bar{x}, \bar{y})$. However, unlike the case where the Income Monotonicity assumption holds, the convergence of x_t to $\bar{x}$ is no longer monotonic, but damped oscillatory about $\bar{x}$.

In order to produce any local bifurcation the characteristic root λ_1 of J must pass through -1. If $\lambda_1 = -1$, then $-\lambda_2 = \gamma$ must hold. However, this condition implies $-1 + \lambda_2 = \gamma + \eta + 1$, or $-1 - \gamma = \gamma + \eta + 1$; solving for γ yields the condition

$$\gamma = -1 - \frac{1}{2}\eta < 0. \tag{3.3}$$

Equation (3.3) is equivalent to the condition on the data

$$\left[\left(1 - \frac{1}{H}\right)\bar{x} + \frac{1}{2}\delta u'(\bar{y})/u''(\bar{y})\right] f''(\bar{x}) + \frac{1}{\delta} + 1 = 0. \tag{3.4}$$

Notice that equation (3.4) cannot hold if $H = 1$ since $u' > 0$, $u'' < 0$, and $f'' < 0$. Thus $H > 1$ is necessary for the occurrence of any local bifurcation and in particular the flip bifurcation.[14]

Suppose that the agents find themselves in a model where capital and labor are not too substitutable so that $g'(\bar{x}) < 0$ holds. If the magnitude of $g'(\bar{x})$ is sufficiently large, then a small increment in the capital stock reduces capital income and total income for the capital holding agent even though the total wage bill rises. Since consumption is a normal good, intuition suggests that this agent will want to reduce consumption in light of the income reduction. But how far will consumption fall? The agent's *intertemporal tolerance* function, $A(y) \equiv -u'(y)/u''(y)$, offers a measure of sensitivity of u to consumption variation.[15] A low value of the intertemporal tolerance coefficient signifies the desire to smooth consumption. In the presence of an income fall due to a small rise in the capital stock, the consumer will try to balance the income drop with a consumption change - the lower the intertemporal tolerance, then the less the agent cuts current consumption forcing the impact of the income change to be absorbed as a fall in capital holdings. This reduction in capital in turn induces an increase in next periods capital and total incomes again because of the weak substitution possibilities of capital for labor. The agent consumes more and accumulates capital thereby setting up the potential for an oscillating total capital stock to follow as a result of the corresponding drop in capital income in the following period. The degree of discounting future rewards also enters the agent's calculations regarding smoothing of oscillating consumption profiles.

Equation (3.4) may be rewritten in the following manner

$$g'(\bar{x}) = -1 + \frac{\delta}{2} A(\bar{y}) f''(\bar{x}). \tag{3.4A}$$

[14]If there is only one household, then the model is equivalent to the standard Ramsey optimal growth model with a one sector technology. Optimal capital sequences are always monotonic in that framework.

[15]The intertemporal tolerance is the reciprocal of the Arrow-Pratt measure of absolute risk aversion. This is of course a measure of the degree of concavity of u. Hence the tolerance function also serves the same purpose. The *intertemporal elasticity of substitution*, $A(y)/y$, is a more commonly used measure of sensitivity to intertemporal consumption variation.

This equation expresses the precise conditions that must be satisfied by the data for the agent's intertemporal tolerance, discount factor and the degree of substitutability of capital and labor in the production function (the curvature coefficient $f''(\bar{x})$ and the change in total income, $g'(\bar{x})$) to balance in just the right way for the oscillations in consumption and capital to become undamped and pass into a cyclic pattern.[16] The fluctuations that result are perfectly foreseen, but not smoothed owing to the ways in which these factors interact. While economic intuition suggests the presence of the various ingredients found in (3.4), the discussion shows that to have guessed the exact form of equation (3.4) defining the bifurcation set would have been quite difficult. At this time a purely intuitive derivation of (3.4) is still missing.

Economic intuition suggests the elasticity of substitution at the steady state, $\sigma_f(\bar{x})$, is the crucial technological parameter in the bifurcation condition. However $\sigma_f(\bar{x})$ is hidden in (3.4). We may again rewrite (3.4) to emphasize the place of the elasticity of substitution and record the following expression:

$$\alpha - \frac{1}{\beta}\frac{\delta}{2}A(\bar{y}) = \sigma_f(\bar{x})\frac{f(\bar{x})}{f(\bar{x}) - \delta^{-1}\bar{x}}, \tag{3.4B}$$

where $\beta = 1 + 1/\delta$ and $\alpha = (1 - 1/H)/\beta$. Equation (3.4B) expresses the exact balance between the technological and preference parameters that must obtain for a local bifurcation of the steady state into a two period cycle. Suppose that $\sigma_f(\bar{x})$ is very close to zero-capital and labor are used in approximately fixed proportions. Then (3.4B) holds for $\sigma_f(\bar{x})$ near zero provided $A(\bar{y})$ is chosen appropriately close to $2/\delta$. Hence cycles may appear for all sufficiently small values of the elasticity of substitution at the steady state provided the intertemporal tolerance is also sufficiently small.

3.2 The Center Manifold

The matrix J has a single eigenvalue equal -1 in the case where the data satisfies the condition (3.4). Hence J is nonhyperbolic.

[16]The units of measurement in (3.4A) are output per unit of capital.

Local analysis of the dynamics cannot proceed further on the basis of linear analysis alone. However, the dimensionality of (3.1) may be further reduced to a one dimensional system whose orbits lie on an invariant manifold of the original dynamical system. This invariant manifold is called the *center manifold.* The existence of a center manifold is guaranteed under general conditions; see Kelly (1967) for details.[17]

The economically interesting parameter in the Ramsey equilibrium model is the elasticity of substitution at the equilibrium point, $\sigma_f(\bar{x})$.[18] To meet our goal of obtaining a bifurcation result we will introduce a one parameter family of production functions $f(\bullet, \mu)$ for $|\mu| < \mu^*$, for some μ^*. We emphasize the parameter μ is strictly increasing rational function of the value of the elasticity of substitution at the steady state; the use of our parameter μ is technically more convenient (see Lemma 4.3). Each production function $f(\bullet, \mu)$ is a function of x (where $f(x, 0) = f(x)$ for all $x \in (0, \infty)$), satisfies (3.4), and will be shown to yield a flip bifurcation. For these production functions the following Center Manifold Theorem will play a fundamental role. In the next section we will introduce a scalar parameter perturbation of the production function in a neighborhood of $\bar{x}$ that alters the value of the elasticity of substitution at that point and allows us to check whether or not the conditions for a flip bifurcation obtain at the fixed point.

The introduction of a family of production functions depending on a parameter μ leads to the one parameter family of functions $f(x, \mu)$, $g(x, \mu)$, and $F(x, y, \mu)$. Here $f(\bullet, 0)$ is the original production function satisfying our basic axioms and in particular equation (3.4) at $\bar{x}$, $g(\bullet, \mu)$ and $F(\bullet, \bullet, \mu)$ agree with the original functions g and F for all values of x and y evaluated at the parameter value $\mu = 0$. The corresponding dynamical system if the map Y defined by

$$Y : \begin{pmatrix} x \\ y \end{pmatrix} \mapsto \begin{pmatrix} x_v \\ y_v \end{pmatrix} = \begin{pmatrix} g(x, \mu) - F(x, y, \mu) \\ F(x, y, \mu) \end{pmatrix}.$$

[17]Other useful references on center manifold theory may be found in Carr (1981) and Ruelle (1989).

[18]See Becker and Foias (1987) for the underlying intuition.

To analyze the family of dynamical systems depending on the parameter μ we must consider the expanded dynamical system defined by the map also denoted by Y:[19]

$$Y: \begin{pmatrix} x \\ y \\ \mu \end{pmatrix} \mapsto \begin{pmatrix} x_v \\ y_v \\ \mu_v \end{pmatrix} = \begin{pmatrix} g(x,\mu) - F(x,y,\mu) \\ F(x,y,\mu) \\ \mu \end{pmatrix}. \tag{3.5}$$

The differential of Y evaluated at $(\bar{x}, \bar{y}, 0)$ is expressed by the matrix

$$M = \begin{pmatrix} g'(\bar{x},0) - F_x(\bar{x},\bar{y},0) & -F_y(\bar{x},\bar{y},0) & g_\mu(\bar{x},0) - F_\mu(\bar{x},\bar{y},0) \\ F_x(\bar{x},\bar{y},0) & F_y(\bar{x},\bar{y},0) & F_\mu(\bar{x},\bar{y},0) \\ 0 & 0 & 1 \end{pmatrix}$$

The definition of the one-parameter class of functions f, g, and F implies that the derivatives f_x, g_x and F_x evaluated at $(\bar{x}, \bar{y}, 0)$ equal the derivatives of the original functions f, g, and F evaluated at $(\bar{x}, \bar{y})$. The characteristic roots of M are also denoted by λ_1, λ_2, and λ_3 where λ_1 and λ_2 are the eigenvalues of J and $\lambda_3 = 1$.

We will find a center manifold for (3.5) under the supplementary hypothesis that the parametric production function defined on $\Re_+ \times M$ where M denotes the parameter space is $C^{(4)}$ in a neighborhood of $(\bar{x}, 0)$; later we will show how to perturb the original $C^{(2)}$ production function, defined only over $\Re_+$, to satisfy this additional smoothness condition in a neighborhood of $\bar{x}$ and imbed that function in a family of production functions with the desired smoothness in x and μ. We will also require that $u \in C^{(4)}$ in a neighborhood of $\bar{y}$.

Center manifold theorem: *There exists a center manifold φ defined on a neighborhood of $(\bar{x}, 0)$ for $Y \in C^{(4)}$ with a unique Taylor expansion at $(\bar{x}, 0)$.*[20] *More precisely, for some ε, $\mu^* > 0$, there exists a $C^{(3)}$ function $\varphi : (\bar{x} - \varepsilon, \bar{x} + \varepsilon) \times (-\mu^*, \mu^*) \to \Re$ with $\varphi(\bar{x}, 0) = \bar{y}$, $\varphi_x(\bar{x}, 0) = \eta/2$ and such that $|x - \bar{x}| < \varepsilon$, $|\mu| < \mu^*$ and $(x_v, y_v, \mu_v) = Y(x, \varphi(x, \mu), \mu)$ implies $y_v = \varphi(x_v, y_v) = \varphi(x_v, \mu)$.*

---

[19] This should not lead to any confusion since the analysis applies only to the map.
[20] This Taylor series might not converge.

Proof: The existence of φ follows from the general result in Kelly (1967). We note that

$$F(x, y, \mu) = y_v = \varphi(x_v, \mu_v) = \varphi(g(x, \mu) - F(x, y, \mu), \mu);$$
$$y = \varphi(x, \mu). \tag{3.6}$$

The invariant subspaces of the matrix M are of the form $Y = \eta/(1 - \lambda_i)X$ for $i = 1, 2$ and $Y = X$ corresponding to $\lambda_3 = 1$. The subspace tangent to the graph of $\varphi(\bullet, 0)$ taken at the point $\bar{x}$ is the invariant subspace belonging to $\lambda_1 = -1$. It follows from this tangency that $\varphi_x(\bar{x}, 0) = \eta/(1 - \lambda_1) = \eta/2$. Q.E.D.

The dynamics on the center manifold may be represented by the two dimensional $C^{(3)}$ map ψ defined by

$$(x, \mu) \mapsto (\psi(x, \mu), \mu) = (g(x, \mu) - F(x, \varphi(x, \mu), \mu), \mu). \tag{3.7}$$

Moreover, $\psi(\bar{x}, 0) = \bar{x}$ and

$$\psi_x(\bar{x}, 0) = \gamma - F_x(\bar{x}, \bar{y}, 0) - F_y(\bar{x}, \bar{y}, 0)\varphi_x(\bar{x}, 0)$$
$$= \frac{\gamma(1 - \lambda_1) - \lambda_1 \eta}{1 - \lambda_1} = \frac{2\gamma + 1}{2} = -1. \tag{3.8}$$

The dynamics of the aggregate capital stock are represented as iterates of the map ψ for a constant μ value. We note that for a fixed value of μ the dynamical system is one dimensional. In particular, for $\mu = 0$ we have dynamics corresponding to the original specification of the model - this is the one dimensional dynamical system governing the equilibrium trajectories of (3.1). Moreover, the uniqueness of the Taylor expansion of φ implies $\psi_x(\bar{x}, 0) = -1.$[21]

We record below the conditions for a flip bifurcation. We recall that there are two generic possibilities depending on the signs of a coefficient computed from the derivatives of $\psi(\bullet, \mu)$ up to order three and evaluated at $(\bar{x}, 0)$. We will be interested in the case where $\psi(\bullet, \mu)$ has an attracting 2-cycle orbit for $\mu < 0$. We also note that near the fixed point $\bar{x}$, there is the family of fixed points

[21] This follows from the Implicit Function theorem.

$x(\mu) = \psi(x(\mu), \mu)$ for μ near 0 with $\bar{x} \equiv x(0)$ since $\psi_x(x(0), 0) \equiv \partial\psi/\partial x|_{(\bar{x},0)} = -1$.

Flip bifurcation theorem: *Assume*

$$\psi_{xx}(\bar{x}, 0)\psi_{\mu}(\bar{x}, 0) + 2\psi_{x\mu}(\bar{x}, 0) \neq 0; \tag{3.9}$$

$$a_* \equiv -2\psi_{xxx}(\bar{x}, 0) - 3(\psi_{xx}(\bar{x}, 0))^2 \neq 0. \tag{3.10}$$

Then $x(\mu)$ can be extended to a $C^{(1)}$-function on $[-\mu_*, 0]$ for some $\mu_* > 0$, $\psi(x(\mu), \mu) \equiv x(\mu)$, and $x(\mu)$ is unstable for $\mu < 0$. Moreover, there exist $x_{\pm}(\mu)$ defined for μ in

$$(0, \mu_*] \text{ if } a_* > 0;$$
$$(-\mu_*, 0] \text{ if } a_* < 0,$$

(with μ_* small enough) such that

$$x_-(\mu) < x(\mu) < x_+(\mu), \;\; x_{\pm}(\mu) \to x(0)$$

for $\mu \to 0$, and

$$\psi(x_{\pm}(\mu), \mu) = x_{\mp}(\mu).$$

The 2-cycle $\{x_-(\mu), x_+(\mu)\}$ is asymptotically stable if $a_* < 0$ and unstable in the other case.

Proof: See Guckenheimer and Holmes (1983, p. 158). Q.E.D.

We refer to the conditions on the derivatives of ψ in (3.8) and (3.9) along with $a_* < 0$ as the *Conditions for a Stable Flip Bifurcation*, abbreviated *CFB*, at $(\bar{x}, \bar{y})$. The strategy for checking the bifurcation conditions has several steps. First, we show that original $C^{(2)}$ production function may be perturbed to a $C^{(4)}$ production function without altering equation (3.4). Moreover, we will choose this perturbation so that its fourth derivative, $f''''(\bar{x}) >> 1$.[22] This property of $f''''(\bar{x})$ will insure that the nondegeneracy condition

[22] That is, the fourth derivative is very much larger than 1.

(3.9) holds and $a_* < 0$. Finally, we will show how to imbed the original production function in a one parameter family of maps that yields a sufficiently smooth dynamical system in both the (x, μ) variables for the center manifold reduction to obtain and govern the Ramsey equilibrium dynamics.

4. Perturbations of the Production Function

We show in this section how to perturb the original production function and imbed it in a family of production functions for which a one-parameter family of maps ψ will be defined which is suitable for a bifurcation theory.

4.1 Manifolds of Production Functions

Let $\wp$ denote the class of all production functions satisfying (A3); $\wp$ is an infinite dimensional manifold in the Fréchet space $C^{(2)}(\Re_{++})$ endowed with the topology of uniform convergence of f, f', f'' on compact intervals.[23] We are interested in the subsets of $\wp$ for which (i) $\bar{x}$ is a candidate for a steady state, and (ii) those functions for which (3.4) holds given the first household's utility function, discount factor and the capital stock $\bar{x}$. For notational simplicity we set $k = \bar{x}$ in this section.

Define $\wp_k = \{f \in \wp : \delta f'(k) = 1\}$ and endow this space with topology of uniform convergence of f, f', f'' on any compact interval in $\Re_{++}$. The set $\wp_k$ is an infinite dimensional manifold.[24] We take $\wp_k$ to be the space of production functions in which we will construct the perturbations giving rise to the one parameter family of maps ψ. Define $\mathcal{Q}_k$ to be the subset of $\wp_k$ of determined by the condition (3.4). Elements of $\mathcal{Q}_k$ yield production functions which we will perturb and induce a stable flip bifurcation. Put differently, the set $\mathcal{Q}_k$ will be our bifurcation set.

---

[23]The tangent spaces are given by the space $C_0^{(2)}(\Re_+)$ of twice continuously differentiable real-valued functions defined on $\Re_+$ having compact support. This space is a strict inductive limit of Banach spaces. Since this structure does not play any economic role in our analysis we will leave all the functional-geometric details to the mathematically minded reader.

[24]Indeed, $\wp_k$ is a submanifold of $\wp$ of codimension 1, that is defined by one equation. For example, a surface in $\Re^3$ is a codimension 1 submanifold. Hence, a codimension 1 submanifold in $\wp$ signifies the submanifold is a "rich" class of functions

Before developing the specific forms of the perturbation we show that $\mathcal{Q}_k$ is a submanifold of codimension 1 in $\wp_k$ for the case $u(y) = y^{1-\alpha}$, $\alpha \in (0,1)$. The restriction (3.4) is then seen to be a joint condition on the preference of household one, the number of agents, as well as the technology.

Tolerance Lemma: *Assume f is a production function such that there exists $\bar{x} > 0$ such that $f'(\bar{x}) > 1$ and $(xf'(x))'|_{x=\bar{x}} < 0$. Let $\delta = 1/f'(\bar{x})$ and $u(y) = y^{1-\alpha}$ with $\alpha \in (0,1)$.*

i) Let H be large enough, or $\alpha \in [1/2, 1)$; then $f \in \mathcal{Q}_k$ only if

$$\frac{1}{2}\bar{x}f''(\bar{x}) + f'(\bar{x}) < 0, \tag{C}$$

that is $(x^2 f'(x))' < 0$ evaluated at $\bar{x}$.

ii) Conversely, if the following Condition (D)

$$-f'(\bar{x})^2 < \bar{x}f''(\bar{x}) + f'(\bar{x}) < -f'(\bar{x}) \tag{D}$$

holds, then there exists $\alpha \in (1/2, 1)$ such that $f \in \mathcal{Q}_k$ for all H large enough.[25]

Proof: The intertemporal tolerance is given by $A(y) = y/\alpha$. The budget balance for the capital holding agent is

$$\bar{x} + \bar{y} = \left(1 - \frac{1}{H}\right)\delta^{-1}\bar{x} + \frac{f(\bar{x})}{H}.$$

Solving for consumption yields

$$\bar{y} = \left[\left(1 - \frac{1}{H}\right)\delta^{-1} - 1\right]\bar{x} + \frac{f(\bar{x})}{H}.$$

Substituting this expression into (3.4) and simplifying leads to the condition that must be satisfied

$$\left[\left(1 - \frac{1}{H}\right)\bar{x}\left(1 - \frac{1}{2\alpha}\right) + \frac{\delta}{2\alpha}\left(\bar{x} - \frac{f'(\bar{x})}{H}\right)\right]$$
$$f''(\bar{x}) + \delta^{-1} + 1 = 0. \tag{3.4C}$$

[25] Notice that the second inequality in Condition (D) implies (C).

We first show that (C) is a necessary condition for $f \in \mathcal{Q}_k$. If (3.4C) holds, then $\delta^{-1} = f'(\bar{x})$ implies

$$\left[\left(1 - \frac{1}{H}\right)\bar{x}\left(1 - \frac{1}{2\alpha}\right) + \frac{\delta}{2\alpha}\left(\bar{x} - \frac{f(\bar{x})}{H}\right)\right]$$
$$f''(\bar{x}) + f'(\bar{x})$$
$$(1+\delta) = 0.$$

Rearranging yields

$$\frac{1}{2}\bar{x}f''(\bar{x}) + f'(\bar{x}) = \frac{1}{1+\delta}$$
$$\left\{\frac{1-\alpha}{2\alpha}\left(1 - \frac{1}{f'(\bar{x})}\right)\bar{x} - \frac{1-2\alpha}{2\alpha H}\bar{x} + \frac{\delta f(\bar{x})}{2\alpha H}\right\}$$
$$f''(\bar{x}). \tag{3.4D}$$

For H very large or $\alpha > \frac{1}{2}$, the right side of (3.4D) is negative as $f''(\bar{x}) < 0$. Therefore, the condition (C) is necessary for (3.4C) to have a solution. This completes the proof of (i).

Conversely, we show that the condition (D) is sufficient for $-f'(\bar{x})(1+\delta)/f''(\bar{x})$ to lie in the interval formed by varying $\beta \equiv 1/\alpha$ between 1 and 2 according to the function

$$G(\beta, H) = \left[\left(1 - \frac{1}{H}\right)\bar{x}\left(1 - \frac{\beta}{2}\right) + \frac{\delta\beta}{2}\left(\bar{x} - \frac{f(\bar{x})}{H}\right)\right].$$

The endpoints of the interval $G(1,H) = \frac{1}{2}\left(1 - \frac{1}{H}\right)\bar{x} + \frac{\delta}{2}\left(\bar{x} - \frac{f(\bar{x})}{H}\right)$ and $G(2,H) = \delta\left(\bar{x} - \frac{f(\bar{x})}{H}\right)$. Since $0 < \delta < 1$, taking H sufficiently large implies $G(1,H) > G(2,H)$. Let $H \to \infty$ and define $G(1,\infty) = (1+\delta)\bar{x}/2$ and $G(2,\infty) = \delta\bar{x} = \bar{x}/f'(\bar{x})$.

It suffices to show that $-f'(\bar{x})(1+\delta)/f''(\bar{x}) \in (G(2,\infty), G(1,\infty))$. This is equivalent to showing

$$\frac{\bar{x}}{(1+\delta)f'(\bar{x})} < -\frac{f'(\bar{x})}{f''(\bar{x})} < \frac{\bar{x}}{2}.$$

The right inequality is equivalent to Condition (C); the left inequality if equivalent to the first inequality in Condition (D).

Hence (D) is sufficient for (3.4C) to have a solution provided the number of households H is sufficiently large and the tolerance coefficient α is appropriately chosen in $\left(\frac{1}{2}, 1\right)$. This proves statement (ii). Q.E.D.

Now if $f \in \wp_k$ then by using Lemma 4.3 below we can modify f on an arbitrarily small neighborhood of $\bar{x}$ to make $f \in \wp_k$ satisfy

$$f''(\bar{x}) = \frac{1}{2\bar{x}}\frac{1}{\delta}\left(1 + \frac{3}{2\delta}\right).$$

Then Condition (D) holds, i.e.

$$\begin{aligned}-f'(\bar{x})^2 &= -\frac{1}{\delta^2} < \bar{x}f''(\bar{x}) + f'(\bar{x}) \\ &= -\frac{1}{\delta^2} - \frac{1}{2\delta} < -\frac{1}{\delta} = -f'(\bar{x}).\end{aligned}$$

Therefore $f \in \mathcal{Q}_k$ for an appropriate $\alpha \in \left(\frac{1}{2}, 1\right)$ provided H is large enough. The restrictions (C) and (D) are essentially additional curvature conditions on the production function.

Existence Lemma: *Suppose $u(y) = y^{1-\alpha}, \delta$, and $\bar{x} = k$ are given as above. Then there exists f satisfying the assumptions of the Tolerance Lemma as well as Condition (D). Moreover for H large enough $\mathcal{Q}_k$ is a submanifold of $\wp_k$ of codimension 1.*

The following argument can be given a rigorous functional analysis justification using the tangent space $C_0^{(2)}(\Re_+)$. We only present the formal calculations in our proof-the computations will make the validity of the claim clear. The essential point of the proof is to show that there are no singularities in the set $\mathcal{Q}_k$ that would prevent it from having a manifold structure.

Proof of Existence Lemma: The maintained assumptions of the Lemma imply that $\bar{y} = g(\bar{x}) - \bar{x} = \left(\left(1 - \frac{1}{H}\right)\frac{1}{\delta} - 1\right)\bar{x} + \frac{f(\bar{x})}{H}$. Define a functional $B(f)$ according to the formula

$$B(f) = \left[\left(1 - \frac{1}{H}\right)\bar{x} + \frac{1}{2}\delta v(Y(f))\right]\delta''_{\bar{x}}(f) + \frac{1}{\delta} + 1,$$

where $\bar{y} = Y(f)$, $v = u'/u''$, and $\delta''_{\bar{x}}(f) = f''(\bar{x})$. Our bifurcation condition is that the production function satisfy $B(f) = 0$. In order for $\mathcal{Q}_k$ to be a codimension 1 submanifold in $\wp_k$, we need the differential of B at any point f in $\mathcal{Q}_k$ to be surjective. As B is real-valued, this is equivalent to $B'(f) \neq 0$ on the tangent space $T_f(\wp_k) = \{h \in C^2(\Re_{++}),\ h'(\bar{x}) = 0\}$ to $\wp_k$ at f. For a function $h \in C^2(\Re_{++})$, $h'(\bar{x}) = 0$, it can be shown that

$$B'(f)(h) = \left[\left(1 - \frac{1}{H}\right)\bar{x} + \frac{1}{2}\delta v(Y(f))\right] h''(\bar{x}) + \frac{\delta}{2}v'(Y(f))f''(\bar{x})h(\bar{x})/H.$$

For the given utility function, the condition $B'(f) = 0$ is equivalent to $w(\bar{y}) = 0$ and $v'(\bar{y}) = 0$ where $w(y) = \left(1 - \frac{1}{H}\right)\bar{x} + \delta v(y)/2$. Since $f \in \mathcal{Q}_k$

$$w(\bar{y}) = \left(B(f) - \left(\frac{1}{\delta} + 1\right)\right) \Big/ f''(\bar{x}) = -\left(\frac{1}{\delta} + 1\right) \Big/ f''(\bar{x}) \neq 0.$$

Therefore, $B'(f) \neq 0$. Q.E.D.

Notice that the Tolerance Lemma uses the intertemporal tolerance in a crucial manner - a high α value yields the low tolerance for intertemporal consumption variation needed to insure the bifurcation condition holds. In the Existence Lemma δ is given a *priori* and may be taken arbitrarily close to 1. The Tolerance and Existence Lemma also highlight the difference in the origin of competitive cycles in the heterogeneous agent incomplete market framework compared to the multisector representative agent case. Cyclic equilibrium examples in the later models (e.g. Benhabib and Nishimura (1985) and Nishimura and Yano (1994)) frequently postulate $u(y) = y$ implying a very high tolerance for consumption variation over time. Cycles are then produced by construction of an appropriate technology for a given discount factor. Our approach to periodic solutions emphasizes the way in which the elasticity of substitution and the intertemporal tolerance coefficient interact along the lines of the intuition given earlier for the bifurcation condition. Further, the Tolerance Lemma gives an important role to the size of the economy. A large number of heterogeneous wage earners without capital contribute to the potential for swings in the capital holding agent's income with variation in the capital stock.

4.2 Approximation of the Production Function

Approximation of production functions in $\wp_k$ and $\mathcal{Q}_k$ are developed in a series of lemmas. The proofs may be found in the Appendix. Some technical definitions and results needed for the statements of the lemmas are reviewed below.

A $C^{(\infty)}$ map $\Phi : \Re \rightarrow [0,1]$ is said to be a *bump function* provided

(i) $\Phi(x) = 1$ *for* $|x| \leq 1$,
(ii) $\Phi(x) = 0$ *for* $|x| \geq 2$,
(iii) $\Phi(x) = \Phi(-x)$ *for all* $x \in \Re$.

Abraham, Marsden and Ratiu (1988, p. 273) give the details on the construction of the standard example of a bump function. We topologize $C^{(\infty)}$ with the weak or compact-open topology. This topology may be described in terms of a countable family of seminorms and may be made into a Fréchet space.[26] Let $f^{(m)}$ denote the mth*-order derivative* of f.[27] A sequence $\{f_j\}_{j=1}^{\infty}$ of production functions converges to f in the metric for this space if and only if all of the derivatives $f_j^{(m)}$ converge uniformly to the derivatives $f^{(m)}$ on every compact interval, for each m. The $C^{(r)}$*-topology* is defined in an analogous fashion with the mth*-order* derivatives (each $m \leq r$) converging uniformly on each compact interval.

Lemma 4.1: *For any* $f \in \wp_k$ *there exists a sequence* $\{f_j\}_{j=1}^{\infty} \in \wp_k$ *converging to* f *in* $\wp_k$ *such that for all* j, $f_j(k) = f(k)$, $f_j'(k) = f'(k)$, $f_j''(k) = f''(k)$, *and such that the restriction* $f_j|I_j$ *of* f_j *to some adequate open interval* I_j *containing* k *satisfies* $f_j|I_j \in C^{(4)}(I_j)$.

---

[26]See Aubin (1977, p. 52) for details of this construction.

[27]We retain the use of the notation f', f'' etc. for the unperturbed production (and income) functions when the meaning is clear.

Corollary 4.1: *If in Lemma 4.1, $f \in \mathcal{Q}_k$ then the approximating sequence $\{f_j\}_{j=1}^{\infty}$ provided by that lemma is in $\mathcal{Q}_k$.*

This lemma implies that the originally specified production function may be taken to be of class $C^{(4)}$ at least in a small interval about k without changing the steady state capital and consumption or disturbing the bifurcation condition (3.4), yet maintaining the no arbitrage and balance condition (2.8). The perturbation of f given by this lemma tells us that the $C^{(4)}$ property is a generic characteristic for the production function (at least locally). Hence we may assume that in the sequel the original production function is already taken to be $C^{(4)}$ at least in an interval containing k. Formally:

A4: $f \in C^{(4)}(0, \infty)$.

A production function is assumed to satisfy (A3) and (A4) for the remainder of the chapter. Lemma 4.1 also applies to the felicity function. Hence we assume that u is $C^{(4)}$ for the remainder of the chapter.

Lemma 4.2: *Let $\alpha > 0$ and β be given numbers and let $\bar{x}$ satisfy $\delta f'(\bar{x}) = 1$. Then there exists an $\xi > 0$ such that*

$$f_1(x) = f(x) + \frac{1}{24}\Phi\left(\frac{x - \bar{x}}{\xi}\right)(\beta - f^{(4)}(\bar{x}))(x - \bar{x})^4 \qquad (4.1)$$

satisfies

$$f_1^{(1)} > 0,\ f_1^{(1)}(0+) = \infty,\ f_1^{(2)} < 0; \qquad (4.2)$$

$$f_1^{(m)}(\bar{x}) = f^{(m)}(\bar{x})\ (0 \leq m \leq 3)\ \textit{and}\ f_1^{(4)}(\bar{x}) = \beta; \qquad (4.3)$$

$$|f_1^{(m)}(x) - f^{(m)}(x)| \leq \alpha\ \textit{for all}\ x > 0,\ \textit{all}\ 0 \leq m \leq 3. \qquad (4.4)$$

Lemma 4.2 implies that the production function may be perturbed in such a manner to keep the outputs and first three derivatives as close as desired while yielding any arbitrarily chosen value of the fourth derivative of the perturbed function. This implies that generically we can assume $f^{(4)}$ is very much larger than 1. We will use this property of the production function to insure that nondegeneracy condition (3.10) obtains. The lemma says something more important, namely that for $\beta > 0$ the value of $f^{(4)}$ may be taken as a large positive number. We will show a_* is negative in the case $\beta > 0$; this will imply the bifurcation is supercritical (i.e., it is stable). However, we could also study the case $\beta > 0$ and produce a subcritical bifurcation where the two period cycle is unstable. The existence of a cycle does not depend on the sign of β, but the sign does determine the stability of the cycle.

The next lemma introduces the one parameter family of production functions that will be used to defined the one parameter family of maps ψ_μ. The basic idea is to induce a small change in the curvature of the production function at $\overline{x}$ in order to alter the elasticity of substitution. The single parameter μ yields the disturbance of the curvature of f necessary to yield a bifurcation result.

Lemma 4.3: *With the same notation as Lemma 4.2, for* $|\mu| < 1$ *set*

$$f_\mu(x) = f(x) + \frac{\mu}{2}\Phi\left(\frac{x-\overline{x}}{\xi}\right)(x-\overline{x})^2, \tag{4.5}$$

where $\xi = \overline{x}/3$. *Then* f_μ *satisfies*

$$f_\mu^{(1)} > 0,\ f_\mu^{(1)}(0+) = \infty,\ f_\mu^{(2)} < 0; \tag{4.6}$$

$$f_\mu^{(m)}(\overline{x}) = f^{(m)}(\overline{x})(m = 0, 1, 3, 4) \text{ and}$$
$$f_\mu^{(2)}(\overline{x}) = f''(\overline{x}) + \mu; \tag{4.7}$$

$$|f_1^{(m)}(x) - f^{(m)}(x)| \le \alpha \text{ for all } x > 0,\ m = 0, 1 \tag{4.8}$$

provided $|\mu|$ *is small enough.*

Lemma 4.1-4.3 allow us to introduce a family of production functions $\{f_\mu\}$ for μ sufficiently small such that $\lambda_{1\mu} = -1$ and equation (3.4) holds for $\mu = 0$. The dynamical system is defined by the map $x \mapsto \psi(x, \mu)$. Notice that for $\mu < 0$ that $f(\bullet, \mu)$ is "more concave" than $f(\bullet, 0)$ near the point $\bar{x}$ and "less concave" for $\mu > 0$.

5. The Local Bifurcation of Ramsey Equilibrium

5.1 The Main Theorem

The following theorem is the main result of our chapter.

Local bifurcation theorem: *Any* $f \in \mathcal{Q}_k$ *is the limit in* $C^{(2)}(0, \infty)$ *of functions* $f_j \in \wp_k$ *for which there exist nonstationary stable periodic Ramsey equilibria. Precisely*

$$x^1_{2t-1} = k_1(f_j) \neq k_2(f_j) = x^1_{2t},\ x^h_t = 0\ (h = 2, ..., H;\\ t = 1, 2, ...).$$

Moreover, $k_1(f_j), k_2(f_j) \to k$ *as* $f_j \to f$ *in the* $C^{(2)}$*-topology.*

The perturbations of the production function and imbedding of f in a family of production functions such that $\lambda_{1\mu} = -1$ and (3.4) obtain at $\mu = 0$ were the subject of Section 4. The remaining details of the proof of this theorem depend on verifying (CFB). The computations of the derivatives of ψ supporting the verification of (CFB) are tedious. We report the final outcome of those calculations in the following subsection. The computations will show that (3.9) and (3.10) hold. Further, $2\psi_{xxx}(\bar{x}, 0) + 3(\psi_{xx}(\bar{x}, 0))^2$ has the form: $cf^{(4)}(\bar{x})+$ *terms depending on* $f(\bar{x})$, $f^{(m)}(\bar{x})$ where c is a positive constant and $1 \leq m \leq 3$. By taking $f^{(4)}(\bar{x}) >> 1$ we can be sure that $\alpha_* < 0$.

5.2 Computation of the Bifurcation Conditions

The derivatives of order up to three must be computed in order to check the conditions for a stable flip bifurcation hold. These computations are tedious and we only report the final outcome.

We record the one-parameter families of maps that enter into our calculations of the conditions for a stable flip bifurcation at the fixed point $(\bar{x}, \bar{y})$. The parameter μ is used to define the maps:

$$f(x, \mu) = f(x) + \frac{\mu}{2}\Phi\left(\frac{x - \bar{x}}{\xi}\right)(x - \bar{x})^2; \tag{5.1}$$

$$\begin{aligned} g(x, \mu) = g(x) + \left(1 - \frac{1}{H}\right) \\ \left[\mu\Phi\left(\frac{x - \bar{x}}{\xi}\right)(x - \bar{x}) + \frac{\mu}{2\xi}\Phi'\left(\frac{x - \bar{x}}{\xi}\right)(x - \bar{x})^2\right] x \\ + \frac{1}{H}\left[\frac{\mu}{2}\Phi\left(\frac{x - \bar{x}}{\xi}\right)(x - \bar{x})^2\right]. \end{aligned} \tag{5.2}$$

The function $F(x, y, \mu)$ is defined implicitly by the perturbed form of the no arbitrage condition:

$$\begin{aligned} u'(y) = \\ \delta\left[f_x(x, \mu) + \mu\Phi\left(\frac{x - \bar{x}}{\xi}\right)(x - \bar{x}) + \frac{\mu}{2\xi}\Phi'\left(\frac{x - \bar{x}}{\xi}\right)(x - \bar{x})^2\right] \\ u'(F(x, y, \mu)). \end{aligned} \tag{5.3}$$

The center manifold is defined by the map $\varphi(x, \mu)$ and the corresponding map $\psi(x, \mu)$ is defined so that the following equations hold:

$$\psi(x, \mu) = g(x, \mu) - F(x, y, \mu); \tag{5.4}$$

$$\begin{aligned} F(x, y, \mu) &= \varphi(g(x, \mu) - F(x, \varphi(x, \mu)), \mu) \\ &= \varphi(\psi(x, \mu), \mu). \end{aligned} \tag{5.5}$$

We will use the notation $\psi_x, \psi_{xx}, \psi_{xxx}$ to denote the first, second and third order partial derivatives of functions such as ψ. We also reserve the notation $\bar{\psi}_x$ for $\frac{\partial\psi}{\partial x}(\bar{x}, 0)$, etc. However, since we only report here the final results of the calculations we will drop the overbars when the meaning is clear to ease typesetting.

The verification of the bifurcation conditions in equations (3.9) and (3.10) requires computations of the partial derivatives of ψ taken with respect to x of up to order 3, the first order partial of ψ taken with respect to μ, and the mixed second order partial of ψ. We only report the values of derivatives actually needed for the final determination of the coefficients entering equations (3.9) and (3.10).

The derivatives of $f(x,\mu)$ evaluated at $(\bar{x},0)$ are: $f_x = f'$, $f_{xx} = f'' + \mu$, $f_{xxx} = f'''$, $f_{xxxx} = f''''$, $f_\mu = 0$, $f_{\mu x} = 0$, and $f_{xx\mu} = 1$. The derivatives of $g(x,\mu)$ evaluated at $(\overline{x},0)$ are: $g_x = g' \equiv \gamma$, $g_{x\mu} = \left(1 - \frac{1}{H}\right)\bar{x}$, $g_{xx} = g''$, $g_{xxxx} = g''''$, and $g_\mu = 0$.

The first order partial derivatives of F at $(\bar{x},0)$ with $\bar{y} = \varphi(\bar{x},0)$ may be shown to be given by: $F_x = -\eta$, $F_y = 1$, $F_\mu = 0$. We let u' and u'' stand for $u'(\bar{y})$ and $u''(\bar{y})$ respectively in this report on the calculations. The second order partial derivatives of F evaluated at $(\bar{x},0)$ are as follows: $F_{x\mu} = -\delta u'/u''$, $F_{y\mu} = 0$, $F_{yy} = 0$, and

$$F_{xx} = \frac{(f'')^2 \delta u'(2 - \delta(u'u'''/u''^2))}{u''} - f''' \frac{u'}{u''}$$

$$F_{xy} = f''\delta\left(\frac{u'''u'}{u''} - u''\right).$$

We may simplify the representation of these derivatives by introducing constants U_i that depend only on the values of the derivatives of u up to order three evaluated at the fixed point and the constant δ. For example, we will let $U_1 = -u'/u''$; we set the constant $U_0 = \left(1 - \frac{1}{H}\right)\bar{x}$. Thus the previous nonvanishing derivatives have the form:

$$F_x = f''U_1\delta;\; F_{xy} = f''^2 U_2;\; F_{xx} = f''^2 U_3 + f''' U_1.$$

The third order partials of F may also be calculated; we report only the simplified expressions in terms of the U_i as follows:

$$F_{xxx} = -f''''\delta u' + f''' f'' U_4 + f''^3 U_5;$$
$$F_{xxy} = f''' U_6 + f''^2 U_7;\; F_{xyy} = f'' U_8.$$

The derivatives of φ and ψ are expressed in terms of the partials of F, the derivatives of g and the constants η and γ. As before, we introduce constants U_i and only report the final form of the calculations:

$$\varphi_x = \eta/2,\ \varphi_\mu = 0;\ \psi_x = -1,\ \psi_\mu = 0;$$
$$\psi_{xx} = f''U_9 + \frac{f'''}{f''}U_{10};\ \psi_{x\mu} = -U_0/\left(\frac{1}{2}\delta f''U_1 - 1\right).$$

Notice that $U_0 > 0$, $U_1 > 0$ implies $\psi_{x\mu} > 0$ and (3.9) holds.[28] The third order derivative of ψ may be shown to be of the form:

$$\psi_{xxx} = (1 - U_1\delta f'')^{-1}\mathcal{M}$$

$$\text{where } \mathcal{M} \left[\begin{array}{c} f''''U_0 + f'''f''U_{11} + f''^3U_{12} + f''^2U_{13} + f'''^2U_{14} - \\ \left(\frac{f'''U_{15}}{f''^2}\right) - \left(\frac{f'''U_{16}}{f''}\right) \end{array} \right] \tag{5.6}$$

Notice that $U_0 > 0$; by perturbing f so that f'''' is a very large positive number, we can insure that $\psi_{xxx} > 0$. It is now easy to see that the conditions for a stable flip bifurcation hold at the fixed point.

Proof of the local bifurcation theorem: The preceding computations prepare the way for demonstrating the main result.

Proof of the local bifurcation theorem: Suppose $f \in \mathcal{Q}_k$ and define the family of maps f_μ as above. We have shown previously that the conditions for a stable flip bifurcation are met for the corresponding family of maps ψ_μ. Therefore application of the flip bifurcation theorem yields the existence of a 2-cycle equilibrium $x_\pm(\mu)$ defined for μ in some interval $[-\mu_*, 0)$ with $\mu_* > 0$. This 2-cycle is asymptotically stable and $x_\pm(\mu) \to \bar{x}$ as $\mu \to 0$. By choosing $\{\mu_j\}_{j=1}^\infty \subseteq [-\mu_*, 0)$, $\mu_j \to 0$, we see the maps $f_j \equiv f_{\mu j} \to f$ uniformly in $C^{(2)}(0, \infty)$ by Lemma 4.3. Q.E.D.

---

[28] The calculation of the second order derivatives of ψ uses the fact that $\varphi_{xx} = g'' + (F_{xy}F_x - F_{xx})(1 - 2/F_x)$.

A natural question to ask now is the steady state $\bar{x}$ for $\mu = 0$ locally stable? The answer is affirmative; the proof uses the derivatives of ψ calculated in the demonstration of the Local Bifurcation Theorem.

Local stability theorem: *For $\mu = 0$ the fixed point $\bar{x}$ is locally asymptotically stable.*

Proof: First make the standard change of variables by replacing x with $x - \bar{x}$ and $\bar{x}$ by 0. For $\mu = 0$ denote the map $\psi(x, 0)$ by $\psi(x)$. For x near 0 the third order Taylor approximation of ψ may be written as

$$\psi(x) = -x + \frac{\alpha}{2}x^2 + x^3\theta(x)$$

where $\theta(0) > 0$. Then

$$\begin{aligned}
\psi(\psi(x)) &= -\psi(x) + \frac{\alpha}{2}(\psi(x))^2 + (\psi(x))^3\theta(\psi(x)) \\
&= x - \frac{\alpha^2}{2}x^3 - x^3\theta(x) - x^3\theta(\psi(x)) + x^4\theta_1(x) \\
&= x - x^3\left(\frac{\alpha^2}{2} + \theta_2(x)\right) + x^4\theta_1(x)
\end{aligned}$$

where $\theta_2(0) > 0$ and θ_1, θ_2 are continuous at 0. Hence we have $0 \leq \psi(\psi(x)) < x$ for $0 \leq x \leq x_{00}$ with x_{00} small enough and $x < \psi(\psi(x)) \leq 0$ for $-x_{00} \leq x \leq 0$. Thus if $|x_0| \leq x_{00}$ and $x_{n+1} = \psi(\psi(x_n))$ we have that $\{x_n\}_{n=0}^{\infty}$ is monotonic and bounded. Therefore $\lim_n x_n = x_\infty$ exists and $\psi(\psi(x_\infty)) = x_\infty, |x_\infty| < x_{00}$. Obviously x_∞ must be 0. It follows easily that

$$(\psi \circ \psi \circ \ldots \circ \psi)^n(x_0) \to \begin{cases} x_\infty = 0, & \text{if } n \text{ is even} \\ \psi(x_\infty) = 0, & \text{if } n \text{ is odd,} \end{cases}$$

where $(\psi \circ \psi \circ \ldots \circ \psi)^n$ denotes the nth-iterate of ψ. Therefore 0 is asymptotically stable. Q.E.D.

6. Conclusion

One question remaining for future study is a global analysis of the bifurcation problem. This would necessitate treating the production function as the parameter and seem to require new mathematical methods. The exploration of the potential for a chaotic Ramsey equilibrium to emerge from a period doubling scenario or via an alternative route to chaos would be the object of that study. We note the Local Bifurcation Theorem remains valid for f'''' sufficiently negative (i.e. $\beta < 0$) but the emergent cycles are unstable. This would seem to indicate that more complicated dynamics may be exhibited along equilibrium profiles as the elasticity of substitution decreases further.

The Ramsey equilibrium model introduces the elasticity of substitution as a factor in explaining business cycles. The local bifurcation theory also links the magnitude of this elasticity to the capital holding agent's intertemporal tolerance for fluctuations. Our analysis demonstrates the possibility of cycles emerging in equilibrium if the elasticity of substitution at the steady state is already below 1. Moreover, cycles may emerge for all sufficiently small values of the elasticity of substitution at the steady state. It would be interesting to know if more complicated equilibrium trajectories exist. A comparison between business cycle fluctuations as measured with aggregate data and empirical magnitudes of the elasticity of substitution would also be a worthwhile study.[29] Such an investigation would of course be independent of the structure of our model which is based on certain assumptions.

Appendix

Proofs of several technical lemmas presented in Section 4 are collected in this appendix.

[29]We note that Berndt (1976) notes the time series estimates of the elasticity of substitution generally provide lower estimates than 1 for that parameter. Indeed, he notes those estimates tend to be centered in the range 0.1 to 0.2.

Proof of Lemma 4.1:

Let

$$f_j(x) = \Phi\left(\frac{x-k}{\varepsilon_j}\right)\left(f(k) + \frac{x-k}{1!}f'(k) + \frac{(x-k)^2}{2!}f''(k)\right) + \left(1 - \Phi\left(\frac{x-k}{\varepsilon_j}\right)\right)f(x)$$

where Φ is a bump function and the sequence $\{\varepsilon_j\}_{j=1}^{\infty}$ satisfying $k/2 \geq \varepsilon_1 > \varepsilon_2 ... \to 0$ will be further specified later. Then f_j is a polynomial on the interval $(k - \varepsilon_j, k + \varepsilon_j)$ and thus f_j is in $C^{(\infty)}$ in a neighborhood of k and obviously $f_j(k) = f(k)$, $f_j'(k) = f'(k)$, and $f_j''(k) = f''(k)$. Take

$$\infty > M \geq \max|\Phi|, \max|\Phi'|, \max|\Phi''| \text{ and}$$
$$\eta_j = \max\{|f''(x) - f''(k)| : |x-k| \leq 2\varepsilon_j\}.$$

It is clear that

$$\begin{aligned}
|f_j(x) - f(x)| &= \left|\Phi\left(\frac{x-k}{\varepsilon_j}\right)\left[f(k) + \frac{x-k}{1!}f'(k) + \frac{(x-k)^2}{2!}f''(k) - f(x)\right]\right| \\
&\leq M\frac{(2\varepsilon_j)^2}{2}\eta_j = 2\varepsilon_j^2 M\eta_j \to 0; \\
|f_j'(x) - f'(x)| &= \left|\Phi\left(\frac{x-k}{\varepsilon_j}\right)[f'(k) + (x-k)f''(k) - f'(x)] \right. \\
&\quad \left. + \frac{1}{\varepsilon_j}\Phi'\left(\frac{x-k}{\varepsilon_j}\right)\left[f(k) + \frac{x-k}{1!}f'(k) + \frac{(x-k)^2}{2!}f''(k) - f(x)\right]\right| \\
&\leq M2\varepsilon_j\eta_j + \frac{1}{\varepsilon_j}M\frac{(2\varepsilon_j)^2}{2}\eta_j = 4\varepsilon_j M\eta_j \to 0
\end{aligned}$$

and

$$
\begin{aligned}
|f_j''(x) - f''(x)| = &\left|\Phi\left(\frac{x-k}{\varepsilon_j}\right)[f''(k) - f''(x)]\right. \\
&+ 2\varepsilon_j \Phi'\left(\frac{x-k}{\varepsilon_j}\right)[f'(k) + (x-k)f''(k) - f'(x)] \\
&+ \frac{1}{\varepsilon_j^2}\Phi''\left(\frac{x-k}{\varepsilon_j}\right)\left[f(k) + \frac{x-k}{1!}f'(k)\right. \\
&\left.\left.+\frac{(x-k)^2}{2!}f''(k) - f(x)\right]\right| \\
&\leq M\eta_j + 2\frac{1}{\varepsilon_j}M2\varepsilon_j\eta_j + \frac{1}{\varepsilon_j^2}M\frac{(2\varepsilon_j)^2}{2}\varepsilon_j^2\eta_j \\
&= 5M\eta_j \to 0
\end{aligned}
$$

for any $x > 0$, i.e. $f_j(x) \to f(x)$, $f_j'(x) \to f'(x)$, and $f_j''(x) \to f''(x)$ uniformly on $(0, \infty)$. Moreover, if we choose ε_1 to be small enough for

$$(1 + 5M)\eta_1 < |f''(k)|/2, 6\varepsilon_1 M\eta_1 < f'(k)/2$$

to hold, then for all $x > 0$

$$
\begin{aligned}
&f_j''(x) = f''(x) < 0, \text{ for } |x-k| \geq 2\varepsilon_j \text{ and} \\
&f_j''(x) \leq f''(x) + 5M\eta_j \leq f''(k) + (1+5M)\eta_j < 0, \text{ for} \\
&|x-k| \leq 2\varepsilon_j; \\
&f_j'(x) = f'(x) > 0, \text{ for } |x-k| \geq 2\varepsilon_j \text{ and} \\
&f_j'(x) \geq f'(x) - 4\varepsilon_j M\eta_j \geq f'(k) - 6\varepsilon_j M\eta_j > 0, \text{ for} \\
&|x-k| \leq 2\varepsilon_j
\end{aligned}
$$

since $\eta_j \leq \eta_1$, for all $j = 1, 2, \ldots$. Thus $f_j \in \wp_k$ for all $j = 1, 2, \ldots$. Q.E.D.

Proof of Lemma 4.2:

Define f_1 by (4.1) for some $\xi \in (0, 1]$ and observe that

$$f_1(x) = f(x) \text{ for } |x - \bar{x}| \geq 2\xi, \; x > 0. \tag{A-1}$$

Choosing a constant C such that

$$
\begin{aligned}
&|\Phi^{(m)}(x)| \le C \text{ for } 1 \le m \le 3,\ x \in \Re \text{ and} \\
&|\beta - f^{(4)}(\bar{x})| \le C
\end{aligned} \tag{A-2}
$$

we have

$$
|f_1(x) - f(x)| \le \frac{1}{24} C\xi^4 = C_0\xi^4 \text{ for } x > 0; \tag{A-3}
$$

$$
|f_1'(x) - f'(x)| \le \frac{1}{24} C^2\xi^3 + \frac{1}{6} C\xi^3 = C_1\xi^3 \text{ for } x > 0; \tag{A-4}
$$

$$
\begin{aligned}
|f_1''(x) - f''(x)| &\le \frac{1}{24} C^2\xi^2 + \frac{1}{3} C^2\xi^2 + \frac{1}{2} C^2\xi^2 = C_2\xi^2 \\
&\text{for } x > 0;
\end{aligned} \tag{A-5}
$$

$$
\begin{aligned}
|f_1'''(x) - f'''(x)| &\le \frac{1}{24} C^2\xi + \frac{1}{2} C^2\xi + \frac{3}{2} C^2\xi + C\xi \\
&= C_3\xi \text{ for } x > 0
\end{aligned} \tag{A-6}
$$

Now there exists $\alpha_1 > 0$ such that

$$
f'(x) \ge \alpha_1,\ \ f''(x) \le -\alpha_1 \text{ for } |x - \bar{x}| \le 2\xi \tag{A-7}
$$

so that if

$$
C_1\xi^3, C_2\xi^2 < \alpha_1, \text{ and } C_0\xi^4, C_1\xi^3, C_2\xi^2, C_3\xi < \alpha, \tag{A-8}
$$

then we obtain (4.1)-(4.4) by referring to (A-1). Q.E.D.

Proof of Lemma 4.3:

Equation (4.7) is obvious. The analog of (A-3) is

$$
|f_\mu(x) - f(x)| \le 2\mu C\xi^2.
$$

The analog of (A-4) is

$$
|f_\mu'(x) - f'(x)| \le \mu\xi C + 2\mu\xi C = 3\mu\xi C.
$$

The analog of (A-5) is

$$f''_\mu(x) - f''(x)| \leq \mu C + 4\mu C + 2\mu C = 7\mu C.$$

For (4.8) we need

$$2\mu C\xi^2,\ 3\mu\xi C < \alpha.$$

For (4.6) we need

$$3\mu\xi C < \alpha_1,\ 7\mu C < \alpha_1.$$

Therefore if

$$\mu < \min\left\{\frac{\alpha}{2C\xi^2}, \frac{\alpha}{3C\xi}. \frac{\alpha_1}{3\xi C}, \frac{\alpha_1}{7C}\right\},$$

then all the conditions (4.6)-(4.8) are satisfied. Q.E.D.

6

On the Structure of Ramsey Equilibrium: Cycles, Indeterminacy and Sunspots

Gerhard Sorger

1. Introduction

In this chapter we demonstrate the possibility of periodicity and indeterminacy of perfect foresight equilibrium under borrowing constraints in a dynamic one-sector economy with infinitely-lived heterogeneous households. As a consequence of indeterminacy we can also prove the existence of sunspot equilibria in this economy. The Ramsey model we consider is basically the same as the one used by Becker (1980), Becker and Foias (1987, 1994), and Becker et al. (1991). It differs from standard neoclassical growth models in two aspects, namely, that households are heterogeneous and that markets are incomplete which prevents households from using their discounted future income to finance present consumption.

Our model has already been analyzed in the papers mentioned above. In Becker et al. (1991) existence of a perfect foresight equilibrium is demonstrated under quite general assumptions on the intertemporal preferences of households. The other papers focus on utility functionals that are additively separable over time. In Becker (1980) an existence and uniqueness result for stationary perfect foresight equilibria is proved whereas in Becker and Foias (1987) non-stationary perfect foresight equilibria are characterized. In particular, it is shown in Becker and Foias (1987) that all equilibrium paths must converge (eventually monotonically) to the unique stationary perfect foresight equilibrium if the total capital income is an increasing function of capital. If this capital income monotonicity condition is not satisfied then deterministic equilibrium cycles with period 2 can occur as has already been shown in Becker and Foias (1987) by an example. In Becker and Foias (1994), the authors discuss in more detail the issue of equilibrium cycles and demonstrate by a systematic approach that the stationary perfect foresight Ramsey equilibrium may bifurcate into stable period-2-cycles if the elasticity of substitution between capital and labor varies.

In the present chapter we add three new observations to the above catalogue of results. First we show that periodic paths of periods different from 2 can also occur as perfect foresight Ramsey equilibria. More specifically, we construct for every odd number $p \geq 3$ a Ramsey economy which has a period-p-cycle as a perfect foresight equilibrium. This shows that quite complicated dynamics are possible in this model when the capital income monotonicity condition does not hold. The result is also in sharp contrast to one-sector models with homogeneous households for which it has already been known for some time that all equilibria must be monotonic. We restrict ourselves to odd periods because the proof for this case is already rather long and cumbersome. However, there is nothing special about the odd periods and we are pretty much convinced that a similar construction can be used to find economies with deterministic equilibrium cycles of even periods. Our examples provide also another interesting insight, namely, that perfect foresight Ramsey equilibrium is not necessarily unique when the capital income monotonicity condition fails. In fact, the periodic equilibria mentioned above coexist with the unique stationary equilibrium. In other words, even if the initial endowments are chosen in such a way that the stationary Ramsey equilibrium is feasible a non-stationary (periodic) equilibrium cannot be ruled out.

Then we strengthen the non-uniqueness result by constructing an example for which the (periodic) Ramsey equilibrium is not even locally unique. In any sufficiently small neighborhood of the periodic perfect foresight Ramsey equilibrium there are uncountably many other equilibria. We prove this by showing that the forward dynamics describing perfect foresight equilibria under the so-called turnpike property has a locally asymptotically stable cycle of period two. Demonstrating indeterminacy in this example involves already some messy algebra so that we refrain from fully characterizing the conditions under which it occurs in the general model. It is worth mentioning that the existence of periodic equilibria is necessary for our approach since the stationary perfect foresight Ramsey equilibrium can never be a stable fixed point of the forward dynamics.

The indeterminacy result is a first hint to the fact that the borrowing constraints together with the heterogeneity of households make the capital accumulation model with infinitely-lived households exhibit dynamic structures that are similar to those of perfect foresight equilibria in overlapping generations models with finitely-lived households. As a further example of this similarity we establish the existence of non-trivial sunspot equilibria in the Ramsey model. The proof of this result is based on the same idea that has already been used to show that indeterminacy is a sufficient condition for the existence of sunspot equilibria in models with overlapping generations (see, e.g., Peck (1988)), models that are mathematically equivalent to OLG models (see Woodford (1986b)), and optimal growth models with externalities (see Spear (1991)). Again we restrict ourselves to the presentation of an example because a detailed characterization seems to require conditions which can hardly be given any economic interpretation.

In the following section we briefly summarize the dynamic equilibrium model under consideration and define perfect foresight Ramsey equilibria. We also state a few preliminary results which characterize such equilibria. In Section 3 we prove the existence of perfect foresight equilibria of arbitrary long (odd) periods. As a by-product of this proof we obtain an example of non-uniqueness of perfect foresight Ramsey equilibrium. In Section 4 we demonstrate the possibility of indeterminacy and in Section 5 we deal with the existence of sunspot equilibria. Section 6 concludes the chapter and gives some directions for future research.

2. Model Formulation and Preliminary Results

We start by summarizing the capital accumulation model analyzed in this chapter and by defining perfect foresight Ramsey equilibria. Since the model is identical to the one of (Becker and Foias (1987)) and (Becker and Foias (1994)) (see also Becker (1980) and Becker et al (1991)) we shall be brief and refer the reader to those articles for more detailed descriptions. In Section 5 below we shall introduce extrinsic uncertainty (sunspots) and define rational expectations Ramsey equilibria in this more general setting.

The economy consists of $H \geq 1$ heterogeneous households indexed by $h = 1, 2, \ldots, H$ and a production sector producing a single commodity which can be used for consumption and investment at each time $t = 1, 2, \ldots$. Inputs have to precede outputs by one period and capital is the only variable factor of production. The technology is described by a neoclassical production function f. Let us denote by $1+r_t$ the rental factor of capital in period t. In a perfect foresight Ramsey equilibrium firms take the sequence of competitive rental prices $\{1+r_t\}$ as given and solve the following sequence of profit maximization problems:

$$P_t(f) : \sup_{K_{t-1} \geq 0} \left[f(K_{t-1}) - (1+r_t)K_{t-1} \right], \; t = 1, 2, \ldots$$

Household h is described by the triple (u^h, δ^h, k^h) where u^h is its one-period utility function, δ^h is its constant discount factor, and k^h is the initial endowment available to household h at the beginning of period one. We denote by c_t^h, x_t^h, and w_t the consumption of household h in period t, the capital stock of household h at the end of period t, and the wage rate in period t, respectively. In a perfect foresight equilibrium households are assumed to correctly anticipate the sequence of factor returns $\{1+r_t, w_t\}$. Taking these as competitive prices household h solves the following infinite-horizon maximization problem:

$$P(h) : \begin{cases} \sup_{\{c_t^h, x_{t-1}^h\}} & \sum_{t=1}^{\infty} (\delta^h)^{t-1} u^h(c_t^h) \\ \text{s. t.} & c_t^h + x_t^h = w_t + (1+r_t) x_{t-1}^h \; t = 1, 2, \ldots \\ & c_t^h \geq 0, x_{t-1}^h \geq 0 \; t = 1, 2, \ldots \\ & x_0^h = k^h \end{cases}$$

We would like to emphasize the non-negativity constraint for capital holdings, $x_{t-1}^h \geq 0$. This borrowing constraint reflects the incomplete market structure of the Ramsey model. Together with the heterogeneity of households it makes possible the occurrence of complicated non-linear dynamic phenomena in this model.

We postulate the following assumptions:

A1: *The production function f is continuous on $[0, \infty)$, twice continuously differentiable on $(0, \infty)$, and satisfies for all $K > 0$ the conditions $f(0) = 0$, $f'(K) > 0$, $f''(K) < 0$, $\lim_{K \to 0} f'(K) = \infty$, and $\lim_{K \to \infty} f'(K) = 0$.*

A2: *For each h, the utility function u^h is continuous on $[0, \infty)$, twice continuously differentiable on $(0, \infty)$, and satisfies for all $c > 0$ the conditions $(\partial/\partial c) u^h(c) > 0$, $(\partial^2/\partial c^2) u^h(c) < 0$, and $\lim_{c \to 0} (\partial/\partial c) u^h(c) = \infty$.*

A3: *Households are ordered according to increasing impatience, that is, the discount factors satisfy $1 > \delta^1 > \delta^2 > \ldots > \delta^H > 0$.*

A4: *Endowments k^h are non-negative and the total endowment $K = \sum_{h=1}^{H} k^h$ is strictly positive.*

This concludes the description of the economy $\mathcal{E} = (f, \{u^h, \delta^h, k^h\}, h = 1, 2, \ldots, H)$.

A sequence $\{1 + r_t, w_t, K_{t-1}, c_t^h, x_{t-1}^h\}$ constitutes a perfect foresight Ramsey equilibrium for $\mathcal{E}$ if the following conditions hold:

PFE1: *The sequence $\{1 + r_t, w_t\}$ is non-negative and non-zero.*

PFE2: *For each $t \geq 1$, K_{t-1} solves $P_t(f)$ given $1 + r_t$.*

PFE3: *For each $h = 1, 2, \ldots, H$, $\{c_t^h, x_{t-1}^h\}$ solves $P(h)$ given $\{1 + r_t, w_t\}$.*

PFE4: *The capital market is in equilibrium, i.e.,* $\sum_{h=1}^{H} x_{t-1}^h = K_{t-1}$ *for all* $t \geq 1$.

PFE5: *The output market is in equilibrium, i.e.,* $\sum_{h=1}^{H}(c_t^h + x_t^h) = f(K_{t-1})$ *for all* $t \geq 1$.

It is easily seen that the equilibrium conditions PFE1 - PFE5 imply that the labor market clearing condition $Hw_t = f(K_{t-1}) - (1+r_t)K_{t-1}$ is satisfied for all $t \geq 1$. In the following lemma we state a set of sufficient equilibrium conditions for our model.

Lemma 2.1: *Let* $\mathcal{E} = (f, \{u^h, \delta^h, k^h\}, h = 1, 2, \ldots, H)$ *be a given economy. Assume that there exists a sequence* $\{K_{t-1}, c_t^h, x_{t-1}^h\}$ *such that the following conditions are satisfied for all* $h = 1, 2, \ldots, H$ *and all* $t \geq 1$:

a) $c_t^h > 0,\ x_{t-1}^h \geq 0,\ K_{t-1} > 0$

b) $x_0^h = k^h$

c) $(\partial/\partial c)u^h(c_t^h) \geq \delta^h f'(K_t)(\partial/\partial c)u^h(c_{t+1}^h)$ *with equality whenever* $x_t^h > 0$

d) $c_t^h + x_t^h = (1/H)\left[f(K_{t-1}) - f'(K_{t-1})K_{t-1}\right] + f'(K_{t-1})x_{t-1}^h$

e) $\sum_{h=1}^{H} x_{t-1}^h = K_{t-1}$

f) $\lim_{t\to\infty}(\delta^h)^{t-1}[(\partial/\partial c)u^h(c_t^h)]x_t^h = 0$

Then it follows that $\{1+r_t, w_t, K_{t-1}, c_t^h, x_{t-1}^h\}$ *qualifies as a perfect foresight Ramsey equilibrium for* $\mathcal{E}$ *where the sequence of factor prices* $\{1+r_t, w_t\}$ *is given by*

$$1 + r_t = f'(K_{t-1} \quad \textit{and} \quad w_t = \frac{1}{H}\left[f(K_{t-1}) - f'(K_{t-1})K_{t-1}\right]. \qquad (2.1)$$

Proof: Condition PFE1 is satisfied because of (2.1) and assumption A1. For all $K \geq 0$ we have

$$[f(K_{t-1}) - (1+r_t)K_{t-1}] - [f(K) - (1+r_t)K] \geq$$
$$f'(K_{t-1})(K_{t-1} - K) - (1+r_t)(K_{t-1} - K) = 0$$

which shows that condition PFE2 is satisfied. The sequence $\{c_t^h, x_{t-1}^h\}$ is feasible for problem $P(h)$ because of assumptions a), b), and d). Now consider any other sequence $\{c_t, x_{t-1}\}$ satisfying the constraints of problem $P(h)$. We have

$$\begin{aligned}
&\lim_{T\to\infty}\sum_{t=1}^{T}(\delta^h)^{t-1}\left[u^h(c_t^h)-u^h(c_t)\right]\\
&\geq \lim_{T\to\infty}\sum_{t=1}^{T}(\delta^h)^{t-1}\frac{\partial}{\partial c}u^h(c_t^h)(c_t^h-c_t)\\
&= \lim_{T\to\infty}\sum_{t=1}^{T}(\delta^h)^{t-1}\frac{\partial}{\partial c}u^h(c_t^h)\left[f'(K_{t-1})(x_{t-1}^h-x_{t-1})-x_t^h+x_t\right]\\
&= \lim_{T\to\infty}\left\{\begin{array}{c}\sum_{t=1}^{T-1}(\delta^h)^{t-1}(x_t^h-x_t)\left[\begin{array}{c}\delta^h f'(K_t)\frac{\partial}{\partial c}u^h(c_{t+1}^h)-\\ \frac{\partial}{\partial c}u^h(c_t^h)\end{array}\right]\\ -(\delta^h)^{T-1}(x_T^h-x_T)\frac{\partial}{\partial c}u^h(c_T^h)\end{array}\right\}\\
&\geq \lim_{T\to\infty}\left\{\begin{array}{c}\sum_{t=1}^{T-1}(\delta^h)^{t-1}x_t^h\left[\begin{array}{c}\delta^h f'(K_t)\frac{\partial}{\partial c}u^h(c_{t+1}^h)\\ -\frac{\partial}{\partial c}u^h(c_t^h)\end{array}\right]-\\ (\delta^h)^{T-1}x_T^h\frac{\partial}{\partial c}u^h(c_T^h)\end{array}\right\}=0
\end{aligned}$$

This shows that PFE3 is satisfied. Condition PFE4 is identical to assumption e) and condition PFE5 follows immediately from assumptions d) and e). Q.E.D.

We say that a perfect foresight Ramsey equilibrium $\{1+r_t, w_t, K_{t-1}, c_t^h, x_{t-1}^h\}$ is p-periodic if p is the smallest positive integer such that for all $h = 1, 2, \ldots, H$ and all $t \geq 1$ it holds that $r_t = r_{t+p}$, $w_t = w_{t+p}$, $K_{t-1} = K_{t-1+p}$, $c_t^h = c_{t+p}^h$, and $x_{t-1}^h = x_{t-1+p}^h$. An equilibrium which is 1-periodic is called a stationary perfect foresight Ramsey equilibrium for $\mathcal{E}$. The main result of Becker (1980) is the following proposition (the "if" part of which is a simple corollary of Lemma 2.1).

Proposition 2.1: *Let $\mathcal{E}$ be an economy satisfying* A1 - A4. *Then $\mathcal{E}$ admits a stationary Ramsey equilibrium if and only if $k^1 = \bar{K}$ and $k^h = 0$ for $h = 2, \ldots, H$, where $\bar{K}$ is the unique* *capital stock satisfying* $\delta^1 f'(\bar{K}) = 1$. *This equilibrium is given by*

$$1 + r_t = \frac{1}{\delta^1}, \; K_{t-1} = x^1_{t-1} = \bar{K}, \; c^1_t = \frac{1}{H}\left[f(\bar{K}) + (H-1)f'(\bar{K})\bar{K}\right] - \bar{K}$$

$$c^h_t = w_t = \frac{1}{H}\left[f(\bar{K}) - f'(\bar{K})\bar{K}\right], \; x^h_{t-1} = 0 \; h = 2, 3, \ldots, H.$$

Probably the most interesting consequence of this proposition is that in a stationary Ramsey equilibrium only the most patient household holds capital whereas the other ones just consume their wage income without saving anything. Becker and Foias (1987) call a situation in which the entire capital stock of the economy is held by household one the zero capital state and they show that this state is reached infinitely often along every perfect foresight Ramsey equilibrium (recurrence property). On the other hand, it is not difficult to construct examples in which the impatient households $h = 2, 3, \ldots, H$ hold positive capital stocks in infinitely many periods and these positive capital stocks are uniformly bounded away from zero. This shows that without additional assumptions on technology and/or preferences the so-called turnpike property does not obtain. Here, the turnpike property is defined as in Becker and Foias (1987) by requiring that the zero capital state is reached in finite time and maintained thereafter. A sufficient condition for the turnpike property to hold along every perfect foresight Ramsey equilibrium is that

$$\frac{d}{dK}\left[f'(K)K\right] > 0 \tag{2.2}$$

for all $K > 0$. Since the term in brackets on the left hand side is the total capital income earned by all households we call (2.2) the capital income monotonicity condition. Becker and Foias (1987) prove that under this condition every perfect foresight Ramsey equilibrium converges eventually monotonically to the unique stationary Ramsey equilibrium and the turnpike property obtains. If

(2.2) does not hold for all $K > 0$ then equilibrium cycles of period two can emerge as has been shown in Becker and Foias (1987) and Becker and Foias (1994).

The purpose of the present chapter is to show that even more complicated dynamical patterns can occur and that Ramsey equilibrium is, in general, not uniquely determined. In order to do this we shall construct various examples of Ramsey economies. To keep the analytical burden as small as possible we restrict ourselves to economies with $H = 2$ households. The following result presents sufficient equilibrium conditions for this special case.

Lemma 2.2: *Let f and u be functions satisfying* A1 *and* A2, *respectively, and let δ, ϵ_1, and ϵ_2 be positive real numbers with $\delta \in (0, 1)$. Finally, assume that there exists a sequence $\{K_{t-1}, c_t\}$ such that the following conditions hold true for all $t \geq 1$:*

a) $\epsilon_1 < c_t$, $\epsilon_1 < K_t < \epsilon_2$

b) $u'(c_t) = \delta f'(K_t) u'(c_{t+1})$

c) $c_t = (1/2)\,[f(K_{t-1}) + f'(K_{t-1})K_{t-1}] - K_t$

For any utility function u^2 satisfying A2 *and for any sufficiently small discount factor $\delta^2 > 0$ it holds that the economy $\mathcal{E} = (f, \{u^h, \delta^h, k^h\}, h = 1, 2)$ defined by $u^1 = u$, $\delta^1 = \delta$, $k^1 = K_0$, and $k^2 = 0$ has a perfect foresight Ramsey equilibrium $\{1 + r_t, w_t, K_{t-1}, c_t^h, x_{t-1}^h\}$ given by*

$$\begin{aligned} 1 + r_t = f'(K_{t-1})\,,\ c_t^1 = c_t\,,\ c_t^2 = w_t = \\ \frac{1}{2}\,[f(K_{t-1}) - f'(K_{t-1})K_{t-1}]\,,\ x_{t-1}^1 \\ = K_{t-1}\,,\ x_{t-1}^2 = 0 \end{aligned} \tag{2.3}$$

Proof: We choose f, u^1, δ^1, k^1, and k^2 as specified in the lemma and show that for any function u^2 satisfying assumption A2 the conditions of Lemma 2.1 hold for the economy $\mathcal{E} = (f, \{u^h, \delta^h, k^h\}, h = 1, 2)$ provided the discount factor δ^2 of household $h = 2$ is sufficiently small. Conditions b), d), and e) of Lemma 2.1 follow immediately from the assumptions of the present lemma and from the definitions of x_{t-1}^h and c_t^h. From assumption a) of the present lemma, from assumption A1, and from (2.3) it follows that condition a) of Lemma 2.1 is satisfied. Condition c) of Lemma 2.1 for $h = 1$ follows from assumption b) in the present lemma. To verify condition c) of Lemma 2.1 for $h = 2$ we first observe that the definition of c_t^2 in (2.3) implies that there are positive constants ϵ_3 and ϵ_4 such that $\epsilon_3 \leq c_t^2 \leq \epsilon_4$ for all $t \geq 1$. From this observation it follows that we can satisfy condition b) of Lemma 2.1 for $h = 2$ for any function u^2 satisfying A2 by choosing δ^2 smaller than the supremum of the function

$$g(K, c, \tilde{c}) = \frac{(\partial/\partial c)u^2(c)}{f'(K)(\partial/\partial c)u^2(\tilde{c})}$$

over the set $(K, c, \tilde{c}) \in [\epsilon_1, \epsilon_2] \times [\epsilon_3, \epsilon_4]^2$. Since this set is compact and g is continuous and positive, δ^2 can be chosen positive. Finally, condition f) of Lemma 2.1 follows from A2 and A3 and from assumption a) in the present lemma. Q.E.D.

3. Non-uniqueness and Cycles

From Becker and Foias (1987) and Becker and Foias (1994) we know that a perfect foresight Ramsey equilibrium may exhibit cycles of period two. The main result of the present section is that other periods are also possible and that the equilibrium is not necessarily unique. We summarize these results in the following theorem.

Theorem 3.1: *Let $p \geq 3$ be any odd integer. There exists an economy $\mathcal{E}$ satisfying* A1 - A4 *which admits (at least) two different perfect foresight Ramsey equilibria one of which is p-periodic whereas the other one is stationary.*

The proof of this result is rather long and technical and will be split up in a series of lemmas. Before we do this, however, let us add a few remarks on the above theorem.

First of all, as in every economic model, it is a slightly disturbing result that equilibrium may not be unique because this implies that preferences, endowments, and technology alone may not suffice to determine the allocation of resources in the economy. Moreover, non-uniqueness renders sensitivity analysis and characterizations of equilibria less attractive unless they are valid for all equilibria like the turnpike theorem in Becker and Foias (1987, p. 183). The situation becomes even worse if equilibria turn out to be indeterminate, that is, not even locally unique. We shall give an example of indeterminacy in Section 4 below.

As a second remark let us mention that we have not dealt with the stability issue and we do not know whether the periodic equilibria constructed in Theorem 3.1 are stable or not. However, since stable 2-periodic Ramsey equilibria have been found by Becker and Foias (1994) by means of the flip bifurcation theorem it does not sound unlikely that periodic equilibria of other periods might also be stable.

Finally, the reader might be wondering why we state the theorem only for odd periods p. The reason for this restriction was that the proof for this case is already quite long and that we did not want to overload the chapter with too many technicalities. We do believe, however, that a similar construction as the one used in the case of odd periods would also work for even periods.

The remainder of this section is devoted to the proof of Theorem 3.1. Since it is quite long we would like to give a brief outline of the main strategy behind our construction. The general idea is to apply Lemma 2.2 so that we have to specify p-periodic sequences of capital stocks K_{t-1} and consumption rates c_t. This is done in equations (3.17)-(3.18) below by using the auxiliary variables L_i and d_j from (3.1)-(3.3) and (3.12)-(3.13), respectively. The reason why we specify the sequences in this particular way is that it allows to construct functions f and $u = u^1$ satisfying Assumptions A1 and A2 as well as the two conditions b) and c) from Lemma 2.2. The latter two conditions involve the values of f and its first derivative f' at the capital stocks $K_0, K_1, \dots, K_{p-1}$ and the values of u' at the consumption rates $c_1, c_2, \dots, c_p$. We specify these values in Lemma 3.2 and Lemma 3.4, respectively. The auxiliary results from Lemma 3.1 and Lemma 3.3 as well as Lemma 7.1 from the Appendix are used to show that one can "fill up" the graphs of f and u between the points $K_0, K_1, \dots, K_{p-1}$ and $c_1, c_2, \dots, c_p$ in such a way that Assumptions A1 and A2 are satisfied. Finally, Lemma 3.5 verifies that conditions b) and c) from Lemma 2.2 hold.

Let us recall that $p \geq 3$ is an odd integer and define $q = (p+1)/2$. We choose any real number $\gamma \in (0, 1/3)$ and set $g = \gamma^{-q}$. Finally, we define

$$L_0 = 0\,, \; L_1 = 1 \tag{3.1}$$

$$L_i = 2\left(\frac{2}{1+\gamma}\right)^{i-2} \quad i = 2, 3, \dots, q \tag{3.2}$$

$$L_i = 2\left(\frac{2}{1+\gamma}\right)^{q-2}\left(\frac{2+\gamma}{1+3\gamma}\right)^{i-q} \quad i = q+1, q+2, \dots, p \tag{3.3}$$

Note that $\gamma \in (0, 1/3)$ implies that $0 < L_{i-1} < L_i$ for all $i = 1, 2, \dots, p$. The numbers L_i will turn out to be the capital stocks along the equilibrium path. We shall need the following technical result.

Lemma 3.1: *There exists $\bar{\delta} \in (0,1)$ such that for all $\delta \in (0, \bar{\delta})$ the following conditions are satisfied:*

$$\begin{aligned}(L_2 - L_1)\frac{g\gamma^2}{\delta} &< L_1\frac{g\gamma}{\delta} - 2L_q + 2L_p - L_2\frac{g\gamma^2}{\delta} \\ &< (L_2 - L_1)\frac{g\gamma}{\delta}\end{aligned} \tag{3.4}$$

$$2\delta(L_{q+1} - L_{q-1}) < \frac{1}{2}\left[L_q(3+\gamma) - L_{q+1}(1+3\gamma)\right] \tag{3.5}$$

$$\begin{aligned}2\delta(L_{p+3-i} - L_{p+2-i}) &< g\gamma^{i-1}\left[L_{i-1}(1+\gamma) - 2L_i\gamma\right] \\ i &= 3, 4, \dots, q\end{aligned} \tag{3.6}$$

$$\begin{aligned}2\delta(L_{p+2-i} - L_{p+1-i}) &< \frac{g\gamma^{i-1}}{2} \\ \left[L_{i-1}(3+\gamma) - L_i(1+3\gamma)\right] \; i &= q+1, q+2, \dots, p\end{aligned} \tag{3.7}$$

Proof: Using the definition of L_i from (3.1)-(3.3) and simplifying we see that (3.4) is equivalent to

$$\frac{g\gamma}{\delta}(3\gamma-1) < 4\left(\frac{2}{1+\gamma}\right)^{q-2}\left[\left(\frac{2+\gamma}{1+3\gamma}\right)^{p-q}-1\right] < 2\frac{g\gamma^2}{\delta}.$$

Since the middle term of this inequality is a constant independent of δ and since $g > 0$ and $\gamma \in (0, 1/3)$ it is clear that the inequality holds for δ sufficiently small. This proves (3.4). From (3.3) we see that $L_{i-1}(3+\gamma) - L_i(1+3\gamma) = L_{i-1} > 0$ holds for all $i = q+1, q+2, \dots, p$ so that the right hand sides of (3.5) and (3.7) are positive and independent of δ. Therefore, (3.5) and (3.7) must hold provided that δ is sufficiently small. Finally, by using (3.2) we obtain $L_{i-1}(1+\gamma) - 2L_i\gamma = L_{i-1}(1-\gamma)^2/(1+\gamma) > 0$ for all $i = 3, 4, \dots, q$ which shows that the right hand side of (3.6) is also independent of δ and positive. Therefore, (3.6) is satisfied for all sufficiently small δ. Q.E.D.

From now on we shall assume that $\delta \in (0, \bar{\delta})$ is fixed. In the following lemma we specify the production function.

Lemma 3.2: *There exists a function f satisfying* A1 *such that the following conditions hold:*

$$f(L_i) > 2, \; f'(L_i) = \frac{g\gamma^i}{\delta} \quad i = 1, 2, \dots, p \tag{3.8}$$

$$f(L_i) = f(L_1) + L_1\frac{g\gamma}{\delta} - 2L_q + 2L_{p+2-i} - L_i\frac{g\gamma^i}{\delta} \quad i = 2, 3, \dots, q \tag{3.9}$$

$$f(L_i) = f(L_{i-1}) + (L_i - L_{i-1})\frac{g\gamma^{i-1}(1+\gamma)}{2\delta} \quad i = q+1, q+2, \dots, p \tag{3.10}$$

Proof: Choose $f(L_1) > \max\{2, L_1 g\gamma/\delta\}$ arbitrary. Together with (3.9) and (3.10) this determines the values $f(L_i)$ for all $i =$

$1, 2, \ldots, p$. According to Lemma 7.1 in the Appendix it suffices to verify the inequality

$$f(L_{i-1}) + f'(L_i)(L_i - L_{i-1}) < f(L_i) < f(L_{i-1}) + f'(L_{i-1})(L_i - L_{i-1}) \qquad (3.11)$$

for all $i = 1, 2, \ldots, p$ where $f(L_0) = 0$ and $f'(L_0) = \infty$. We have to consider four cases:

Case 1: $i = 1$. By substituting from (3.1) and (3.8) we see that, for $i = 1$, (3.11) is given by $L_1 g\gamma/\delta < f(L_1) < \infty$ which holds trivially by our choice of $f(L_1)$.

Case 2: $i = 2$. For $i = 2$, (3.11) is given by

$$f(L_1) + (L_2 - L_1)\frac{g\gamma^2}{\delta} < f(L_2) < f(L_1) + (L_2 - L_1)\frac{g\gamma}{\delta}.$$

Substituting for $f(L_2)$ from (3.9) we see that this inequality is equivalent to (3.4) so that it holds according to Lemma 3.1.

Case 3: $i = 3, 4, \ldots, q$. Using the definitions of $f'(L_i)$ and $f(L_i)$ from (3.8) and (3.9), respectively, in (3.11) we obtain after cancelling identical terms

$$2L_{p+3-i} - L_{i-1}\frac{g\gamma^{i-1}}{\delta} + (L_i - L_{i-1})\frac{g\gamma^i}{\delta} < 2L_{p+2-i} - L_i\frac{g\gamma^i}{\delta} < 2L_{p+3-i} - L_{i-1}\frac{g\gamma^{i-1}}{\delta} + (L_i - L_{i-1})\frac{g\gamma^{i-1}}{\delta}.$$

This is easily seen to be equivalent to

$$g\gamma^{i-1}\left[2L_{i-1} - L_i(1+\gamma)\right] < 2\delta(L_{p+3-i} - L_{p+2-i}) < g\gamma^{i-1}\left[L_{i-1}(1+\gamma) - 2L_i\gamma\right].$$

The leftmost term is identically equal to zero because of the definition of L_i in (3.2). The middle term is positive because the numbers L_i form a strictly increasing sequence. Consequently, the left inequality is satisfied. The right inequality follows from (3.6) in Lemma 3.1.

Case 4: $i = q+1, q+2, \ldots, p$. In this case (3.8) and (3.10) imply that (3.11) is given by

$$f(L_{i-1}) + (L_i - L_{i-1})\frac{g\gamma^i}{\delta} < f(L_{i-1}) + (L_i - L_{i-1})\frac{g\gamma^{i-1}(1+\gamma)}{2\delta}$$
$$< f(L_{i-1}) + (L_i - L_{i-1})\frac{g\gamma^{i-1}}{\delta}.$$

Since this is equivalent to $\gamma < (1+\gamma)/2 < 1$ it holds true.

The four cases discussed above are exhaustive so that (3.11) is proven. The result follows now from Lemma 7.1 in the Appendix. Q.E.D.

Having specified the production function we now turn to the utility function of the first household. To this end let us define

$$d_i = \frac{1}{2}\left[f(L_{p+1-i}) + L_{p+1-i}\frac{\gamma^{q-i}}{\delta} - 2L_i\right] \quad i = 1, 2, \ldots, q-1 \tag{3.12}$$

$$d_q = \frac{1}{2}\left[f(L_1) + L_1\frac{g\gamma}{\delta} - 2L_q\right] \tag{3.13}$$

The numbers d_i will be used later on as the consumption levels of the first household along the equilibrium path. We need the following technical result.

Lemma 3.3: *The numbers d_i defined above satisfy the following conditions:*

$$0 < d_1 < d_2 < \ldots < d_q \tag{3.14}$$

$$d_q = \frac{1}{2}\left[f(L_{q+1-j}) + L_{q+1-j}\frac{\gamma^{-j+1}}{\delta} - 2L_{q+j}\right] \quad j = 1, 2, \ldots, q-1 \tag{3.15}$$

Proof: To prove (3.14) we proceed in three steps:

Step 1: $0 < d_1$. We can use (3.12) to rewrite this inequality as $0 < f(L_p) + L_p\gamma^{q-1}/\delta - 2L_1$ which holds because of $f(L_p) > 2 = 2L_1$; see (3.1) and (3.8).

Step 2: $d_{j-1} < d_j$ for $j = 2, 3, \dots, q-1$. Using (3.12) we see that this inequality is equivalent to

$$f(L_{p+2-j}) + L_{p+2-j}\frac{\gamma^{q-j+1}}{\delta} - 2L_{j-1} < f(L_{p+1-j}) + L_{p+1-j}\frac{\gamma^{q-j}}{\delta} - 2L_j.$$

We can use the transformation $i = p + 2 - j$ and the identity $g = \gamma^{-q}$ to rewrite this inequality as

$$f(L_i) + L_i\frac{g\gamma^i}{\delta} - 2L_{p+1-i} < f(L_{i-1}) + L_{i-1}\frac{g\gamma^{i-1}}{\delta} - 2L_{p+2-i}$$
$$i = q+2, q+3, \dots, p.$$

Substituting for $f(L_i)$ from (3.10) and collecting terms we obtain (3.7) which holds on account of Lemma 3.1.

Step 3: $d_{q-1} < d_q$. Because of (3.12) and (3.13) this inequality can be written as

$$f(L_{q+1}) + L_{q+1}\frac{\gamma}{\delta} - 2L_{q-1} < f(L_1) + L_1\frac{g\gamma}{\delta} - 2L_q.$$

Substituting for $f(L_{q+1})$ from (3.10) and using $g\gamma^q = 1$ one obtains

$$f(L_q) + (L_{q+1} - L_q)\frac{1+\gamma}{2\delta} + L_{q+1}\frac{\gamma}{\delta} - 2L_{q-1} < f(L_1) + L_1\frac{g\gamma}{\delta} - 2L_q.$$

Substituting again for $f(L_q)$ from (3.9) and rearranging terms one can see that this inequality is equivalent to (3.5) so that it holds according to Lemma 3.1. This completes the proof of (3.14).

Next we use (3.13) and the transformation $i = q + 1 - j$ to rewrite equation (3.15) as follows:

$$f(L_1) + L_1\frac{g\gamma}{\delta} - 2L_q = f(L_i) + L_i\frac{g\gamma^i}{\delta} - 2L_{p+2-i} i = 2, 3, \dots, q$$

Since this is identical to (3.9) it holds according to Lemma 3.2. Q.E.D.

We are now in a position to specify the utility function u of the first household.

Lemma 3.4: *There exists a function u satisfying* A2 *such that*

$$u'(d_i) = g\gamma^i \; i = 1, 2, \ldots, q. \tag{3.16}$$

Proof: Since $\gamma \in (0,1)$ we have $u'(d_{i-1}) > u'(d_i)$ for all $i = 2, 3, \ldots, q$. Let us specify the function u at the points d_i recursively by $u(d_1) > d_1 g\gamma$ and

$$u(d_i) = u(d_{i-1}) + (d_i - d_{i-1})\frac{u'(d_i) + u'(d_{i-1})}{2} =$$

$$u(d_{i-1}) + \frac{g\gamma^{i-1}(1+\gamma)}{2}(d_i - d_{i-1})$$

$$i = 2, 3, \ldots, q.$$

The result follows now from Lemma 7.1 in the Appendix, from (3.14), and from the observation that the conditions in A2 are less restrictive than those in A1. Q.E.D.

We are now ready to specify the sequence $\{K_{t-1}, c_t\}$ which is required in Lemma 2.2. For $t = 1, 2, \ldots, p$ we define it by

$$K_{2i} = L_{q-i}\,, \; c_{2i} = d_{q-i} \; i = 0, 1, \ldots, q-1 \tag{3.17}$$

$$K_{2i-1} = L_{q+i}\,, \; c_{2i-1} = d_q \; i = 1, 2, \ldots, q-1 \tag{3.18}$$

and for $t > p$ we extend the definition by the periodicity conditions $K_{t-1} = K_{t-1-p}$ and $c_t = c_{t-p}$. We have the following result.

Lemma 3.5: *With δ, f, u, and $\{K_{t-1}, c_t\}$ as defined above there exist positive number ϵ_1 and ϵ_2 such that the conditions of Lemma* 2.2 *are satisfied.*

Proof: Because $\{K_{t-1}, c_t\}$ is a strictly positive and periodic sequence one can find positive numbers ϵ_1 and ϵ_2 such that condition

a) of Lemma 2.2 is satisfied. Furthermore, because of periodicity it is sufficient to verify conditions b) and c) of Lemma 2.2 for $t = 1, 2, \ldots, p$. We have to consider three cases.

Case 1: $t = 2i$ for $i = 1, 2, \ldots, q-1$. In this case we obtain from (3.8) and (3.16)-(3.18) that $u'(c_{2i}) = u'(d_{q-i}) = \gamma^{-i}$ and $\delta f'(K_{2i})u'(c_{2i+1}) = \delta f'(L_{q-i})u'(d_q) = \gamma^{-i}$. This shows that condition b) of Lemma 2.2 is satisfied. Moreover, we have $c_{2i} = d_{q-i}$ and

$$\begin{aligned}
&\frac{1}{2}\left[f(K_{2i-1}) + f'(K_{2i-1})K_{2i-1}\right] - K_{2i} \\
&= \frac{1}{2}\left[f(L_{q+i}) + L_{q+i}\frac{\gamma^i}{\delta} - 2L_{q-i}\right].
\end{aligned}$$

From these identities one can easily see by applying the transformation $j = q - i$ that condition c) of Lemma 2.2 is equivalent to (3.12).

Case 2: $t = 2i - 1$ for $i = 1, 2, \ldots, q-1$. In this case we obtain from (3.8) and (3.16)-(3.18) that $u'(c_{2i-1}) = u'(d_q) = 1$ and $\delta f'(K_{2i-1})u'(c_{2i}) = \delta f'(L_{q+i})u'(d_{q-i}) = 1$. This shows that condition b) of Lemma 2.2 is satisfied in this case, too. Furthermore, we have $c_{2i-1} = d_q$ and

$$\begin{aligned}
&\frac{1}{2}\left[f(K_{2i-2}) + f'(K_{2i-2})K_{2i-2}\right] - K_{2i-1} \\
&= \frac{1}{2}\left[f(L_{q+1-i}) + L_{q+1-i}\frac{\gamma^{-i+1}}{\delta} - 2L_{q+i}\right]
\end{aligned}$$

so that condition c) of Lemma 2.2 is equivalent to (3.15) which holds according to Lemma 3.3.

Case 3: $t = p$. Here we have $u'(c_p) = u'(d_q) = 1$ and $\delta f'(K_p)u'(c_{p+1}) = \delta f'(L_q)u'(d_q) = 1$ which proves that condition b) of Lemma 2.2 holds. Finally, we have $c_p = d_q$ and

$$\frac{1}{2}\left[f(K_{p-1}) + f'(K_{p-1})K_{p-1}\right] - K_p = \frac{1}{2}\left[f(L_1) + L_1\frac{g\gamma}{\delta} - 2L_q\right]$$

which shows that condition c) of Lemma 2.2 is identical to (3.13). Q.E.D.

Because of the above result it follows from Lemma 2.2 that there exists an economy $\mathcal{E}$ satisfying $k^1 = K_0$ and $k^2 = 0$ which admits the p-periodic Ramsey equilibrium defined by (2.3), (3.17), and (3.18). Moreover, since $\delta^1 f'(K_0) = \delta f'(L_q) = 1$ we see that the necessary and sufficient conditions for the existence of a stationary Ramsey equilibrium stated in Proposition 2.1 are satisfied with $\bar{K} = K_0$. This completes the proof of Theorem 3.1.

4. Indeterminacy

Theorem 3.1 shows that a perfect foresight Ramsey equilibrium may not be uniquely determined. However, it could still be the case that the equilibria are locally unique, that is, that there are no other equilibria in any sufficiently close neighborhood of any given equilibrium. Local uniqueness is also called determinacy whereas a situation in which equilibria are not even locally unique is referred to as indeterminacy. Of course, indeterminacy is a much stronger property than non-uniqueness and it makes sensitivity analysis impossible even if one restricts oneself to local perturbations of the model. Indeterminacy occurs under quite general assumptions in models with overlapping generations of finitely-lived agents (see, e.g., Woodford (1984)). In the present section we will prove that without additional assumptions indeterminacy can neither be ruled out in the Ramsey model with infinitely-lived households. This result is stated in the following theorem.

Theorem 4.1: *There exists an economy $\mathcal{E}$ satisfying* A1-A4 *which has an indeterminate perfect foresight Ramsey equilibrium.*

Proof: The idea of the proof is rather simple. As in the previous section we restrict ourselves to economies with $H = 2$ households in order to keep the notation simple. Let us start with the conditions stated in Lemma 2.2. Substituting for consumption from condition c) of that lemma into the Euler equation stated as condition b) we obtain the following implicit second order difference equation:

$$\begin{aligned} &u'([f(K_{t-1}) + f'(K_{t-1})K_{t-1}]/2 - K_t) \\ &= \delta f'(K_t)u'([f(K_t) + f'(K_t)K_t]/2 - K_{t+1}) \end{aligned} \tag{4.1}$$

Since $\delta f'(K_t) > 0$ by A1 and since u' is a strictly monotonic function by A2 we can solve the above equation uniquely for K_{t+1}. Let us write the solution as

$$K_{t+1} = g(K_t, K_{t-1}). \tag{4.2}$$

Now assume that there exists a trajectory $\{K_{t-1}\}$ of this difference equation such that for all $t \geq 1$ the capital stocks K_{t-1} are positive, uniformly bounded, and uniformly bounded away from zero. From this trajectory of capital stocks we can compute a corresponding trajectory of consumption rates $\{c_t\}$ by condition c) of Lemma 2.2. If c_t is uniformly bounded away from zero then it follows that these two trajectories satisfy the assumptions of Lemma 2.2 and, therefore, give rise to a perfect foresight Ramsey equilibrium E for an economy $\mathcal{E} = (f, (u, \delta, K_0), (u^2, \delta^2, 0))$ where u^2 is any function satisfying A2 and δ^2 is a sufficiently small positive number. Now assume in addition that the trajectory $\{K_{t-1}\}$ is locally asymptotically stable in the sense that all trajectories of (4.2) which emanate from a nearby initial state $(\tilde{K}_0, \tilde{K}_1)$ converge towards the given trajectory as t approaches infinity. In particular, this has to be true for trajectories starting at points $(\tilde{K}_0, \tilde{K}_1)$ with $\tilde{K}_0 = K_0$ and $\tilde{K}_1 \approx K_1$. By the same argument as above this new trajectory $\{\tilde{K}_{t-1}\}$ together with the corresponding consumption trajectory $\{\tilde{c}_t\}$ will satisfy the conditions of Lemma 2.2 if $\tilde{K}_1$ has been chosen sufficiently close to K_1. Therefore, for every $\tilde{K}_1$ in a small neighborhood of K_1 it follows from Lemma 2.2 that there exists a perfect foresight Ramsey equilibrium $E(\tilde{K}_1)$ for $\mathcal{E}$ which is different from (because $\tilde{K}_1 \neq K_1$) but close to the given equilibrium E. The equilibrium E would therefore be indeterminate.

In view of the above discussion all we have to do in order to prove the existence of an indeterminate perfect foresight Ramsey equilibrium is to find a uniformly bounded and positive solution to (4.2) which is locally asymptotically stable. The first candidate for such a solution that springs to ones mind is the capital stock of the stationary Ramsey equilibrium $\bar{K}$. It can be shown, however, that under assumptions A1 - A4 $\bar{K}$ cannot be a stable fixed point of the difference equation (4.2); at least one of the eigenvalues of the Jacobian matrix

$$J(\bar{K},\bar{K}) = \begin{pmatrix} g_1(\bar{K},\bar{K}) & g_2(\bar{K},\bar{K}) \\ 1 & 0 \end{pmatrix}$$

lies outside the unit circle (see Becker and Foias (1994)). Therefore, we will prove local non-uniqueness of periodic Ramsey equilibria. More specifically, we shall construct models with indeterminate 2-periodic equilibria by choosing utility and production functions u and f in such a way that the difference equation (4.2) has a locally asymptotically stable cycle of period 2. The conditions for the existence of a 2-periodic solution $K_1, K_2, K_1, K_2, \dots$ are

$$g(K_1,K_2) = K_2 \text{ and } g(K_2,K_1) = K_1 \tag{4.3}$$

whereas the stability condition is that the two eigenvalues of the following product of Jacobian matrices

$$J(K_1,K_2)J(K_2,K_1) = \begin{pmatrix} g_1(K_1,K_2) & g_2(K_1,K_2) \\ 1 & 0 \end{pmatrix} \begin{pmatrix} g_1(K_2,K_1) & g_2(K_2,K_1) \\ 1 & 0 \end{pmatrix} \tag{4.4}$$

have absolute values smaller than one. We treat the two conditions of existence and stability of a 2-periodic solution separately in the following two lemmas thereby completing the proof of Theorem 4.1. Q.E.D.

Lemma 4.1: *Let K_1, K_2, δ, and γ be real numbers with*

$$0 < K_1 < K_2 < \frac{1+\gamma+2\delta\gamma}{\gamma(1+\gamma+2\delta)} K_1\,,\ 0 < \gamma < 1\,,$$
$$0 < \delta < 1. \tag{4.5}$$

Then it follows that there exist functions f and u satisfying assumptions A1 *and* A2, *respectively, as well as the following conditions:*

$$f(K_1) > \max\left\{\frac{K_1}{\gamma\delta}, \frac{1}{\delta}[(1+2\delta)K_1 - (1+\gamma)K_2]\right\},$$
$$f(K_2) = f(K_1) + \frac{1}{\delta}(K_2 - K_1) \tag{4.6}$$

$$f'(K_1) = \frac{1}{\gamma\delta}, \; f'(K_2) = \frac{\gamma}{\delta} \tag{4.7}$$

$$u'(c_1) = 1, \; u'(c_2) = \gamma \tag{4.8}$$

where $c_1 = [f(K_2) + f'(K_2)K_2]/2 - K_1$ *and* $c_2 = [f(K_1) + f'(K_1)K_1]/2 - K_2$. *Furthermore, condition* (4.3) *is satisfied so that* $K_1, K_2, K_1, K_2, \ldots$ *is a 2-periodic solution of equation* (4.2).

Proof: We use Lemma 7.1 in the Appendix to show that there exists a function f satisfying A1, (4.6), and (4.7). According to that lemma it is sufficient to verify that $f(K_1) < f(K_2)$, $f'(K_1) > f'(K_2)$, $f(K_1) > f'(K_1)K_1$, and

$$f(K_1) + f'(K_2)(K_2 - K_1) < f(K_2) < f(K_1) + f'(K_1)(K_2 - K_1).$$

The first three inequalities are immediate implications of (4.5)-(4.7) whereas the last one reduces upon substitution from (4.6) and (4.7) to $\gamma < 1 < 1/\gamma$ which holds because of $\gamma \in (0,1)$.

Analogously, we can use Lemma 7.1 to prove that there exists a function u satisfying A2 and (4.8). To this end we have to show $0 < c_1 < c_2$, $u'(c_1) > u'(c_2)$, $u(c_1) > u'(c_1)c_1$, and

$$u(c_1) + u'(c_2)(c_2 - c_1) < u(c_2) < u(c_1) + u'(c_1)(c_2 - c_1). \tag{4.9}$$

From (4.6) and (4.7) together with the definitions of c_1 and c_2 as stated in the lemma we see that $0 < c_1 < c_2$ is equivalent to

$f(K_1) > [(1+2\delta)K_1-(1+\gamma)K_2]/\delta$ and $K_2 < K_1(1+\gamma+2\delta\gamma)/(\gamma+\gamma^2+2\delta\gamma)$. Both of these inequalities hold according to (4.5) and (4.6). The condition $u'(c_1) > u'(c_2)$ follows immediately from (4.8) and $\gamma \in (0,1)$. To satisfy the remaining two conditions we simply choose $u(c_1) > c_1$ arbitrarily and set $u(c_2) = u(c_1) + (1+\gamma)(c_2 - c_1)/2$.

Finally, because (4.2) is equivalent to (4.1), condition (4.3) is equivalent to the equations $u'(c_1) = \delta f'(K_1)u'(c_2)$ and $u'(c_2) = \delta f'(K_2)u'(c_1)$. These equations are easily verified by substituting from (4.7) and (4.8). This completes the proof of the lemma. Q.E.D.

The conditions in the above lemma are stated in terms of the values of the functions f and u as well as their first derivatives (all evaluated at the points K_1 and K_2, respectively, c_1 and c_2). The important thing to note is that we are still free to specify the values of the second derivatives of these functions at the above mentioned points. According to Lemma 7.1 there is no restriction on these values except for their negativity. This property will now be used to verify the stability of the period-2 cycle K_1, K_2.

Lemma 4.2: *There exist capital stocks K_1 and K_2 as well as a discount factor $\delta \in (0,1)$, a production function f satisfying* A1, *and a utility function u satisfying* A2 *such that $K_1, K_2, K_1, K_2, \ldots$ is a locally asymptotically stable 2-periodic trajectory of the difference equation* (4.2).

Proof: The proof is by constructing an example. Because of the previous lemma it suffices to verify (4.5) and to show that all eigenvalues of (4.4) are inside the unit circle. Let us first compute the partial derivatives of the function g which are used in (4.4). From the implicit function theorem and (4.1) and (4.2) one obtains

$$g_1(x,y) = f'(x) + \frac{1}{2}f''(x)x + \frac{u''([f(y)+f'(y)y]/2-x)}{\delta f'(x)u''([f(x)+f'(x)x]/2-g(x,y))} + \frac{f''(x)u'([f(x)+f'(x)x]/2-g(x,y))}{f'(x)u''([f(x)+f'(x)x]/2-g(x,y))}$$

and

$$g_2(x,y) = -\frac{u''([f(y)+f'(y)y]/2 - x)[f'(y)+f''(y)y/2]}{\delta f'(x)u''([f(x)+f'(x)x]/2 - g(x,y))}.$$

Now assume for a moment that (4.5) is satisfied so that we can find f and u such that conditions (4.3) and (4.6)-(4.8) hold. Substituting from these conditions into the above expressions for g_1 and g_2 we obtain:

$$g_1(K_1,K_2) = \frac{1}{\delta\gamma} + \frac{1}{2}f''(K_1)K_1 + \frac{\gamma u''(c_1)}{u''(c_2)} + \frac{\gamma^2\delta f''(K_1)}{u''(c_2)}$$

$$g_2(K_1,K_2) = -\frac{\gamma^2 u''(c_1)}{\delta u''(c_2)} - \frac{\gamma f''(K_2)K_2 u''(c_1)}{2u''(c_2)}$$

$$g_1(K_2,K_1) = \frac{\gamma}{\delta} + \frac{1}{2}f''(K_2)K_2 + \frac{u''(c_2)}{\gamma u''(c_1)} + \frac{\delta f''(K_2)}{\gamma u''(c_1)}$$

$$g_2(K_2,K_1) = -\frac{u''(c_2)}{\gamma^2\delta u''(c_1)} - \frac{f''(K_1)K_1 u''(c_2)}{2\gamma u''(c_1)}$$

The characteristic polynomial of the matrix in (4.4) is given by $\lambda^2 - A\lambda - B = 0$ where

$$A = g_1(K_1,K_2)g_1(K_2,K_1) + g_2(K_1,K_2) + g_2(K_2,K_1)$$

and

$$B = -g_2(K_1,K_2)g_2(K_2,K_1).$$

It is straightforward to check that both roots of the characteristic equation are located inside the unit circle if and only if the inequality

$$-1 < B < \max\{1-A, 1+A\} \tag{4.10}$$

holds. Now let us choose the following numerical values $K_1 = 1$, $K_2 = 2$, $\delta = 1/2$, $\gamma = 1/4$, $f''(1) = -34$, $f''(2) = -1/2$, and $u''(c_1) = u''(c_2) = -1/4$. For these values it is easy to check that (4.5) holds so that the derivation of condition (4.10) was feasible. Moreover, we can evaluate the defining expressions for A and B

to obtain $A = B = 0$. Apparently, these values satisfy (4.10) so that the 2-periodic path $K_1, K_2, K_1, K_2, \ldots$ is in fact locally asymptotically stable. We have therefore completed the proof of the lemma. Q.E.D.

5. Sunspot Equilibria

Up to now we have only considered deterministic perfect foresight Ramsey equilibria for the capital accumulation model formulated in Section 2. It is well known from the literature that there are economic models with no intrinsic uncertainty whatsoever (that is, all the fundamentals of the economy are deterministic) which nevertheless possess stochastic equilibria. These equilibria have been termed sunspot equilibria (see, e.g., Azariadis (1981), Cass and Shell (1983)). In the present section we will show that the Ramsey model of capital accumulation with heterogeneous households and borrowing constraints belongs to the class of economic models for which sunspots matter.

The proof is based on the observation made by various authors that indeterminacy of perfect foresight equilibria is a sufficient condition for the existence of sunspot equilibria (see, e.g., Laitner (1989), Matsuyama (1990), Peck (1988), Spear (1991), Woodford (1984, 1986a, 1986b)). These studies focus on indeterminate stationary equilibria whereas we could only prove indeterminacy of 2-periodic perfect foresight Ramsey equilibria. The general method, however, which was used by those authors to infer the existence of sunspot equilibria from the presence of indeterminacy can be extended to periodic equilibria as well. In order to do this we first have to define clearly what we mean by a stochastic Ramsey equilibrium.

We consider the same model as in Section 2 with the only exception that agents do not have perfect foresight on the (deterministic) sequence of factor prices $\{1 + r_t, w_t\}$. Instead, they believe that this sequence is a realization of a stochastic process. Let us denote by $\mathcal{F}_t$ the σ-algebra generated by the first t observations of this stochastic process so that $1 + r_s$ and w_s are $\mathcal{F}_t$-measurable random variables for all $s = 1, 2, \ldots, t$. Moreover, we denote by

E_t the conditional expectation with respect to $\mathcal{F}_t$. In a rational expectations equilibrium the agents' expectations in period t are assumed to coincide with E_t, the true expectations conditional on the available information. Therefore, both the profit maximization problems of the firms and the utility maximization problems of the households are stochastic optimization problems. They are defined as follows:

$$\tilde{P}_t(f) : \begin{cases} \sup\limits_{K_{t-1}} & E_{t-1}[f(K_{t-1}) - (1+r_t)K_{t-1}] \\ \text{s. t.} & K_{t-1} \geq 0 \\ & K_{t-1} \text{ is } \mathcal{F}_{t-1}\text{-measurable} \end{cases} \qquad t = 1, 2, \ldots$$

$$\tilde{P}(h) : \begin{cases} \sup\limits_{\{c_t^h, x_{t-1}^h\}} & E_0\left[\sum_{t=1}^{\infty} (\delta^h)^{t-1} u^h(c_t^h)\right] \\ \text{s. t.} & c_t^h + x_t^h = w_t + (1+r_t)x_{t-1}^h \quad t = 1, 2, \ldots \\ & c_t^h \geq 0, x_{t-1}^h \geq 0 \qquad t = 1, 2, \ldots \\ & c_t^h \text{ and } x_t^h \text{ are } \mathcal{F}_t\text{-measurable} \\ & x_0^h = k^h \end{cases}$$

A stochastic process $\{1 + r_t, w_t, K_{t-1}, c_t^h, x_{t-1}^h\}$ constitutes a rational expectations Ramsey equilibrium for $\mathcal{E}$ if the following conditions hold:

REE1: *The sequence* $\{1 + r_t, w_t\}$ *is non-negative and non-zero almost surely (a.s.).*

REE2: *For each* $t \geq 1$, K_{t-1} *solves* $\tilde{P}_t(f)$.

REE3: *For each* $h = 1, 2, \ldots, H$, $\{c_t^h, x_{t-1}^h\}$ *solves* $\tilde{P}(h)$.

REE4: *The capital market is in equilibrium, i.e.,* $\sum_{h=1}^H x_{t-1}^h = K_{t-1}$ *for all* $t \geq 1$ *a.s..*

REE5: *The output market is in equilibrium, i.e.* $\sum_{h=1}^H (c_t^h + x_t^h) = f(K_{t-1})$ *for all* $t \geq 1$ *a.s..*

We say that sunspots matter in the economy $\mathcal{E}$ if there exists a rational expectations equilibrium which is a non-trivial stochastic process. The main result of the present section is the following theorem.

Theorem 5.1: *There exists an economy $\mathcal{E}$ satisfying* A1 - A4 *in which sunspots matter.*

Proof: The general idea of the proof has been used by several authors so that we will not present any details but merely outline the approach. The two crucial optimality conditions for the Ramsey model are the first order conditions for the production sector and the most impatient household. In the stochastic setting of a rational expectations equilibrium these conditions can be written as

$$f'(K_{t-1}) - E_{t-1}(1 + r_t) = 0 \tag{5.1}$$

and

$$u'(c_t) - \delta E_t\left[(1 + r_{t+1})u'(c_{t+1})\right] = 0 \tag{5.2}$$

where we have dropped the household superscript 1 from u^1, δ^1, and c_t^1, respectively (compare with Lemma 2.2). Now assume that μ_t and ν_t, $t \geq 1$, are mutually independent and identically distributed random variables with $E\mu_t = E\nu_t = 0$ for all t. Moreover, let $\mathcal{F}_t$ be the σ-algebra generated by $\{(\mu_s, \nu_s) \mid s = 1, 2, \ldots, t\}$. We replace conditions (5.1) and (5.2) by the following equations:

$$f'(K_{t-1}) - (1 + r_t) + \mu_t = 0 \tag{5.3}$$

$$u'(c_t) - \delta(1 + r_{t+1})u'(c_{t+1}) + \nu_{t+1} = 0. \tag{5.4}$$

Because μ_t and ν_t have zero expectations it follows that (5.3) and (5.4) imply (5.1) and (5.2), respectively. Now we can proceed exactly as in the proof of Theorem 4.1 and use the budget equation for household 1 and (5.3) to obtain $c_t = [f(K_{t-1}) + (f'(K_{t-1}) + \mu_t)K_{t-1}]/2 - K_t$. (Here, we have again chosen $H = 2$ in order to be in accordance with the setting of Section 4.) Substituting this and (5.3) into (5.4) we get the following stochastic second order difference equation:

$$\begin{aligned}
&u'([f(K_{t-1}) + (f'(K_{t-1}) + \mu_t)K_{t-1}]/2 - K_t) \\
&= \delta(f'(K_t) + \mu_{t+1})u'([f(K_t) + (f'(K_t) + \mu_{t+1})K_t]/2 - K_{t+1}) \\
&- \nu_{t+1}
\end{aligned}$$

By the same argument as in the proof of Theorem 4.1 one can solve this equation for K_{t+1} in the form

$$K_{t+1} = G(K_t, K_{t-1}, \mu_t, \mu_{t+1}, \nu_{t+1}) \tag{5.5}$$

where $G(x, y, \mu', \mu, \nu)$ is a smooth function satisfying $G(x, y, 0, 0, 0) = g(x, y)$ for all x and y. Now let us consider an economy $\mathcal{E}$ as described in Theorem 4.1 for which $\bar{K}_1, \bar{K}_2, \bar{K}_1, \bar{K}_2, \ldots$ is a 2-periodic indeterminate perfect foresight Ramsey equilibrium. Because of the stability of the cycle and the smoothness of G there exists some real number $\epsilon > 0$ such that for all (x, y, μ', μ, ν) with $|x - \bar{K}_1| < \epsilon$, $|y - \bar{K}_2| < \epsilon$, $|\mu'| < \epsilon$, $|\mu| < \epsilon$, and $|\nu| < \epsilon$ it holds that $|G(x, y, \mu', \mu, \nu) - \bar{K}_2| < \epsilon$ and $|G(y, x, \mu', \mu, \nu) - \bar{K}_1| < \epsilon$. It follows that as long as μ_t and ν_t are supported on the interval $(-\epsilon, \epsilon)$ and the initial values for K_1 and K_2 are chosen sufficiently close to the period-2 cycle equation (5.5) determines a unique stochastic process $\{K_{t-1}\}$ which is adapted to the σ-algebra $\mathcal{F}_{t-1}$ (that is, K_{t-1} is $\mathcal{F}_{t-1}$-measurable for all t). Defining

$$\begin{aligned} 1 + r_t &= f'(K_{t-1}) + \mu_t \,, \; c_t^1 \\ &= \frac{1}{2}[f(K_{t-1}) + (f'(K_{t-1} + \mu_t)K_{t-1}] - K_t \,, \; x_{t-1}^1 = K_{t-1} \\ c_t^2 = w_t &= \frac{1}{2}[f(K_{t-1}) - (f'(K_{t-1}) + \mu_t)K_{t-1}] \,, \; x_{t-1}^2 = 0 \end{aligned}$$

it is straightforward to see that $\{1 + r_t, w_t, K_{t-1}, c_t^h, x_{t-1}^h\}$ qualifies as a rational expectation Ramsey equilibrium for $\mathcal{E}$. This equilibrium will be a non-trivial stochastic process if the distributions of μ_t and ν_t are non-trivial. In this case, it follows that sunspots matter in $\mathcal{E}$ and the proof of the theorem is complete. Q.E.D.

6. Conclusions

In this chapter we have shown that the perfect foresight equilibria in a capital accumulation model with infinitely-lived heterogeneous households and borrowing constraints can be periodic of arbitrary long period, that indeterminate perfect foresight equilibria may exist, and that sunspots do matter. Our results can only be regarded as a preliminary analysis and they give rise to a number of interesting questions for future research.

First of all, one should try to develop sufficient conditions for the occurrence of the three phenomena (cycles, indeterminacy, and sunspot equilibria) in the general model. Here, we have restricted ourselves to presenting examples because the analytical burden seemed to be to great. It would also be interesting to investigate if there are general conditions which ensure (local) uniqueness of Ramsey equilibria. The turnpike property alone (see Becker and Foias (1987) for a definition) is certainly not enough since the periodic and non-unique equilibria of Theorem 3.1 and Theorem 4.1 satisfy this property. Another candidate for a uniqueness condition is the capital income monotonicity condition (2.2) which has been shown to imply the turnpike property and to rule out periodic equilibria (see Becker and Foias (1987)).

Another interesting question concerns the various dynamic patterns that are possible in the Ramsey model under consideration. Are there perfect foresight equilibria which are even more complicated than periodic ones? A positive answer to this question is given in Sorger (1994b) where it is shown that topological chaos can occur as a solution to the equilibrium conditions. This provides quite a remarkable contrast to one-sector capital accumulation models with homogeneous households.

Finally, let us mention that in optimal growth models there is a strong relation between the dynamic complexity of optimal paths of capital accumulation and the time preference rate of the planner (see, e.g., Sorger (1992b)). It would be worthwhile to investigate whether the rate of impatience has a similar smoothing effect on the equilibrium dynamics in the present model. In this regard it may be interesting to point out that the construction used to prove all our results uses "sufficiently small" discount factors for all households.

We would like to conclude the chapter by trying to compare the Ramsey model with other models generating complicated dynamics. Above all, these are models with overlapping generations of finitely lived agents and infinitely lived agent models with externalities. All these models have one point in common, namely, that there are incomplete markets so that it is safe to say that market incompleteness is one possible source of complicated dynamics. However, there are also differences between these three classes of models. For example, in overlapping generations models complicated dynamics can occur irrespective from the assumptions on the production technology. As a matter of fact, there can be all kinds of non-linear dynamic phenomena in overlapping generations models without production and it depends entirely on the properties of the households' preferences whether this is the case or not. In the Ramsey model, on the other hand, complicated dynamics cannot occur under the capital income monotonicity condition as has been shown in Becker and Foias (1987). Since this is a condition on the technology alone, we see that the source of dynamic complexity in the Ramsey model must be different from that in the overlapping generations case.

7. Appendix

Lemma 7.1: *Let* $L_1 < L_2 < \ldots < L_n$, $f_1 < f_2 < \ldots < f_n$, *and* $f_1' > f_2' > \ldots > f_n'$ *be three sets of positive real numbers and let* $f_1'', f_2'', \ldots, f_n''$ *be arbitrary negative real numbers. Moreover, define* $L_0 = 0$, $f_0 = 0$, *and* $f_0' = \infty$. *If the condition*

$$f_{i-1} + f_i'(L_i - L_{i-1}) < f_i < f_{i-1} + f_{i-1}'(L_i - L_{i-1}) \tag{7.1}$$

holds for all $i = 1, 2, \ldots, n$ *then there exists a function* f *satisfying* A1 *such that* $f(L_i) = f_i$ and $f'(L_i) = f_i'$ *for all* $i = 0, 1, \ldots, n$ *and* $f''(L_i) = f_i''$ *for all* $i = 1, 2, \ldots, n$.

Proof: The proof will only be outlined as the result becomes rather intuitive once one has understood the meaning of (7.1). What this condition implies is that the set of points $M = \{(L_i, f_i) \,|\, i = 0, 1, \ldots, n\}$ describes a strictly concave "curve". More specifically, every point (L_i, f_i) is an extreme point of the convex hull of M and the straight line through (L_i, f_i) with slope determined by f_i' is a supporting hyperplane of the convex hull of M which has no point in common with M other than (L_i, f_i) itself. Now choose a sufficiently small number $\epsilon > 0$ and define f on the intervals $(L_i - \epsilon, L_i + \epsilon)$, $i = 1, 2, \ldots, n$, as the unique quadratic polynomial $q_i(L)$ which satisfies $q_i(L_i) = f_i$, $q_i'(L_i) = f_i'$, and $q_i''(L_i) = f_i''$. On the interval $[0, \epsilon)$ define f as $f(L) = \sqrt{L}$. Provided that ϵ has been chosen sufficiently small it should be clear that one can "fill up" the graph of f on the remaining parts of the non-negative real line in such a way that the conditions in A1 are satisfied. Q.E.D.

7

Robust Ergodic Chaos in Discounted Dynamic Optimization Models

Mukul Majumdar and Tapan Mitra

1. Introduction

It is by now well-known that a variety of models in economics gives rise to discrete time, *non-linear* processes of the form

$$x_{t+1} = h(x_t) \tag{1.1}$$

where the function h satisfies the Li-Yorke condition for "chaotic" or "complex" behavior. Besides the *relative abundance* of examples of chaos, yet another theme has rightly been stressed: *quite simple models* of economic theory may lead to such examples.

While formal results on chaotic processes have been derived primarily in the last ten years or so, interest in non-linear difference equations can be traced back at least to Samuelson's *Foundations* (pp. 302-310). Interestingly enough, Samuelson also speculated on a theory of *comparative dynamics*, which will "include the theory of comparative statics as a special case", but "will cover a much richer terrain". The central notion of *comparative dynamics* is that "we change something, and investigate the effect of this change on the whole motion or behavior of the economic system". The changes Samuelson talked about included (i) changes in initial conditions and (ii) changes in some parameter affecting the system. To this effect we can explicitly introduce a "parameter" μ and study a family of dynamic processes

$$x_{t+1} = h_\mu(x_t). \tag{1.2}$$

Now, admitting the possibility or even pervasiveness of complicated behavior, one can ask a natural question of interest to mathematical modeling of social phenomena: "to what extent is chaotic behavior robust?" If a process like (1.1) displays complexity due to some accidental matching of parameters, then clearly one need not take such behavior seriously. More generally, given a particular definition of complicated behavior, if the family (1.2) displays such behavior for a "negligible" set of parameter values, one can argue that the "typical" model is still "well-behaved".

In this chapter we continue the investigation of the properties of optimal programs in discounted dynamic optimization models studied earlier (Majumdar and Mitra, Chapter 3). Our focus is primarily on *ergodic chaos.* We attempt to throw light on the issue of its robustness by using a parametric variation approach in the spirit of (1.2). A family of discounted dynamic optimization problems is identified in which the one period return function depends on current consumption, capital input and some taste parameter μ. It is shown that this family of optimization problems gives rise to the dynamical system represented by the "quadratic family": that is, for each value of the parameter μ, the optimal transition function is given by $h_\mu(x) = \mu x(1-x)$.

Using this result, it is shown that *the set of parameter values generating optimal programs that display ergodic chaos has positive Lebesgue measure.* The basic mathematical result that we apply is due to Jakobson (1981) [see also Benedicks and Carleson (1985) and Carleson (1991)].[1]

By using Jakobson's theorem one can also throw light on the issue of *sensitive dependence* of the processes (1.2) on initial conditions in the strong sense of Guckenheimer (1979). Roughly speaking, for a given value of μ, the process (1.2) possesses sensitive dependence on initialconditions if there is some positive number such that in every neighborhood of an initial x there exists a point such that its image under a finite number of iterations

[1]The question of whether the family of dynamic optimization problems displaying robust ergodic chaos is itself non-negligible in an appropriate class of dynamic optimization problems is an important one. We do not pursue this question in this paper. It is worth noting, though, that Jakobson's result actually holds for some families of maps "close to" the quadratic family in the C^3 metric. (See Jakobson (1981, p. 40)). This suggests that the answer to the above question might be in the affirmative in a "sufficiently smooth" class of dynamic optimization problems.

separates from that of x by that positive number. Here, again, the size of such a set of initial conditions can be an issue: if it is of Lebesgue measure zero, it is "negligible". The definition of Guckenheimer requires that such sensitivity holds on a set of initial conditions with positive Lebesgue measure. Jakobson's result also implies that *the set of parameter values generating optimal programs that display Guckenheimer's sensitive dependence has positive Lebesgue measure.*

This quadratic family of maps has been the focus of attention of "bifurcation analysis", which examines, roughly speaking, how the limit sets (asymptotic behavior) of a family of dynamic processes (like (1.2)) change as the parameter μ changes. Given this literature (emphasizing especially the "period-doubling route to chaos") our family of dynamic optimization problems allows us to also explore - in the spirit of Samuelsonian comparative dynamics - how the long-run behavior of optimal programs undergoes qualitative changes as the taste parameter μ varies.

In Section 2 we collect some definitions. The family of dynamic optimization problems is formally set up in Section 3. The main result on robustness of ergodic chaos and sensitive dependence on initial conditions is presented in Section 4. The steps needed to show that the dynamical system generated by this family of dynamic optimization problems is given by the quadratic family involve some detailed calculations and verifications (more refined than those in Majumdar and Mitra, Chapter 3). These are relegated to an appendix.

It should perhaps be stressed that while we use an explicit parametric variation in the spirit of (1.2) in this chapter, our earlier result on robust topological chaos rested on varying the one period return and production functions on an open set in the function spaces (and the discount factor on an open interval). We note that while the question of robustness of topological chaos has been discussed in a few other environments, we do not know of any attempt to characterize robust ergodic chaos or sensitive dependence.

2. Technical Preliminaries

Consider a probability space $(X, \sum, v)$ and a $\sum$-measurable map $h : X \to X$; v is *invariant under* h if $v(E) = v(h^{-1}(E))$ for all $E \in \sum$; v is *ergodic* if "$E \in \sum$, $h^{-1}(E) = E$" implies "$v(E) = 0$ or $v(E) = 1$". Let X be a closed interval $[\alpha, \beta]$ of the real line with $\alpha < \beta$. We say that $h : X \to X$ exhibits *ergodic chaos* if there is an ergodic invariant measure that is absolutely continuous with respect to the Lebesgue measure. It is known, for example, that

$$h(x) \equiv 4x(1 - x)$$

exhibits ergodic chaos.

If $x \in X$, the sequence $\tau(x) \equiv (h^j(x))_{j=0}^{\infty}$ is called the *trajectory* from (the initial condition) x. The *orbit* from x is the set $\gamma(x) \equiv \{y : y = h^j(x) \text{ for some } j \geq 0\}$. The asymptotic behavior of a trajectory from x is described by the *limit set*, which is defined as the set of all limit points of $\gamma(x)$, and is denoted by $\omega(x)$.

A point $x \in X$ is a *fixed point* of h if $h(x) = x$. A point $x \in X$ is called *periodic* if there is $k \geq 1$ such that $h^k(x) = x$. The smallest such k, call it $\mathbf{k}$, is the *period* of x. Note that if $x \in X$ is a periodic point, then $\omega(h^j(x)) = \gamma(x)$ for every $j = 0, 1, \ldots$. A periodic point $\bar{x} \in X$ is *stable* if there is an open interval R (in X) containing $\bar{x}$, such that $\omega(x) = \gamma(\bar{x})$ for all $x \in R$. [In this case we also say that the periodic orbit $\gamma(\bar{x})$ is stable.]

Consider the iterations

$$h^0(x) = x \text{ and } h^k(x) = h[h^{k-1}(x)] \text{ for } k \geq 1. \tag{2.1}$$

The process

$$x_{t+1} = h(x_t) \tag{2.2}$$

has *sensitive dependence on initial conditions* (Guckenheimer (1979)) if there are a set $\Phi \subset X$ of positive Lebesgue measure and $\varepsilon > 0$ such that given any $x \in \Phi$ and any neighborhood U of x, there is $y \in U$ and $n \geq 1$ such that $|h^n(x) - h^n(y)| > \varepsilon$.

To address the question of robustness, let (h_μ) be the "quadratic" family of maps, where $\mu \in A \equiv [1, 4]$ and

$h_\mu : [0,1] \to [0,1]$ is defined as:

$$h_\mu(x) \equiv \mu x(1-x). \tag{2.3}$$

Let $\Lambda \equiv \{\mu \in A : h_\mu$ exhibits ergodic chaos$\}$. The basic result of Jakobson characterizes this set Λ: it is uncountable *and* non-negligible. More precisely, let us state:

Jakobson's Theorem (Theorem B and remarks on p. 40 of Jakobson [1981]): *The set Λ, such that for each $\mu \in \Lambda$ the map h_μ exhibits ergodic chaos, has positive Lebesgue measure; also, h_μ has sensitive dependence in the sense of Guckenheimer.*

The following results from bifurcation analysis summarizes the period-doubling route to chaos for the family of maps (2.3).

Bifurcation Theorem: *There is a sequence $(\mu_k)_{k=0}^{\infty}$ of bifurcation values of μ, such that*

(i) $\mu_0 = 1$, $\mu_k < \mu_{k+1}$ for $k = 0, 1, 2, 3, \ldots$

(ii) $\mu_k \to \mu_\infty$ as $k \to \infty$, where $\mu_\infty \backsimeq 3.569946$.

(iii) For $\mu \in (\mu_k, \mu_{k+1})$, there is a stable periodic point $x(\mu)$, with period 2^k, such that for almost every $x \in X$, $\omega(x) = \gamma(x(\mu))$.

3. A Family of Dynamic Optimization Problems

We consider a *family* of intertemporal economies, indexed by a parameter μ, where $\mu \in A \equiv [1,4]$. Each economy in this family has the same *production function*, $\mathbf{f}$: $\Re_+ \to \Re_+$ and the same *discount factor* $\delta \in (0,1)$. The economies in this family differ in the specification of their *welfare functions*, $w : \Re_+^2 \times A \to \Re_+$ [depending on the parameter value of $\mu \in A$ that is picked].

The following assumptions on $\mathbf{f}$ are used:

F1: $\mathbf{f}$*(0) = 0.*

F2: $\mathbf{f}$ *is non-decreasing, continuous and concave on $\Re_+$.*

F3: *There is* $\mathbf{K} > 0$, *such that $f(x) < x$ for all $x > \mathbf{K}$, and* $\mathbf{f}(x) > x$ *for all* $0 < x < \mathbf{K}$.

A *program* from $x \geq 0$ is a sequence $\{x_t\}$ satisfying

$$x_0 = x,\ 0 \le x_t \le \mathbf{f}(x_{t-1}) \text{ for } t \ge 1.$$

The *consumption* sequence $\{c_t\}$ generated by a program $\{x_t\}$ is given by

$$c_t = \mathbf{f}(x_{t-1}) - x_t \ (\ge 0) \text{ for } t \ge 1.$$

It is standard to verify that for any program $\{x_t\}$ from $x \ge 0$, we have x_t, $c_{t+1} \le \mathbf{K}(x) = \max(\mathbf{K}, x)$ for $t \ge 0$.

Given any $\mu \in A$, the following assumptions on $w(\bullet; \mu)$ are used:

W1: *$w(x, c, \mu)$ is non-decreasing in x given c, and non-decreasing in c, given x.*

W2: *$w(x, c, \mu)$ is continuous on $\Re^2_+$.*

W3: *$w(x, c, \mu)$ is concave on $\Re^2_+$.*

In defining "optimality" of a program, we note that the notion has to be economy specific. Since we can keep track of the economies by simply noting its μ value, we find it convenient to refer to the appropriate notion of optimality as μ-optimality.

Given any $\mu \in A$, a program $\{\hat{x}_t\}$ from $x \ge 0$ is *μ-optimal* if

$$\sum_{t=0}^{\infty} \delta^t w(\hat{x}_t, \hat{c}_{t+1}, \mu) \ge \sum_{t=0}^{\infty} \delta^t w(x_t, c_{t+1}, \mu)$$

for every program $\{x_t\}$ from x.

Define a set $\Omega \subset \Re^2_+$ by

$$\Omega = \{(x, z) \in \Re^2_+ : z \le \mathbf{f}(x)\}$$

For much of our discussion of μ-optimal programs, what is crucial is the behavior of $w(\bullet; \mu)$ on Ω (rather than on $\Re^2_+$). We now proceed to assume:

W4: *Given any $\mu \in A$, $w(x, c, \mu)$ is strictly increasing and strictly concave in c given x, on the set Ω.*

Standard arguments ensure that given any $\mu \in A$, there is a μ-optimal program from every $x \geq 0$. Assumptions (F.2), (W.3), and (W.4) ensure that a μ-optimal program is unique.

Given any $\mu \in A$, we define a *value function*, $V : \Re_+ \times A \to \Re$ by

$$V(x, \mu) = \sum_{t=0}^{\infty} \delta^t w(\hat{x}_t, \hat{c}_{t+1}, \mu)$$

where $\{\hat{x}_t\}$ is the μ-optimal program from $x \geq 0$.

Since there is a unique μ-optimal program from every $x \geq 0$, one can define an *optimal transition function* $h : \Re_+ \times A \to \Re_+$ by

$$h_\mu(x) = \hat{x}_1$$

where $\{\hat{x}_t\}$ is the μ-optimal program from $x \geq 0$. It is easily checked that this definition also implies that for all $t \geq 0$, we have

$$\hat{x}_{t+1} = h_\mu(\hat{x}_t).$$

4. A Family Displaying Robust Ergodic Chaos and Sensitive Dependence

Consider now a family of economies, where $\mathbf{f}$, δ and w are specified as follows:

$$\mathbf{f}(x) = \begin{cases} (16/3)x - 8x^2 + (16/3)x^4 & \text{for } x \in [0, 0.5] \\ 1 & \text{for } x \geq 0.5 \end{cases}$$
$$\delta = 0.0025 \qquad (4.1)$$

The function w is specified in a more involved fashion. To ease the writing, denote $L \equiv 98$, $a \equiv 425$. Also, denote by I the closed interval [0,1], and define the function $\theta : I \times A \to I$ by

$$\theta(x, \mu) = \mu x(1 - x) \text{ for } x \in I, \ \mu \in A$$

and $u : I^2 \times A \to \Re$ by

$$u(x, z, \mu) = ax - 0.5Lx^2 + z\theta(x, \mu) - 0.5z^2 - \delta[az - 0.5Lz^2 + 0.5\theta(z, \mu)^2] \tag{4.2}$$

Define a set $\mathbf{D} \subset I^2$ by

$$\mathbf{D} = \{(x, c) \in I \times \Re_+ : c \leq \mathbf{f}(x)\}$$

and a function $w : \mathbf{D} \times A \to \Re_+$ by

$$w(x, c, \mu) = u(x, \mathbf{f}(x) - c, \mu) \text{ for } (x, c) \in \mathbf{D} \text{ and } \mu \in A. \tag{4.3}$$

We now extend the definition of $w(\bullet; \mu)$ to the domain Ω. For $(x, c) \in \Omega$ with $x > 1$ [so that $\mathbf{f}(x) = 1$, and $c \leq 1$], define

$$w(x, c, \mu) = w(1, c, \mu). \tag{4.4}$$

Finally, we extend the definition of $w(\bullet; \mu)$ to the domain $\Re^2_+$. For $(x, c) \in \Re^2_+$ with $c > \mathbf{f}(x)$, define

$$w(x, c, \mu) = w(x, \mathbf{f}(x), \mu). \tag{4.5}$$

It can be checked (see Appendix) that for the above specifications, $\mathbf{f}$ satisfies (F.1)-(F.3), and given any $\mu \in A$, $w(\bullet; \mu)$ satisfies (W.1)-(W.4).

We observe that $w(x, c, \mu) \geq w(0, 0, \mu)$ [by (W.1)] $= u(0, \mathbf{f}(0) - 0, \mu) = u(0, 0, \mu) = 0$, for all $(x, c) \in \Re^2_+$. Thus $w(\bullet; \mu)$ maps from $\Re^2_+$ to $\Re_+$. Also, for all $(x, c) \in \Re^2_+$, $w(x, c, \mu) \leq w(1, c, \mu) \leq w(1, \mathbf{f}(1), \mu) = w(1, 1, \mu)$.

We verify (in Appendix) the following:

Theorem 4.1: *The optimal transition functions for the family of economies* $(\mathbf{f}, w(\bullet; \mu), \delta)$ *are given by*

$$h_\mu(x) = \mu x(1 - x) \text{ for all } x \in I. \tag{4.6}$$

Hence, by Jakobson's Theorem, the family exhibits robust ergodic chaos and sensitive dependence on initial conditions.

Using (4.6) of Theorem 4.1, and applying the Bifurcation Theorem, we can conclude that as the taste parameter μ changes, for instance, from $1 < \mu < \mu_1$ to $\mu_1 < \mu < \mu_2$, the long-run behavior of the typical optimal program changes from convergence to a stable fixed point of $h_\mu(x)$ to convergence to a stable periodic point of period two. That is, the long-run dynamic behavior of optimal programs experiences a bifurcation (a distinct qualitative change) as the taste parameter μ crosses the value μ_1. The same is, of course, true of the other bifurcation values μ_k for $k > 1$ as each successive bifurcation value gives rise to a stable periodic point of a period which is double that of the previous bifurcation value.

5. Appendix

5.1 Verification of the Assumptions

The production function
Recall that the production function, $\mathbf{f} : \Re_+ \to \Re_+$ is given by

$$\mathbf{f}(x) = \begin{cases} (16/3)x - 8x^2 + (16/3)x^4 & \text{for } x \in [0, 0.5) \\ 1 & \text{for } x \geq 0.5 \end{cases}$$

Note, that $\mathbf{f}(0) = 0$, and $\mathbf{f}(x) \to 1$ as $x \uparrow (1/2)$, so that $\mathbf{f}$ is clearly continuous on $\Re_+$. Next, note that for $0 \leq x < 0.5$, $\mathbf{f}$ is C^2, and

$$\mathbf{f}'(x) = (16/3) - 16x + (64/3)x^3$$
$$\mathbf{f}''(x) = -16 + 64x^2$$

As $x \uparrow (1/2)$, $\mathbf{f}'(x) \to 0$. Thus, we can conclude that $\mathbf{f}$ is C^1 on $\Re_+$. As $x \uparrow (1/2)$, $\mathbf{f}''(x) \to 0$, so repeating the above argument, $\mathbf{f}$ is C^2 on $\Re_+$. It is clear that $\mathbf{f}''(x) < 0$ for $0 \leq x \leq 0.5$, and so $\mathbf{f}'(x) > 0$ for $0 \leq x < 0.5$ [since $\mathbf{f}'(1/2) = 0$]. Thus, $\mathbf{f}$ is non-decreasing and concave on $\Re_+$.

To verify (F.3), note that the concavity of $\mathbf{f}$ and $\mathbf{f}(0) = 0$ ensure that $[\mathbf{f}(x)/x]$ is non-increasing for $x > 0$. Since $\mathbf{f}(1/2) > (1/2)$, we have $\mathbf{f}(x) > x$ for $0 < x \leq (1/2)$. And for $(1/2) < x < 1$, $\mathbf{f}(x) = 1$, so $\mathbf{f}(x) > x$ for $(1/2) < x < 1$ as well. Thus, defining $\mathbf{K} = 1$, we have $\mathbf{f}(x) > x$ for $0 < x < \mathbf{K}$, and $\mathbf{f}(x) = 1 < x$ for all $x > \mathbf{K}$.

We now note an important property of $\mathbf{f}$, viz,

$$\mathbf{f}(x) \geq \theta(x, \mu) \text{ for all } x \in I,\ \mu \in A$$

Since $\mathbf{f}(x) = 1$ for $x \in [0.5, 1]$, and $\theta(x, \mu) \in I$ for all $x \in I$, $\mu \in A$, we have $\mathbf{f}(x) \geq \theta(x, \mu)$ for $x \in [0.5, 1]$ and $\mu \in A$.

We now examine the case where $x \in [0, 0.5)$ and yet, $\mu \in A$. Clearly $\theta(0, \mu) = 0$ and $\mathbf{f}(0) = 0$, so $\mathbf{f}(x) \geq \theta(x, \mu)$ for $x = 0$, $\mu \in A$. For $0 < x < 0.5$ define

$$\beta(x, \mu) = [\mathbf{f}(x) - \theta(x, \mu)]/x$$

Then $\beta(x, \mu) = ((16/3) - \mu) - (8 - \mu)x + (16/3)x^3$ and $D_1\beta(x, \mu) = -(8 - \mu) + 16x^2 < 0$ for $0 < x < 0.5$, and $\beta(\bullet, \mu)$ is a decreasing function on $(0, 0.5)$. As $x \uparrow (1/2)$, $\beta(x, \mu) \to 2 - (\mu/2) \geq 0$. Thus, $\beta(x, \mu) > 0$ for $0 < x < (1/2)$. This means $\mathbf{f}(x) > \theta(x, \mu)$ for $0 < x < (1/2)$.

The welfare function

Recall that $u : I^2 \times A \to \Re$ was defined by

$$u(x, z, \mu) = ax - 0.5Lx^2 + z\theta(x, \mu) - 0.5z^2 - \delta[az - 0.5Lz^2 + 0.5\theta(z, \mu)^2]$$

We can compute the following derivatives:

$$\begin{aligned}
D_1u(x, z, \mu) &= a - Lx + z\mu(1 - 2x) \\
D_2u(x, z, \mu) &= \mu x(1 - x) - z - \delta a + \delta Lz - \\
&\qquad \delta\mu^2 z + 3\delta\mu^2 z^2 - 2\delta\mu^2 z^3 \\
D_{11}u(x, z, \mu) &= -L - 2\mu z \\
D_{12}u(x, z, \mu) &= \mu(1 - 2x) = D_{21}u(x, z, \mu) \\
D_{22}u(x, z, \mu) &= -1 + \delta L - \delta\mu^2 + 6\delta\mu^2 z - 6\delta\mu^2 z^2
\end{aligned}$$

Recall that $\mathbf{D} \subset I^2$ was defined by

$$D = \{(x, c) \in I \times \Re_+ : c \leq \mathbf{f}(x)$$

and $w : \mathbf{D} \times A \to \Re$ by

$$w(x, c, \mu) = u(x, \mathbf{f}(x) - c, \mu) \text{ for } (x, c) \in \mathbf{D} \text{ and } \mu \in A$$

We check, first, that $w(\bullet, \mu)$ satisfies (W.1)-(W.4) on $\mathbf{D}$, given any $\mu \in A$. Clearly $w(\bullet, \mu)$ is C^2 on $\mathbf{D}$, and we check below that $w(\bullet, \mu)$ is strictly increasing in x (given c) and in c (given x). It is strictly concave in (x, c) on $\mathbf{D}$. The first partials of w are:

$$D_2 w(x, c, \mu) = -D_2 u(x, \mathbf{f}(x) - c, \mu)$$
$$D_1 w(x, c, \mu) = D_1 u(x, \mathbf{f}(x) - c, \mu) + D_2 u(x, \mathbf{f}(x) - c, \mu)\mathbf{f}'(x)$$

$$\begin{aligned}\text{Now, } D_2 u(x, \mathbf{f}(x) - c, \mu) &\leq (\mu/4) - \delta a + \\ &\qquad \delta z(L - \mu^2) - z + 3\delta\mu^2 \\ &\leq (\mu/4) - \delta a + 3\delta\mu^2 \\ [\text{since } \delta(L - \mu^2) < 1] \\ &\leq (\mu/4) - \delta[a - 3\mu^2] \\ &< 0 \ [\text{since } \delta[a - 3\mu^2] > (\mu/4)]\end{aligned}$$

Thus, $D_2 w(x, c, \mu) > 0$ for $(x, c) \in \mathbf{D}$ and $\mu \in A$. Also, $D_1 u(x, z, \mu) \geq a - L - \mu \geq 323$ and

$$\begin{aligned}D_2 u(x, z, \mu) &\geq -\delta a - z + \delta z(L - \mu^2) + \delta\mu^2 z^2(3 - 2z) \\ &\geq -\delta a - z \ (\text{since } L > \mu^2) \\ &\geq -2.0625 > -3\end{aligned}$$

and $\mathbf{f}'(x) \leq \mathbf{f}'(0) = (16/3)$. Thus, $D_2 u(x, z, \mu)\mathbf{f}'(x) \geq -16$, and so $D_1 w(x, c, \mu) \geq 323 - 16 > 0$. Using the signs of the first partials of w, we conclude that $w(\bullet, \mu)$ is strictly increasing in x (given c) and in c (given x) on $\mathbf{D}$.

The second partials of w are:

$$\begin{aligned}
D_{22}w(x,c,\mu) &= D_{22}u(x,\mathbf{f}(x)-c,\mu) \\
D_{21}w(x,c,\mu) &= -D_{21}u(x,\mathbf{f}(x)-c,\mu) - \\
&\quad D_{22}u(x,\mathbf{f}(x)-c,\mu)\mathbf{f}'(x) \\
&= D_{12}w(x,c,\mu) \\
D_{11}w(x,c,\mu) &= D_{11}u(x,\mathbf{f}(x)-c,\mu) + \\
&\quad D_{12}u(x,\mathbf{f}(x)-c,\mu)\mathbf{f}'(x) \\
&\quad + [D_{21}u(x,\mathbf{f}(x)-c,\mu) + \\
&\quad D_{22}u(x,\mathbf{f}(x)-c,\mu)\mathbf{f}'(x)]\mathbf{f}'(x) \\
&\quad + D_2u(x,\mathbf{f}(x)-c,\mu)\mathbf{f}''(x)
\end{aligned}$$

Now, $D_{22}u(x,z,\mu) \leq -1+\delta L + 6\delta\mu^2 z \leq -1+\delta(L+96) < -0.5 < 0$, so we have

$$D_{22}w(x,c,\mu) < 0$$

Also,

$$\begin{aligned}
(D_{22}w)(D_{11}w) - (D_{12}w)^2 &= (D_{22}u)(D_{11}u) + 2(D_{12}u)(D_{22}u)\mathbf{f}'(x) \\
&\quad + (D_{22}u)^2\mathbf{f}'(x)^2 + \\
&\quad (D_{22}u)(D_2u)\mathbf{f}''(x) - (D_{12}u)^2 \\
&\quad - (D_{22}u)^2\mathbf{f}'(x)^2 - \\
&\quad 2(D_{12}u)(D_{22}u)\mathbf{f}'(x) \\
&= (D_{22}u)(D_{11}u) - (D_{12}u)^2 + \\
&\quad (D_{22}u)(D_2u)\mathbf{f}''(x)
\end{aligned}$$

Now,

$$D_{22}u(x,z,\mu) \geq -1+\delta[L-\mu^2] > -1$$

and $0 > D_2u(x,z,\mu) > -2.0625$, as we checked above. Further, $\mathbf{f}''(x) = -16+64x^2$ for $0 \leq x \leq (1/2)$, so $0 \geq \mathbf{f}''(x) \geq -16$ for $x \in I$. These estimates imply that

$$(D_2u)(D_{22}u)\mathbf{f}''(x) > -33$$

We also have $D_{11}u(x,z,\mu) \leq -L < 0$ and $D_{22}u(x,z,\mu) < -0.5 < 0$, while $[D_{12}u(x,z,\mu)]^2 = \mu^2(1-2x)^2 \leq \mu^2 \leq 16$. These estimates imply that

$$(D_{11}u)(D_{22}u) - (D_{12}u)^2 > (L/2) - 16 = 33$$

Thus, $(D_{22}w)(D_{11}w) - (D_{12}w)^2 > 33 - 33 = 0$ and $w(\bullet, \mu)$ is strictly concave on the set $\mathbf{D}$.

We have now checked that $w(\bullet,\mu)$ satisfies (W.1)-(W.4) on $\mathbf{D}$, given any $\mu \in A$. Furthermore, $w(\bullet,\mu)$ is C^2 on $\mathbf{D}$.

Recall that $\Omega \subset \Re^2_+$ was defined by

$$\Omega = \{(x,z) \in \Re^2_+ : z \leq \mathbf{f}(x)\}$$

and w was extended from $\mathbf{D}$ to Ω by defining for $(x,c) \in \Omega$ with $x > 1$ [so that $\mathbf{f}(x) = 1$, and $c \leq 1$], and $\mu \in A$,

$$w(x,c,\mu) = w(1,c,\mu)$$

We now check that $w(\bullet,\mu)$ satisfies (W.1)-(W.4) on Ω, given any $\mu \in A$. Note that $w(\bullet,\mu)$ satisfies (W.1) and (W.2) on Ω. To check that $w(\bullet,\mu)$ satisfies (W.3), let (x,c) and $(\bar{x},\bar{c})$ belong to Ω, and let $0 < \lambda < 1$. Then $w(\lambda(x,c) + (1-\lambda)(\bar{x},\bar{c}),\mu) = w(\lambda x + (1-\lambda)\bar{x},\ \lambda c + (1-\lambda)\bar{c},\mu)$. Now, $[\lambda x + (1-\lambda)\bar{x},\ \lambda c + (1-\lambda)\bar{c}] \in \mathbf{D}$ if $[\lambda x + (1-\lambda)\bar{x}] \leq 1$. If x and $\bar{x}$ are both ≤ 1, then concavity of w follows from concavity of w on $\mathbf{D}$. So consider, without loss of generality, that $x > 1$ while $\bar{x} < 1$. Then $w(\lambda x + (1-\lambda)\bar{x},\ \lambda c + (1-\lambda)\bar{c},\mu) \geq w(\lambda + (1-\lambda)\bar{x},\ \lambda c + (1-\lambda)\bar{c},\mu)$ [using (W.2)] $\geq \lambda w(1,c,\mu) + (1-\lambda)w(\bar{x},\bar{c},\mu)$ [since $(1,c)$ and $(\bar{x},c)$ belong to $\mathbf{D}$] $= \lambda w(x,c,\mu) + (1-\lambda)w(\bar{x},\bar{c},\mu)$. If $[\lambda x + (1-\lambda)\bar{x}] > 1$, then $w(\lambda x t(1-\lambda)\bar{x},\ \lambda c t(1-\lambda)\bar{c},\mu) = w(1, \lambda c + (1-\lambda)\bar{c},\mu)$ $\geq \lambda w(1,c,\mu) + (1-\lambda)w(1,\bar{c},\mu)$ [since $(1,c) \in \mathbf{D}$ and $(1,\bar{c}) \in \mathbf{D}$] $\geq \lambda w(x,c,\mu) + (1-\lambda)w(\bar{x},\bar{c},\mu)$. Thus, $w(\bullet,\mu)$ satisfies (W.3) on Ω.

To check (W.4), let $x \geq 0$ be given. Then if $x \leq 1$, $w(x,c,\mu)$ is strictly increasing and strictly concave in c. If $x > 1$, $w(x,c,\mu) = w(1,c,\mu)$, which is strictly concave and strictly increasing in c.

Recall that $w(\bullet,\mu)$ was extended from Ω to $\Re^2_+$ by defining for $(x,c) \in \Re^2_+$ with $c > \mathbf{f}(x)$, and $\mu \in A$,

$$w(x,c,\mu) = w(x,\mathbf{f}(x),\mu)$$

We now check that $w(\bullet, \mu)$ satisfies (W.1)-(W.4), given any $\mu \in A$. Note that $w(\bullet, \mu)$ clearly satisfies (W.1), (W.2) on $\Re^2_+$. To check (W.3), let (x, c) and $(\overline{x}, \overline{c})$ be in $\Re^2_+$ and let $0 < \lambda < 1$. Then

$$\begin{aligned} & w(\lambda(x, c) + (1-\lambda)(x, \bar{c}), \mu) \\ &= w(\lambda x + (1-\lambda)\overline{x},\ \lambda c + (1-\lambda)\bar{c}, \mu) \end{aligned}$$

Denote $\min[c, \mathbf{f}(x)]$ by $G(x, c)$; $\min[\bar{c}, \mathbf{f}(\bar{x})]$ by $G(\bar{x}, \bar{c})$. Then, we have $w(\lambda x + (1-\lambda)\bar{x},\ \lambda c + (1-\lambda)\bar{c}, \mu) \geq w(\lambda x + (1-\lambda)\bar{x}, \lambda G(x, c) + (1-\lambda)G(\bar{x}, \bar{c}), \mu)$. Also, $\lambda G(x, c) + (1-\lambda)G(\bar{x}, \bar{c}) \leq \lambda \mathbf{f}(x) + (1-\lambda)\mathbf{f}(\bar{x}) \leq \mathbf{f}(\lambda x + (1-\lambda)\bar{x}$. Further, $G(x, c) \leq \mathbf{f}(x)$ and $G(\bar{x}, \bar{c}) \leq \mathbf{f}(\bar{x})$. Thus, $(x, G(x, c)), (\bar{x}, G(\bar{x}, \bar{c}))$ and $[\lambda x + (1-\lambda)\bar{x}, \lambda G(x, c) + (1-\lambda)G(\bar{x}, \bar{c})]$ all belong to Ω and so we have

$$\begin{aligned} & w(\lambda x + (1-\lambda)\bar{x},\ \lambda c + (1-\lambda)\bar{c}, \mu) \\ &\geq \lambda w(x, G(x, c), \mu) + (1-\lambda) w(\bar{x}, G(\bar{x}, \bar{c}), \mu). \end{aligned}$$

If $\min(c, \mathbf{f}(x)) = c$, then $w(x, G(x, c), \mu) = w(x, c, \mu)$; if $\min(c, \mathbf{f}(x)) \neq c$, then $c > \mathbf{f}(x)$, and $w(x, G(x, c), \mu) = w(x, \mathbf{f}(x), \mu) = w(x, c, \mu)$; thus, in either case, $w(x, G(x, c), \mu) = w(x, c, \mu)$.
Similarly, $w(\bar{x}, G(\bar{x}, \bar{c}), \mu) = w(\bar{x}, \bar{c}, \mu)$. Hence, $\lambda w(x, G(x, c), \mu) + (1-\lambda) w(\bar{x}, G(\bar{x}, \bar{c}), \mu) = \lambda w(x, c, \mu) + (1-\lambda) w(\bar{x}, \bar{c}, \mu)$, completing our demonstration of the concavity of $w(\bullet, \mu)$ on $\Re^2_+$.

5.2 Verification of Optimal Transition Function

The verification of the optimal transition function is obtained by following closely the technique of Boldrin-Montrucchio (1986).

Define $\phi : I^2 \times A \to \Re$ by

$$\phi(x, z, \mu) = ax - 0.5Lx^2 + z\mu x(1-x) - 0.5z^2$$

We can compute the following derivatives:

$$\begin{aligned} D_1\phi(x, z, \mu) &= a - Lx + z\mu(1-2x) \\ D_2\phi(x, z, \mu) &= \mu x(1-x) - z \\ D_{11}\phi(x, z, \mu) &= -L - 2\mu z \\ D_{12}\phi(x, z, \mu) &= \mu(1-2x) = D_{21}\phi(x, z, \mu) \\ D_{22}\phi(x, z) &= -1 \end{aligned}$$

Looking at the Hessian matrix of $\phi(\bullet, \mu)$, we note that

$$D_{11}\phi(x, z, \mu) < 0 \text{ [since } L > 0]$$

and

$$(D_{11}\phi)(D_{22}\phi) - (D_{12}\phi)^2 = L + 2\mu z - [\mu(1-2x)]^2 > 0$$

$$[\text{since } |1-2x| \leq 1 \text{ and } |\mu| \leq 4, \text{ and } L > 16]$$

Thus $\phi(\bullet, \mu)$ is strictly concave on I^2.

Result 5.1: *Given any $x \in I$, $\theta(x, \mu) \equiv \mu x(1-x)$ solves the problem:*

$$\left.\begin{array}{ll} \text{Max} & \phi(x, z, \mu) \\ \text{Subject to} & z \in I \end{array}\right\} (P)$$

Furthermore, this solution is unique.

Proof: Clearly, $\theta(x, \mu) \in I$ for all $x \in I$ and $\mu \in A$. For any $z \in I$ with $z \neq \theta(x, \mu)$ we have by the strict concavity of $\phi(\bullet, \mu)$ on I^2,

$$\phi(x, z, \mu) - \phi(x, \theta(x, \mu), \mu < D_2\phi(x, \theta(x, \mu), \mu)(z - \theta(x, \mu)) = 0$$

Thus, $\theta(x, \mu)$ uniquely solves problem (P).

Define $\psi : I \times A \to I$ by

$$\psi(x, \mu) = \phi(x, \theta(x, \mu), \mu) \text{ for } (x, \mu) \in I \times A$$

Then, by the definition of ϕ and θ,

$$\psi(x, \mu) = ax - 0.5Lx^2 + \theta(x, \mu)^2 - 0.5\theta(x, \mu)^2 = ax - 0.5Lx^2 + 0.5\theta(x, \mu)^2$$

Note that by the definition of $u(x, z, \mu)$,

$$u(x, z, \mu) = ax - 0.5Lx^2 + z\theta(x, \mu) - 0.5z^2 - \delta[az - 0.5Lz^2 + 0.5\theta(z, \mu)^2]$$
$$= \phi(x, z, \mu) - \delta\psi(z, \mu)$$

for $(x, z) \in I^2$ and $\mu \in A$.

Result 5.2: *Given any $x \in I$,*
(i) $\theta(x, \mu)$ solves the problem:

$$\left.\begin{array}{ll} \text{Max} & u(x, z, \mu) + \delta\psi(z, \mu) \\ \text{Subject to} & z \in I \end{array}\right\} (Q)$$

and
(ii) $\psi(x, \mu) = \underset{z \in I}{\text{Max}} \, [u(x, z, \mu) + \delta\psi(z, \mu)]$

Proof: We have $u(x, z, \mu) + \delta\psi(z, \mu) = \phi(x, z, \mu)$ by definition of u. Thus (i) follows from Result 1.
Using (i), we get

$$\begin{aligned} \underset{z \in I}{\text{Max}} \, [u(x, z, \mu) + \delta\psi(z, \mu)] &= u(x, \theta(x, \mu), \mu) + \delta\psi(\theta(x, \mu), \mu) \\ &= \phi(x, \theta(x, \mu), \mu) \\ &\quad \text{[by definition of } u] \\ &= \psi(x, \mu) \text{ [by definition of } \psi] \end{aligned}$$

This proves (ii).

Consider a sequence $\{\hat{x}_t\}_0^\infty$ defined by $\hat{x}_0 = x \in I$, $\hat{x}_{t+1} = \theta(\hat{x}_t, \mu)$ for $t \geq 0$. Notice that $\hat{x}_t \in I$ for $t \geq 0$. Also, for $t \geq 0$,

$$\begin{aligned} \psi(\hat{x}_t, \mu) &= \phi(\hat{x}_t, \theta(\hat{x}_t, \mu), \mu) \\ &= \phi(\hat{x}_t, \hat{x}_{t+1}, \mu) \\ &= u(\hat{x}_t, \hat{x}_{t+1}, \mu) + \delta\psi(\hat{x}_{t+1}, \mu) \end{aligned}$$

Thus, iterating on this relation,

$$\psi(x, \mu) = \sum_{t=0}^{T} \delta^t u(\hat{x}_t, \hat{x}_{t+1}, \mu) + \delta^T \psi(\hat{x}_{T+1}, \mu)$$

Since $\psi(\bullet, \mu)$ is bounded on I and $u(\bullet, \mu)$ is bounded on I^2, and $0 < \delta < 1$, we get

$$\psi(x, \mu) = \sum_{t=0}^{\infty} \delta^t u(\hat{x}_t, \hat{x}_{t+1}, \mu) \tag{5.1}$$

Consider, next, any sequence $\{x_t\}_0^\infty$ defined by $x_0 = x \in I$, $(x_t, x_{t+1}) \in I^2$ for $t \geq 0$.

Then, by using Result 5.2,

$$\psi(x_t, \mu) \geq u(x_t, x_{t+1}, \mu) + \delta\psi(x_{t+1}, \mu)$$

Iterating on this relation

$$\psi(x, \mu) \geq \sum_{t=0}^{T} \delta^t u(x_t, x_{t+1}, \mu) + \delta^T \psi(x_{T+1}, \mu)$$

Again, using the boundedness of $u(\bullet, \mu)$ on I^2 and of $\psi(\bullet, \mu)$ on I, and $0 < \delta < 1$,

$$\psi(x, \mu) \geq \sum_{t=0}^{\infty} \delta^t u(x_t, x_{t+1}, \mu) \tag{5.2}$$

Given (5.1) and (5.2), and the strict concavity of $u(\bullet, \mu)$ on I^2, we can conclude

$$\sum_{t=0}^{\infty} \delta^t u(\hat{x}_t, \hat{x}_{t+1}, \mu) = \psi(x, \mu) > \sum_{t=0}^{\infty} \delta^t u(x_t, x_{t+1}, \mu) \tag{5.3}$$

for every $\{x_t\}_0^\infty$ satisfying $x_0 = x$, $(x_t, x_{t+1}) \in I^2$ for $t \geq 0$, and $x_t \neq \hat{x}_t$ for some $t \geq 1$.

Note that $\{\hat{x}_t\}$ is a *program* from $x \in I$, since $\hat{x}_0 = x$, $\hat{x}_t \geq 0$ for $t \geq 1$ and $\hat{x}_t = \theta(\hat{x}_{t-1}, \mu) \leq \mathbf{f}(\hat{x}_{t-1})$ for $t \geq 1$. Also, by definition of w, we have

$$\sum_{0}^{\infty} \delta^t w(\hat{x}_t, \mathbf{f}(\hat{x}_t) - \hat{x}_{t+1}, \mu) = \sum_{0}^{\infty} \delta^t u(\hat{x}_t, \hat{x}_{t+1}, \mu) \tag{5.4}$$

If $\{x_t\}$ is any program from $x \in I$, then $x_0 = x$, $x_t \in I$ for $t \geq 0$, and

$$\sum_0^\infty \delta^t w(x_t, \mathbf{f}(x_t) - x_{t+1}, \mu) = \sum_0^\infty \delta^t u(x_t, x_{t+1}, \mu) \tag{5.5}$$

Using (5.3), (5.4) and (5.5), we conclude that

$$\sum_0^\infty \delta^t w(\hat{x}_t, \widehat{c}_{t+1}, \mu) > \sum_0^\infty \delta^t w(x_t, c_{t+1}, \mu) \tag{5.6}$$

for every program $\{x_t\}$ from x, for which $x_t \neq \hat{x}_t$ for some $t \geq 1$. Thus $\{\hat{x}_t\}$ is the unique μ-optimal program from x. Consequently, the optimal transition function $h_\mu(x)$ satisfies

$$h_\mu(x) = \mu x(1 - x) \text{ for all } x \in I, \mu \in A$$

8

Non-Linear Dynamics and Chaos in Optimal Growth: A Constructive Exposition

Kazuo Nishimura and Makoto Yano

1. Introduction

In the recent literature, it has been demonstrated that optimal capital accumulation may be chaotic; see Boldrin and Montrucchio (1986), and Deneckere and Pelikan (1986).[1] This finding indicates, as Scheinkman (1990) discusses, that the deterministic equilibrium model of a dynamic economy may explain various complex dynamic behaviors of economic variables, and, in fact, search for such explanations has already begun (see Brock, 1986, and Scheinkman and LeBaron, 1989, for example). In the existing literature, however, not much has yet been revealed with respect to the circumstances under which optimal accumulation exhibits complex non-linear dynamics. In particular, it has not yet been known whether or not chaotic optimal accumulation may appear in the case in which future utilities are discounted not so strongly.[2]

The main result of this chapter is to provide a positive answer to this question. For this purpose, we construct a two-sector model in

---

[1]These studies are followed by the recent work of Boldrin and Deneckere (1990) and Majumdar and Mitra (Chapter 3). For earlier treatments of chaotic economic behaviors, see Benhabib and Day (1982), Day (1982) and Grandmont (1985), the models of which are not of optimal growth.

[2]It has been known that if, given a technology and a period-wise preference, the discounting of future utilities is sufficiently weak, the classical turnpike result holds, i.e., the optimal paths converges to a stationary state; see Cass and Shell (1976), Scheinkman (1976), Brock and Scheinkman (1976) and McKenzie (1983). Sorger (1990), moreover, demonstrates the existence of an upper bound of discount factors with which a given trajectory is a solution to an optimal growth model. These results, however, do not exclude the possibility that for any given degree of the discounting of future utilities, there may be a technology and a period-wise preference that result in chaotic optimal dynamics.

which the objective function $\sum_{t=1}^{\infty} \rho^t u_t$; in the discrete time model of this type, the discount factor of future utilities, ρ, is related to the length of an individual period. We present a constructive method, which indicates the way in which an example of an ergodically chaotic optimal path can be constructed for values of the discount factor ρ close to 1. By using that method, we provide an example in which optimal paths are ergodically chaotic in the case in which future utilities are discounted by about 11% (or, to be precise, for the case of $\rho = 0.886$).[3] In the existing examples or ergodically chaotic optimal accumulation, in contrast, discount factors are set around $\rho = 0.01$.[4]

Another interesting finding of this chapter is that chaotic nonlinear dynamics may appear in a two-sector model in which neither factor intensity reversals nor sign changes in the cross partial derivative of a reduced-form utility function occur. In our model, as in the existing two-sector models of chaotic optimal accumulation, we assume that one sector produces the pure consumption good and that the other sector produces the pure capital good. Unlike the existing studies, however, we assume that both sectors have Leontief production functions; in this case, the factor intensity ranking of the two sectors does not change.[5] We choose the production functions in such a way that the consumption good sector is capital intensive. In this case the optimal transition function, which expresses the optimal choice of an end-of-a-period stock level as a function of the beginning-of-a-period stock level, is monotonically decreasing on the interior of the feasible set (Benhabib and Nishimura, 1985).[6] Within the case of interior optimal

[3]See Deneckere and Pelikan (1986) and Boldrin and Montrucchio (1986b).

[4]Boldrin and Montrucchio (1986b) discuss the possibility of chaotic optimal dynamics for values of ρ up to $\rho = 0.25$; see also Boldrin and Deneckere (1990), which focuses on topological chaos. As is well-known, topological chaos is a weaker form of chaotic dynamics (see Section 3 for an explanation).

[5]Deneckere and Pelikan (1986), in contrast, assume that one sector has a Cobb-Douglas production function and that the other sector has a Leontief production function. In this case, a factor intensity reversal occurs.

[6]Benhabib and Nishimura (1985) demonstrates that if and only if the sign of the cross partial derivative of a reduced form utility function is negative, the optimal transition function is downard sloping. (This result can be derived as a special case of the results obtained by Topkis (1978) and Marshall and Olkin (1979).) Because the existence of a

transition functions, therefore, a factor intensity reversal is necessary so as to guarantee the optimal transition function must lie on the boundary of the feasible set, which is increasing, for any level of beginning-of-a-period stock below a certain cut-off point. Once that cut-off point is passed, the optimal transition function lies on the interior of the feasible set and is monotonically decreasing. This forces the optimal transition function to have a non-linear shape similar to a tent map. We utilize this non-linearity in our examples of chaotic optimal accumulation.[7]

This fact implies that, in our setting, an optimal transition function is non-differentiable.[8] We therefore adopt a sufficient condition of ergodic chaos different from that of Deneckere and Pelikan (1986) and Boldrin and Montrucchio (1986). In those studies, on the one hand, optimal growth models are constructed in such a way that their optimal transition function coincides with quadratic function $k_t = ck_{t-1}(1-k_{t-1})$. In our study, on the other hand, the optimal transition function of the model is shown to be expansive and unimodal.[9] Both that quadratic transition function and an expansive-unimodal transition function are known to generate ergodic chaos (Lasota and Yorke, 1973, and Li and Yorke, 1978, and Collet and Eckmann, 1980).

The following part of this chapter is organized as follows. Section 2 introduces our model. Section 3 discusses the mathematical results on ergodic chaos on which our study is based. Section 4 provides a sufficient condition for an optimal transition function to be ergodically chaotic. Section 5 provides our numerical examples. Section 6 makes concluding remarks.

---

capital intensive consumption good sector implies a negative cross partial derivative, the transition function monotone dressing. Because Benhabib and Nishimura (1985) assumes differentiability, the result of that study does not directly apply to our case, which is non-differentiable. However, a more or less similar relationship appears in our setting, as is shown below.

[7]Boldrin and Deneckere (1990) discusses use of the boundary imposed by the limit of capital depreciation.

[8]On the differentiability of an optimal transition function, see Araujo (1991), Boldrin and Montrucchio (1989) and Santos (1991).

[9]See Section 3 for precise definitions.

2. The Two-Sector Leontief Model

The model we consider is of two sectors, each of which have Leontief production functions (*two-sector Leontief model*). The model is very simple. As we will explain below, however, that simplicity suffers to bring out chaotic optimal paths.

Think of two goods C and K. Good C is a pure consumption good. Each sector uses both good K and labor L as input. Good K is not consumed. Good K input must be made one period prior to the period in which output is produced. In this sense, we may call good K a capital good. Labor input is made in the same period as output is produced. Sectors C and K have Leontief production functions as follows:

$$c_t = \min\{K_{Ct-1}, L_{Ct}/\alpha\}; \tag{2.1}$$

$$y_t = \mu \min\{K_{Kt-1}, L_{Kt}/\beta\}. \tag{2.2}$$

For $i = C, K$, $K_{it-1} \geq 0$ is sector i's capital input in period $t-1$, and $L_{it} \geq 0$ is sector i's labor input in period t. Moreover, $c_t \geq 0$ and $y_t \geq 0$ are outputs in period t. For the sake of simplicity, assume that every period a level of each sector's capital input is freely adjustable. Thus, be denoting k_{t-1}, the aggregate capital input,

$$k_{t-1} = K_{Ct-1} + K_{Kt-1} \geq 0. \tag{2.3}$$

Denote by n_t the aggregate labor input. Then,

$$n_t = L_{Ct} + L_{Kt} \geq 0. \tag{2.4}$$

Assume that labor supply is inelastic and time independent. Goods are normalized so that the labor endowment $\bar{n}$ is equal to 1. Thus,

$$n_t \leq \bar{n} = 1. \tag{2.5}$$

Assume that the consumers' preference is represented by a linear utility function

$$u(c_t) = c_t \geq 0 \tag{2.6}$$

for $c_t \geq 0$ in each period. The *set of feasible input/output combinations* is

$$F = \{(k, n, y, c) \in R_+^4 |\ (2.1) \text{ through } (2.6) \text{ are satisfied}\}. \quad (2.7)$$

The optimal growth model is described by

$$\max_{\{k_{t-1}, n_t, k_t, c_t\}} \sum_{t=1}^{\infty} \rho^t u(c_t)$$
$$\text{s.t. } k_0 \leq \bar{k}, k_t \leq y_t \text{ and } (k_{t-1}, n_t, y_t, c_t) \in F \text{ for } t = 1, 2, ..., \quad (2.8)$$

where discount factor ρ satisfies

$$0 < \rho < 1. \quad (2.9)$$

With respect to parameters ρ, μ, β, and a, assume the following:

$$\mu > 1/\rho; \quad (2.10)$$

$$\beta > \alpha > 0. \quad (2.11)$$

As (2.2) indicates, condition (2.10) implies that if labor input is not binding, the marginal product of capital input in the capital good sector is larger than 1 plus the long-run interest rate. Condition (2.11) implies that the consumption good sector is capital intensive.

An optimal growth model is often expressed with a reduced form utility function, $v(k, y)$, which associates each pair of beginning-of-a-period and end-of-a-period stock levels, (k, y), with the maximum utility that the economy can achieve in a period. That is,

$$v(k, y) = \max_{(n,c)} \text{ s.t. } (k, n, y, c) \in F. \quad (2.12)$$

By using this reduced form function, define the value function

$$V(y) = \max \sum_{t=1}^{\infty} \rho^t v(k_{t-1}, k_t) \text{ s.t. } k_0 = y \quad (2.13)$$

and the *optimal transition correspondence*

$$\begin{aligned} H(k) = \{h \in R | v(k,h) + V(h) \\ \geq v(k,y) + V(y) \text{ for any } y \geq 0 \\ \text{such that } v(k,y) \text{ is well defined.}\}. \end{aligned} \tag{2.14}$$

In the case in which $H(k)$ is a singleton for every k, we call $H(k)$ an *optimal transition function.*

Let $H^t(k) = H(H^{t-1}(k))$, where $H^0(k) = k$. It is easy to demonstrate that if (k_{t-1}, n_t, y_t, c_t), $t = 1, 2, ...$, is an optimal path from $\bar{k}$ solving the optimization problem (2.8), then

$$k_t \in H^t(\bar{k}) \tag{2.15}$$

for any t. By a routine method, the following can be established.

Proposition 2.1: *For each $k \in R$, the optimal transition correspondence, $H(k)$ is non-empty.*

It is useful to derive the explicit form of the reduced form utility function, $v(k,y)$. To this end, fix $c \geq 0$, and construct the indifference curve on the $k - y$ plane that corresponds to utility level $u = c$. Let $k = K_K + K_C$ and $n = L_K + L_C$. By (2.1) and (2.2),

$$y = -\frac{\alpha}{\beta - \alpha}\mu k + \frac{1}{\beta - \alpha}\mu n \geq 0 \tag{2.16}$$

and

$$c = \frac{\beta}{\beta - \alpha}k - \frac{1}{\beta - \alpha}n \geq 0. \tag{2.17}$$

In Figure 1, line PQ illustrates equation (2.16) for $n = 1$; if an only if $n \leq \bar{n} = 1$, (k,y) lies below or one line PQ. Therefore, if (k,y) lies either below or on line PQ, the relationship between c and (k,y) can be obtained by eliminating n from (2.16) and (2.17), i.e., $c = k - y/\mu$. If (k,y) lies above line PQ, the n satisfying (2.16) is $n > 1$. In this case, in order to produce the maximum amount of the consumption good, $n = \bar{n} = 1$ must hold. Thus, by setting $n = 1$ in and eliminating k from (2.16) and (2.17), $c = (1 - \beta y/\mu)/\alpha$.

With these considerations, it can be demonstrated that

$$v(k,y) = \begin{cases} k - \frac{1}{\mu}y & \text{if } y \leq -\frac{\alpha}{\beta-\alpha}\mu k + \frac{1}{\beta-\alpha}\mu \\ \frac{1}{\alpha} - \frac{\beta/\alpha}{\mu}y & \text{if } y \geq -\frac{\alpha}{\beta-\alpha}\mu k + \frac{1}{\beta-\alpha}\mu \end{cases} \tag{2.18}$$

By (2.18) and the non-negativity conditions, the domain of v is

$$D = \{(k,y) \in R_+^2 | y \leq \mu k \text{ if } k \leq 1/\beta \text{ and } y \leq \mu/\beta \text{ if } k \geq 1/\beta\}. \tag{2.19}$$

In Figure 1, D is the region below segment OP and half line PZ. The kinked line OPZ is the indifference curve with $u = c = 0$. Any indifference curve with $u > 0$ is parallel to the kinked line OPZ with the kink on line PQ: $y = -\frac{\alpha}{\beta-\alpha}\mu k + \frac{1}{\beta-\alpha}$.

Because we deal with Leontief production functions, it might appear easier to handle the present model than those of the existing studies. In fact, however, the choice of Leontief production functions creates an extra difficulty.

As (2.18) indicates, the graph of our reduced form utility function consists of flat faces; these faces are called non-trivial von Neumann facets (McKenzie, 1983 and 1986). Under the existence of non-trivial von-Neumann facets, the optimal level of end-of-a-period stock may not be determined uniquely. Therefore, the optimal transition correspondence may not be determined uniquely. Therefore, the optimal transition correspondence may not be written as a function. The existing studies on chaotic optimal dynamics, in contrast, deal with the case in which the existence of an optimal transition function can be taken for granted. Although in that setting, for example, the existence of a periodic optimal path of period 3 automatically implies the existence of a chaotic optimal transition function in the sense of Li and Yorke, it does not in the present setting without guaranteeing that the optimal transition correspondence is actually a function.

Our reduced form utility function is not differentiable on segment PQ. Thus, we cannot express an optimal path by the Euler equation, which plays a major role in the construction of a chaotic optimal path in the existing studies. We cannot use, moreover, the theorem of Benhabib and Nishimura (1985) that relates the sign of the cross partial derivative of $v(k, y)$ to the sign of the slope of the optimal transition function.

In our model, in short, difficulties arise from the non-strict concavity and non-differentiability of the reduced form utility function. We overcome these difficulties by deriving an optimal transition function directly.

3. Chaos: Definitions

This section briefly explains the mathematical result on chaotic dynamics on which this study is based. For the sake of discussion, let I be a non-degenerate interval with endpoints a and b, $a < b$. Call function f a *transition function* if it maps I into itself. Transition function $f : I \to I$ generates a *trajectory of states* $f^t(x)$, $t = 0, 1, 2, \ldots$, for any initial point $x \in I$, where $f^t(x) = f(f^{t-1}(x))$ and $f^0(x) = x$. A transition function, $f : I \to I$, is *expansive* if it is continuous and piecewise twice continuously differentiable and if there is $g > 0$ such that

$$|f'(x)| \geq 1 + g \tag{3.1}$$

for every $x \in I$ at which f has a derivative (Lasota and Yorke, 1973). A transition function, $f : I \to I$ is *unimodal* if it is continuous and if there is $c \in I$, $a < c < b$, such that f is strictly increasing on the interval between a and c and strictly decreasing on the interval between c and b (Collet and Eckmann, 1980). Point $x \in I$ is a *cyclical point* of transition function $f : I \to I$ if there is $t \geq 1$ such that $f^t(x) = x$. In particular, if $f^t(x) = x$ and $f^\tau(x) \neq x$ for $\tau = 1, 2, \ldots, t-1$, x is a *cyclical point of period* t. Call a trajectory starting from a cyclical point a cyclical trajectory.

In the existing literature on chaotic dynamics, various concepts of chaos have been constructed. Correspondingly, various sufficient conditions of chaos have been provided. In this study, among those conditions, we adopt that based on expansiveness, which generates the so-called ergodic chaos. Perhaps the most well-known sufficient condition, which generates the so-called topological chaos, is the cyclicity of period 3 (Li and Yorke, 1975). As is well-known, however, the set in which chaotic behaviors are observable under that condition is of measure zero if there is at least one locally stable cyclical path; in other words, even if a transition function is shown to generate a cyclical path of period 3, topological chaos may be generically non-existent (Collet and Eckmann, 1980, and Grandmont, 1985). In contrast, the existence of ergodic chaos is guaranteed almost everywhere.

Although the existing sufficient conditions of chaotic dynamics differ in specifics from one to another, they typically guarantee such features as (i) the wild oscillations of individual trajectories themselves and (ii) the extreme sensitivity of a trajectory to a small perturbation of a state on that trajectory. In what follows, we explain the property of an expansive and unimodal transition function in these two respects.

Lasota and Yorke (1973) introduce expansiveness and characterize the oscillations of individual trajectories generated by an expansive transition function. Li and Yorke (1978) demonstrates that if a transition function is expansive and unimodal, there exists a unique absolutely continuous measure with respect to which the transition function is ergodic and measure preserving (see Day and Pianigiani, 1991, for a summary). This implies that along almost every trajectory that the transition function generates, the relative frequency of the state visiting any given interval is eventually equal to the area below the density function of the measure on that interval. In other words, those trajectories are ergodic (the time average can be expressed by a space average). For the sake of precise statement, denote by χ_B the characteristic function; $\chi_B(x) = 1$ if $x \in B$ and $\chi_B(x) = 0$ if $x \notin B$.

Definition 3.1: A transition function $f : I \to I$ is *chaotic in the sense of ergodic oscillations* if there is a unique probability measure v that is absolutely continuous and satisfies that for almost every $x \in I$,

$$\lim_{t\to\infty} t^{-1} \sum_{\tau=0}^{t-1} \chi_B(f^\tau(x)) = v(B) \tag{3.2}$$

for any Borel set $B \subset I$.

In the definition of ergodic oscillations, the absolute continuity of measure v is important. The mere existence of a probability measure that satisfies (3.2) does not imply the ergodic oscillations of trajectories. Consider, for example, a transition function that has a unique globally stable fixed point and is flat around the fixed point. Then, all the trajectories reach the fixed point within finite periods. Although such a transition function produces no chaotic trajectory, the probability measure that assigns 1 to that fixed point satisfies (3.2). Note that, in this case, the measure is not absolutely continuous.

Chaos in the sense of ergodic oscillations is primarily concerned with the oscillations of individual trajectories. With respect to the sensitivity of a trajectory to a perturbation of a state, an expansive transition function has the property described below.

Definition 3.2: A transition function$f : I \to I$ is *chaotic in the sense of geometric sensitivity* if there is $g > 0$ such that for any t, there is $\in > 0$ such that $|\Delta x| < \in$ implies

$$|f^\tau(x + \Delta x) - f^\tau(x)| \geq (1 + g)^\tau |\Delta x| \tag{3.3}$$

for $\tau = 1, 2, ..., t$.

The above definition does not imply that the geometric magnification of the effect of a small perturbation lasts indefinitely, which is impossible because f is defined on a bounded interval. Note that geometric sensitivity implies that there is no locally stable cyclical path.

Proposition 3.1: *If a transition function $f : I \to I$ is expansive and unimodal, it is chaotic in the sense of ergodic oscillations and geometric sensitivity.*

Proof: That f is of ergodic oscillations from Li and Yorke (1978). With respect to geometric sensitivity, take an arbitrary point $x \in I$, and consider a small perturbation of x by Δx. Define $f_\tau = f^\tau(x)$ and $\Delta f_\tau = f^\tau(x + \Delta x) - f^\tau(x)$. Then, formally,

$$\left|\frac{f^t(x+\Delta x) - f^t(x)}{\Delta x}\right|$$
$$= \left|\frac{f(f_{t-1} + \Delta f_{t-1}) - f(f_{t-1})}{\Delta f_{t-1}}\right| \left|\frac{f(f_{t-2} + \Delta f_{t-2}) - f(f_{t-2})}{\Delta f_{t-2}}\right|$$
$$\dots \left|\frac{f(x - \Delta x) - f(x)}{\Delta x}\right| \tag{3.4}$$

For any given t, by the continuity of f, it is possible to choose $|\Delta x|$ sufficiently small so that $f_\tau + \Delta f_\tau$ and f_τ belong to a closed interval $[x_{i-1}, x_i]$. Because, in that case, it holds that for any $\tau = 0, 1, ..., t-1$,

$$\left|\frac{f(f_\tau + \Delta f_\tau) - f(f_\tau)}{\Delta f_\tau}\right| \geq 1 + g$$

by (3.1), there is a small $\in > 0$ such that $|\Delta x| < \in$ implies

$$|f^\tau(x + \Delta x) - f^\tau(x)| \geq (1+g)^\tau |\Delta x| \tag{3.5}$$

for any $\tau = 1, 2, ..., t$. Q.E.D.

In what follows, we demonstrate that under certain conditions, the two-sector Leontief model has an expansive-unimodal optimal transition function. In that case, the optimal paths almost always make ergodic oscillations and are always geometrically sensitive to any small initial perturbations for a certain span of periods.

4. Characterization of Chaotic Optimal Transition Functions

We will construct conditions under which the following transition function is the optimal transition function of our model.

$$h(k) = \begin{cases} \mu k & \text{if } 0 \leq k \leq 1/\beta \\ \frac{\alpha}{\beta-\alpha}\mu k + \frac{1}{\beta-\alpha}\mu & \text{if } 1/\beta \leq k \leq \mu/\beta \end{cases} \tag{4.1}$$

The domain of transition function h is the interval

$$I \equiv \{k \in R | 0 \leq k \leq \mu/\beta\}; \tag{4.2}$$

because, as Figure 1 indicates, a level of capital that is higher than μ/β cannot be reproduced, we may restrict our attention to interval I. The graph of transition function $h(k)$ consists of segments OP and PR, where R corresponds the value of h when $k = \mu/\beta$, i.e., $R = (\mu/\beta, h(\mu/\beta))$. Since $\mu > 1$ by (2.9) and (2.10), transition function h is expansive and unimodal under the assumption that

$$\frac{\alpha}{\beta - \alpha}\mu > 1. \tag{4.3}$$

Before starting formal discussions, it may be useful first to explain diagrammatically the idea behind our construction of the optimal transition function. To this end, denote by $\partial_2 v(k, y)$ and $\partial V(k)$, respectively, the partial subgradient of $v(k, y)$ with respect to y and the subgradient of V;

$$\begin{aligned} \partial_2 v(k, y) = \{-p \geq 0 | v(k, y) + py \geq v(\kappa, \eta) + p\eta \\ \text{for all } (\kappa, \eta) \in D; \end{aligned} \tag{4.4}$$

$$\partial V(y) = \{p \geq 0 | V(y) - p2y \geq V(\eta) - p\eta \text{ for all } \eta \geq 0\}. \tag{4.5}$$

Since v and V are concave, continuous and of free disposal, sets $\partial_2 v(k, y)$ and $\partial V(y)$ are well-defined for any $(k, y) \in D$ and for any $y \geq 0$, respectively.

For each given k, $-\partial_2 v(k, y)$ may be thought of as a correspondence that associates y with a closed interval in R. Because $-\partial_2 v(k, y)$ captures the loss in present-period utility caused by a marginal increase in the capital good output, y, we may call the graph of this correspondence a *marginal cost curve* and denote it by $MC_{|k}$. Similarly, $\rho\partial V(y)$, too, is a correspondence that associates y with a closed interval in R. It captures the present value of the gain in future-period utilities due to a marginal increase in the capital good output, y. Thus, we may call the graph of $\rho\partial V(y)$ a *marginal value curve* and denote it by MV. By definition, for each $k \in I$, optimal transition function $H(k)$ is determined at the intersection of MV and $MC_{|k}$ curves; i.e.,

$$H(k) = \{y \geq 0 | [-\partial_2 v(k, y)] \cap [\rho \partial V(y)] \neq \phi\}, \tag{4.6}$$

where $cX = \{z \in R | z - cx \text{ and } x \in X\}$ for any subset X of R and any number c.

Figures 2 and 3 illustrate typical relationships between curves MV and $MC_{|k}$. As (2.18) indicates, the shape of $MC_{|k}$ depends on k. If $0 < k \leq 1/\beta$,

$$-\partial v_2(k, y) = \begin{cases} \{1/\mu\} & \text{if } y < \mu k \\ \{p | p \geq 1/\mu\} & \text{if } y = \mu k \end{cases} \tag{4.7}$$

which is illustrated in Figure 2. If $1/\beta < k \leq \mu/\beta$,

$$-\partial_2 v(k, y) = \begin{cases} \{1/\mu\} & \text{if } y < -\frac{\alpha}{\beta - \alpha}\mu k + \frac{1}{\beta - \alpha}\mu \\ \{p | 1/\mu \leq p \leq (\beta/\alpha)/\mu\} & \text{if } y = -\frac{\alpha}{\beta - \alpha}\mu k + \frac{1}{\beta - \alpha}\mu \\ \{(\beta/\alpha)/\mu\} & \text{if } -\frac{\alpha}{\beta - \alpha}\mu k + \frac{1}{\beta - \alpha}\mu < y < \mu/\beta \\ \{p | p \geq (\beta/\alpha)/\mu\} & \text{if } y = \mu/\beta \end{cases} \tag{4.8}$$

which is illustrated in Figure 3.

Because curve MV is determined endogenously by the solution to the optimization problem (2.8), its specific shape is generally intractable. Curve MV may have vertical and horizontal segments; it may be, like curve $MC_{|k}$, step-shaped. Whether or not this is the case is, however, not important in our argument. What is important is that by the concavity and monotonicity of V, curve MV is "downward sloping" (or, more precisely, that if $p \in \rho \partial V(y)$, $p' \in \rho \partial V(y')$ and $y < y'$, then $p \geq p'$).

Suppose, for the sake of discussion, that function h is the optimal transition function. Take the case of $0 \leq k \leq 1/\beta$. Then, since

$h(k) = \mu k$, curves MV and $MC_{|k}$ must intersect each other on the vertical part of $MC_{|k}$, as is shown in Figure 2. Instead, take the case of $1/\beta < k \le \mu/\beta$. Then, since $h(k) = -\frac{\alpha}{\beta-\alpha}\mu k + \frac{1}{\beta-\alpha}$, curves MV and $MC_{|k}$ must intersect each other on the first vertical segment of $MC_{|k}$, as is shown in Figure 3. Thus, the intersection of curve MV with the kinked line XYZ must be unique and at $y = \mu/\beta$, and that with the kinked line $XABCZ$ must be unique and at $y = \frac{\alpha/\beta}{\beta-\alpha}\mu^2 + \frac{1}{\beta-\alpha}$.

In what follows, we construct a sufficient condition for this condition to hold because, as is demonstrated below, the converse is also true (see Lemma 4.1). To this end, we focus on the case in which $1/\beta$ is a cyclical point of transition function h; as is noted above, the position of curve MV in the general case is intractable. We first demonstrate that function h is the optimal transition function if the cyclical trajectory from $1/\beta$, which h generates, is the unique optimal path from $1/\beta$ (Lemma 4.1). We then construct a condition under which the cyclical trajectory is actually the unique optimal path from $1/\beta$ (Theorem 4.1).

Lemma 4.1: *If $1/\beta$ is a cyclical point of transition function h (it can be of any periodicity) and if the cyclical path from $1/\beta$ is the unique optimal path from $1/\beta$, then h is the optimal transition function.*

Note: The hypothesis of this lemma is stronger than what is needed. The following proof goes through if it is assumed is that $h(1/\beta)$ and $h(\mu/\beta)$ are the unique optimal choices of an end-of-a-period stock levels for beginning-of-a-period stock levels $1/\beta$ and μ/β, respectively.

Proof: It suffices to demonstrate that

$$H(k) = \{h(k)\} \tag{4.9}$$

for every $k \in I$. The hypothesis of the theorem implies

$$H(1/\beta) = \{\mu/\beta\} \tag{4.10}$$

and

$$H(\mu/\beta) = \{h(\mu/\beta)\}, \tag{4.11}$$

the latter of which follows from $\mu/\beta = h(1/\beta)$. By (4.10), (4.6) implies that if the MV and $MC_{|k=1/\beta}$ curves intersect each other only at $y = \mu/\beta$. Since MV is "downward sloping", this implies

$$\rho\partial V(y) > 1/\mu \text{ if } 0 \leq y < \mu/\beta, \tag{4.12}$$

as Figure 2 indicates. Similarly, by (4.11), (4.6) implies that if the MV and $MC_{|k=\mu/\beta}$ curves intersect each other only at $y = h(\mu/\beta)$. Since the MV curve is "downward sloping", this implies

$$\rho\partial V(y) < (\beta/\alpha)/\mu \text{ if } y > h(\mu/\beta). \tag{4.13}$$

In order to complete the proof, take k such that $0 < k < 1/\beta$. Let $\mu < \mu k$. Since this implies $y < \mu\beta$, by (4.12), $\rho\partial V(y) > 1/\mu$. By (4.7), moreover, $y < \mu k$ implies $-\partial_2 v(k, y) = \{1/\mu\}$. These facts imply $-\partial_2 v(k, y) \cap \partial V(y) = \phi$. Thus, $y \notin H(k)$ for any $y < h(k)$. Since Proposition 1 implies $H(k) \neq \phi$ and since $(k, y) \in D$ implies $y \leq \mu k$, by (4.1), (4.9) holds for k such that $0 < k < 1/\beta$.

Next, take k such that $1/\beta < k < \mu/\beta$. Let $y < h(k)$. Then, as above, it can be shown that

$$y < h(k) \text{ implies } y \notin H(k). \tag{4.14}$$

Let $h(k) < y \leq \mu/\beta$. Since this implies $h(k) > h(\mu/\beta)$, by (4.13), $\rho\partial V(y) < (\beta/\alpha)/\mu$. By (4.8), moreover, $h(k) < y \leq \mu/\beta$ implies $-\partial_2 v(k, y) \geq (\beta/\alpha)/\mu$. These facts imply $-\partial_2 v(k, y) \cap \rho\partial V(y) = \phi$. Thus,

$$h(k) < y \leq \mu/\beta \text{ implies } y \notin H(k). \tag{4.15}$$

Since $H(k) \neq \phi$ by Proposition 1, (4.14) and (4.15) implies (4.9) for k such that $1/\beta < k < \mu/\beta$.

Either if $k = 1/\beta$ or if $k = \mu/\beta$, (4.9) holds by (4.10) and (4.11). If $k = 0$, $(k, y) \in D$ implies $y = 0$. Thus, $H(k) = \{0\}$; (4.9) holds. Q.E.D.

We establish the optimality of the path from $1/\beta$, $h^t(1/\beta)$, under the assumption that $1/\beta$ is a cyclical point of transition function h. For this purpose, we use the value-loss method (McKenzie, 1986). We say that vector (q, p) is a *support price vector of activity* (k, y) if

$$v(k, y) + py - \rho^{-1}qk \geq v(\zeta, \xi) + p\xi - \rho^{-1}q\zeta \tag{4.16}$$

for all $(\zeta, \xi) \in D$; recall D is the domain of v. Given an activity $(k, y) \in D$ and a support price vector (q, p) of that activity, we define the *value loss* of an alternative activity $(\zeta, \xi) \in D$ as follows.

$$\begin{aligned} \Delta(\zeta, \xi; q, p; k, y) = v(k, y) + py - \rho^{-1}qk - \\ [v(\zeta, \xi) + p\xi - \rho^{-1}q\zeta]. \end{aligned} \tag{4.17}$$

Then, by (4.16), $\Delta(\zeta, \xi; q, p; k, y) \geq$ for any $(\zeta, \xi) \in D$.

See Figure 1. Along the cyclical path from $1/\beta$, $k_t = h^t(1/\beta)$, activity (k_{t-1}, k_t) lies either at point P, on segment OP but not at the endpoints, or on PQ but not at the endpoints. We first characterize support price vectors in these cases. Denote

$$\gamma = (\beta - \alpha)/\alpha. \tag{4.18}$$

The next lemma is concerned with the case in which (h_{t-1}, h_t) is at point $P = (1/\beta, \mu/\beta)$.

Lemma 4.2: *Price vector (q_0, q_1) is a support price vector of activity $(1/\beta, \mu/\beta)$ if and only if the following conditions are satisfied.*

$$q_0 \geq 0; \tag{4.19}$$

$$-q_0 + \rho\mu q_1 \geq 0; \tag{4.20}$$

$$\gamma q_0 + \rho\mu q_1 \geq \rho(1 + \gamma). \tag{4.21}$$

Suppose, moreover, that (4.19), (4.20) and (4.21) are satisfied with strict inequality. Then, for $(k, y) \in D$,

$$\begin{aligned} \Delta(k, y; q_0, q_1; 1/\beta, \mu/\beta) > 0 \text{ if and only if} \\ (k, y) \neq (1/\beta, \mu/\beta). \end{aligned} \tag{4.22}$$

Proof: By (2.18), $v(1/\beta, \mu/\beta) = 0$. Thus, point $(v, k, y) = (0, 1/\beta, \mu/\beta)$ lies on the graph of function $v = v(k, y)$ at $(k, y) = (1/\beta, \mu/\beta)$. The plane through point $(0, 1/\beta, \mu/\beta)$ is expressed as

$$v = \rho^{-1} q_0 (k - 1/\beta) - q_1 (y - \mu/\beta) \equiv f(k, y). \tag{4.23}$$

In Figure 1, $P = (1/\beta, \mu/\beta)$, $Q = (1/\alpha, 0)$ and $Z = (1 + 1/\beta, \mu/\beta)$. The corresponding utility levels are $v(1/\beta, \mu/\beta) = 0$ at P, $v(1/\alpha, 0) = 1/\alpha$ at Q, and $v(1 + 1/\beta, \mu/\beta) = 0$ at Z. Let $0' = (0, 0, 0)$, $P' = (0, 1/\beta, \mu/\beta)$, $Q' = (1/\alpha, 1/\alpha, 0)$, $Z' = (0, 1 + 1/\beta, \mu/\beta)$ and $K' = (1/\alpha, 1 + 1/\beta, 0)$ in the $v - k - y$ space. Then, by (2.18), the graph of $v = v(k, y)$ consists of triangle $0'P'Q'$ and the face surrounded by lines $P'Q'$, $P'Z'$ and $Q'K'$. By the concavity of $v = v(k, y)$, therefore, the plane defined by (4.23) supports the graph of $v(k, y)$ at point P' if and only if the rays from P' through $0'$, Q' and Z' all lie below the plane defined by (4.23), i.e., if and only if the following inequalities hold.

$$-\rho^{-1} q_0/\beta + q_1 \mu/\beta = f(0, 0) \geq v(0, 0) = 0; \tag{4.24}$$

$$\rho^{-1} q_0 (1/\alpha - 1/\beta) + q_1 \mu/\beta = f(1/\alpha, 0) \geq v(1/\alpha, 0) = 1/\alpha; \tag{4.25}$$

$$\rho^{-1} q_0 = f(1 + 1/\beta, \mu/\beta) \geq v(1 + 1/\beta, \mu/\beta) = 0. \tag{4.26}$$

These expressions are equivalent to (4.19), (4.20) and (4.21).

Moreover, if the inequality conditions in (4.24), (4.25) and (4.26) are satisfied strictly, then $\Delta(k, y; q_0, q_1; 1/\beta, \mu/\beta) = 0$ if and only if $(k, y) = (1/\beta, \mu/\beta)$. This implies (4.22). Q.E.D.

The next lemma is concerned with the case in which (h_{t-1}, h_t) lies, in Figure 1, on segment PQ but not at the endpoints.

Lemma 4.3: *Suppose that* (k_{t-1}, k_t) *satisfies* $k_t = h(k_{t-1})$ *and* $k_{t-1} > 1/\beta$. *Then,* (q_{t-1}, q_t) *is a support price vector of* (k_{t-1}, k_t) *if and only if there is* q_{t-1} *such that*

$$0 \leq q_{t-1} \leq \rho \tag{4.27}$$

and

$$q_t = [-\gamma q_{t-1} + \rho(1+\gamma)]/(\rho\mu). \tag{4.28}$$

Suppose, moreover, that (4.27) is satisfied with strict inequality. Then, for $(k, y) \in D$,

$$\Delta(k, y; q_{t-1}, q_t; k_{t-1}, k_t) > 0 \text{ if and only if } y \neq h(k) \text{ or } k < 1/\beta. \tag{4.29}$$

Proof: First, we will prove that (q_{t-1}, q_t) is a support price vector of (k_{t-1}, k_t) if and only if there is λ, $0 \leq \lambda \leq 1$, such that

$$(\rho^{-1}q_{t-1}, -q_t) = \lambda\left(1, -\frac{1}{\mu}\right) + (1-\lambda)\left(0, -\frac{\beta/\alpha}{\mu}\right). \tag{4.30}$$

Since $(v, k, y) = (v(k_{t-1}, k_t), k_{t-1}, k_t)$ is on the plane defined by $v = k - \frac{1}{\mu}y$, the concavity of $v(k, y)$ implies that

$$v(k_{t-1}, k_t) + \frac{1}{\mu}k_t - k_{t-1} \geq v(k, y) + \frac{1}{\mu}y - k \tag{4.31}$$

for any $(k, y) \in D$. Since, moreover, $(v, k, y) = (v(k_{t-1}, k_t), k_{t-1}, k_t)$ is on the plane defined by $v = \frac{1}{\alpha} - \frac{\beta-\alpha}{\mu}y$, the concavity of $v(k, y)$ implies

$$v(k_{t-1}, k_t) + \frac{\beta/\alpha}{\mu}k_t - 0 \cdot k_{t-1} \geq v(k, y) + \frac{\beta/\alpha}{\mu}y - 0 \cdot k \tag{4.32}$$

for all $(k, y) \in D$. Let $0 \leq \lambda \leq 1$. Multiply λ to expression (4.31) and $(1-\lambda)$ to expression (4.32). By adding the resulting expressions together, the claim at the beginning of the proof follows. Since $q_{t-1} = \rho\lambda$ by (4.30), there is λ, $0 \leq \lambda \leq 1$, such that (4.32) holds if and only if there is q_{t-1} such that (4.27) and (4.28) hold.

See Figure 1. If (k, y) lies in triangle OPQ, (4.31) holds with equality. If (k, y) lies in the region surrounded by lines PZ, PQ and the horizontal axis of Figure 1, (4.32) holds with equality. Let $0 < \lambda < 1$. Then, by construction, $\Delta(k, y; q_{t-1}, q_t; k_{t-1}, k_t) = 0$

if and only if (k, y) lies on both triangle OPQ and the region surrounded by lines PZ, PQ and the horizontal axis, i.e., (k, y) lies on segment PQ. Since $q_{t-1} = \rho\lambda$ by (4.30), $0 < \lambda < 1$ means $0 < q_{t-1} < \rho$. Thus, (4.29) holds. Q.E.D.

Finally, the next lemma is concerned with in the case in which (k_{t-1}, k_t) lies on segment OP but not at the endpoints.

Lemma 4.4: *Suppose that (k_{t-1}, k_t) satisfies $k_t = h(k_{t-1})$ and $0 < k_{t-1} < 1/\beta$. Then, (q_{t-1}, q_t) is a support price vector of (k_{t-1}, k_t) if there is q_{t-1} such that*

$$q_{t-1} \geq \rho \tag{4.33}$$

and

$$q_t = q_{t-1}/(\rho\mu). \tag{4.34}$$

Suppose, moreover, that (4.33) is satisfied with strict inequality. Then, for $(k, y) \in D$,

$$\Delta(k, y; q_{t-1}, q_t; k_{t-1}, k_t) > 0 \;\; \textit{if and only if} \;\; y \neq h(k) \text{ or } k > 1/\beta. \tag{4.35}$$

Proof: We will first prove that (q_{t-1}, q_t) is a support price vector of (k_{t-1}, k_t) if there is $\varphi \geq 1$ such that

$$(\rho^{-1}q_{t-1}, -q_t) = \varphi\left(1, -\frac{1}{\mu}\right). \tag{4.36}$$

Since $k_t = \mu k_{t-1}$ by the hypothesis of the lemma, by (2.18),

$$\frac{1}{\mu}k_t - k_{t-1} \geq \frac{1}{\mu}y - k \tag{4.37}$$

for all $(k, y) \in D$. Since $(v, k, y) = (0, k_{t-1}, k_t)$ is on the plane defined by $v = k - \frac{1}{\mu}y$, the concavity of $v(k, y)$ implies that

$$v(k_{t-1}, k_t) + \frac{1}{\mu}k_t - k_{t-1} \geq v(k, y) + \frac{1}{\mu}y - k \tag{4.38}$$

for any $(k, y) \in D$. By (4.37) and (4.38), for any $\delta \geq 0$, it holds

$$v(k_{t-1}, k_t) + (1+\delta)\frac{1}{\mu}k_t - (1+\delta)k_{t-1}$$
$$\geq v(k, y) + (1+\delta)\frac{1}{\mu}y - (1+\delta)k \tag{4.39}$$

for any $(k, y) \in D$. By defining $\varphi = 1 + \delta$, the claim at the beginning of the proof follows. Since $q_{t-1} = \rho\varphi$ by (4.36), there is $\varphi \geq 1$ such that (4.36) holds if and only if there is $q_{t-1} > \rho$ such that (3.34) holds.

If and only if (k, y) lies below OP, (4.37) holds with strict inequality. Thus, if $\delta > 0$, $\Delta(k, y; q_{t-1}, q_t; k_{t-1}, k_t) = 0$ if and only if $y = \mu k$. Since $q_{t-1} = \rho\varphi = \rho(1+\delta)$, $\delta > 0$ means $q_{t-1} > \rho$. Thus, (4.35) holds. Q.E.D.

The following theorem provides a sufficient condition that guarantees that function h is the optimal transition function. The sufficient condition consists of two parts: A and B. Part A is simply that $1/\beta$ is a cyclical point of period $N \geq 3$. Part B implies that the cyclical trajectory from $1/\beta$ is the unique optimal path from $1/\beta$.[10]

Theorem 4.1: *Function $h : I \to I$ is the optimal transition function, which is expansive and unimodal, if the following conditions are satisfied.*

A: $1/\beta$ is a cyclical point of period $N \geq 3$.

B: There are prices $q_0, q_1, \ldots, q_N$ such that the following is satisfied:

$$q_N = q_0;$$

$$\begin{cases} q_0 > 0 \\ -q_0 + \rho\mu q_1 > 0 \\ \gamma q_0 + \rho\mu q_1 > \rho(1+\gamma) \end{cases}; \tag{ii}$$

[10] The existence of cyclical optimal paths plays important roles in Deneckere and Pelikan (1986), and Boldrin and Deneckere (1990) as well. Those studies characterize a cyclical optimal path by the Euler equation. In contrast, we characterize it by support prices. This difference is due to the fact we focus on non-interior solutions in a non-differentiable model while the existing studies consider interior solutions in a differentiable model.

(iii) Let $k_t = h^t(1/\beta)$ and $t = 2, 3, ..., N$.

$$a: \text{If } k_{t-1} < 1/\beta, \begin{cases} q_t = q_{t-1}/(\rho\mu) \\ q_{t-1} > \rho \end{cases}.$$

$$b: \text{ If } k_{t-1} > 1/\beta, \begin{cases} q_t = [-\gamma q_{t-1} + \rho(1+\gamma)]/(\rho\mu) \\ 0 < q_{t-1} < \rho \end{cases}.$$

(Note that, given condition A, $k_{t-1} \neq 1/\beta$ for $t = 2, 3, ..., N$.)

Proof: By Lemma 4.1, it suffices to prove that path k_t is the unique optimal path from $1/\beta$. By construction, (k_{t-1}, k_t) lies either at point P, on segment OP but not at its endpoints, or on segment PQ but not at its endpoints. For each $\tau = 0, 1, 2, ...$, define $(q_{N\cdot\tau}, q_{N\cdot\tau+1}, ..., q_{N\cdot\tau+N-1}) = (q_0, q_1, ..., q_{N-1})$. Under conditions B-iii and B-iv, by Lemmas 2, 3 and 4, (q_{t-1}, q_t) is a support price vector of activity (k_{t-1}, k_t) for $t = 1, 2, ...$

Define

$$\Delta_t(k, y) = \Delta(k, y; q_{t-1}, q_t; k_{t-1}, k_t). \tag{4.40}$$

Then, for any $(k, y) \in D$,

$$\Delta_t(k, y) \geq 0, \qquad t = 1, 2, ... \tag{4.41}$$

In order to prove the optimality of path k_t, take an arbitrary alternative path h_t, $t = 0, 1, ...$, such that $(h_{t-1}, h_t) \in D$ and $h_0 = 1/\beta$. Then, by the definition of value losses, for any T,

$$\sum_{t=1}^{T} \rho^t (v(k_{t-1}, k_t) - v(h_{t-1}, h_t))$$

$$= \sum_{t=1}^{T} \rho^t \Delta_t(h_{t-1}, h_t) - \rho^T q_T(h_T - k_T). \tag{4.42}$$

Since $q_T \in \{q_0, q_1, ..., q_{N-1}\}$ by definition, by taking $T \to \infty$,

$$\sum_{t=1}^{\infty} \rho^t v(k_{t-1}, k_t) - \sum_{t=1}^{\infty} \rho^t v(h_{t-1}, h_t) =$$

$$\sum_{t=1}^{\infty} \rho^T \Delta_t(h_{t-1}, h_t) \geq 0 \tag{4.43}$$

for any path h_t, $t = 0, 1, \ldots$, such that $(h_{t-1}, h_t) \in D$ and $h_0 = 1/\beta$. This implies that path k_t is optimal.

In order to prove the uniqueness of path k_t, suppose that there is another optimal path h_t, $t = 0, 1, \ldots$, from $1/\beta$. Then, there is t^* such that $h_{t^*} \neq h(h_{t^*-1})$. Under the hypothesis of the theorem, (4.19), (4.20), (4.21), (4.27) and (4.33) are all satisfied with strict inequality. Therefore, by (4.22), (4.29) and (4.35), $h_{t^*} \neq h(h_{t^*-1})$ implies

$$\Delta_t(h_{t^*-1}, h_{t^*}) > 0. \tag{4.44}$$

This implies, however, by (4.41), (4.43) and (4.44),

$$\sum_{t=1}^{\infty} \rho^t v(k_{t-1}, k_t) - \sum_{t=1}^{\infty} \rho^t v(h_{t-1}, h_t) \geq \Delta_t(h_{t^*-1}, h_{t^*}) > 0, \tag{4.45}$$

which contradicts the optimality of path h_t. Q.E.D.

5. Examples

By using Theorem 4.1, we construct examples of chaotic optimal accumulation for the cases in which the orders of periodicity of point $1/\beta$ are 3, 6, 12 and 24. This exercise demonstrates that, roughly speaking, every time the order of periodicity of point $1/\beta$ is doubled, the discount factor with which we find chaotic optimal accumulation is "square-rooted". That is to say, if the order is 3, $\rho \approx 0.363$; if the order is 6, $\rho \approx 0.61 \approx \sqrt{0.363}$; if the order is 12, $\rho \approx 0.78 \approx \sqrt{0.61}$; if the order is 24, $\rho \approx 0.886 \approx \sqrt{0.78}$. (By $x \approx y$, we mean that x and y are approximately equal.)

Note that the conditions that we have imposed on parameters so far are equivalent to the following:

$$0 < \rho < 1; \tag{5.1}$$

$$\rho\mu > 1; \tag{5.2}$$

$$\mu > \gamma > 0; \tag{5.3}$$

$$\alpha > 0. \tag{5.4}$$

(Recall that $\gamma = (\beta - \alpha)/\alpha$.) In addition, we require

$$(\rho\mu)^3 > \gamma^2. \tag{5.5}$$

5.1 Case of Period 3

We start with the case in which $1/\beta$ is a cyclical point of period 3. Because the basic way of handling the cases of higher orders of periodicity is the same as in the case of period 3, we explain this somewhat in detail. As in the above, let $k_t = h^t(1/\beta)$.

First, we construct a condition under which $1/\beta$ is a cyclical point of period 3. See Figure 1. If $1/\beta$ is a cyclical point of period 3, (k_0, k_1) is at point P, (k_1, k_2) is at point R, and (k_2, k_3) is on segment OP but not at its endpoints; (k_3, k_4) is at point P again. Therefore,

$$k_1 > 1/\beta \text{ and } k_2 < 1/\beta; \tag{5.6}$$

note $k_3 = 1/\beta$ by definition. Thus, by using (4.1),

$$h^3(1/\beta) - 1/\beta = -\frac{1/\alpha}{\gamma(1+\gamma)}(\mu - 1)(\mu^2 - \gamma\mu - \gamma). \tag{5.7}$$

Since $\mu > 1$ by (5.1) and (5.2), by (5.1), $1/\beta$ is a cyclical point of period 3 if and only if $\mu^2 - \gamma\mu - \gamma = 0$, i.e.,

$$\mu = \frac{\gamma + \sqrt{\gamma^2 + 4\gamma}}{2}. \tag{5.8}$$

Next, we will construct a condition under which there are prices q_0, q_1, q_2 and q_3 that satisfy condition B of Theorem 4.1. To this end, take an arbitrary q_0. Choose

$$q_3 = q_0 \tag{5.9}$$

so that condition B-i is satisfied. Since $k_2 < 1/\beta$ by (5.6), choose

$$q_2 = q_3(\rho\mu). \tag{5.10}$$

so that (q_2, q_3) satisfies the equation in condition B-iii-a. Since $k_1 > 1/\beta$ by (5.6), choose

$$q_1 = [-\rho\mu q_2 + \rho(1+\gamma)]/\gamma. \tag{5.11}$$

so that (q_1, q_2) satisfies the equation in condition B-iii-b. If and only if the $q_0, ..., q_3$ constructed in this way satisfies

$$q_2 > \rho \tag{5.12}$$

and

$$0 < q_1 < \rho, \tag{5.13}$$

they satisfy conditions B-i and B-iii of Theorem 4.1.

Condition B-ii of Theorem 4.1 imposes a restriction on the choice of (q_0, q_1). Figure 4 depicts the region of (q_0, q_1) lies on the region above lines X and Y and to the right of the vertical axis. Call this region Γ. Denote by A the intersection of lines X and Y. Then,

$$A = (\rho, 1/\mu). \tag{5.14}$$

Equations (5.9), (5.10) and (5.11) imply

$$(\rho\mu)^2 q_0 + \gamma q_1 = \rho(1+\gamma). \tag{5.15}$$

Call the line defined by this equation line L_3. If the q_0, q_1, q_2 and q_3 constructed above satisfy conditions B of Theorem 4.1, line L_3 must intersect region Γ. Since $(\rho\mu)^3 < \gamma^2$ and $\rho\mu > 1$, $(\rho\mu)^2 < \gamma^2$. Therefore, the slope of line L_3, $-(\rho\mu)^2/\gamma$, is flatter than the slope of line Y, $-\gamma/(\rho\mu)$. Thus, line L_3 intersects region Γ if and only if point A lies below line L_3. The latter condition is equivalent to

$$(\rho\mu - 1)[(\rho\mu)^2 + (\rho\mu) - \gamma] < 0. \tag{5.16}$$

Since $\rho\mu > 1$ by (5.2), (5.16) is equivalent to

$$\rho\mu < \frac{-1 + \sqrt{1+4\gamma}}{2}. \tag{5.17}$$

In short, if there are q_0, q_1, q_2 and q_3 that satisfy conditions B of Theorem 4.1, condition (5.17) must be satisfied.

We now demonstrate that if conditions (5.17) is satisfied, we may construct prices $q_0, ..., q_3$ that satisfy condition B of Theorem 4.1. To this end, choose $q_0 = q_3 = \rho$; condition B-i is satisfied. Choose q_2 by (5.10) and q_1 by (5.11). Since $q_0 = \rho$, and since (q_0, q_1) lies on line L_3, as Figure 4 indicates, $q_1 > 1/\mu$. Thus, (q_0, q_1) lies in region Γ; condition B-ii is satisfied. In order to demonstrate that the $q_0, ..., q_3$ satisfy condition B-iii, it suffices to demonstrate that (5.12) and (5.13) are satisfied. By (5.11), $q_1 = \rho + \rho[1 - (\rho\mu)^2]/\gamma < \rho$. Thus, $q_1 > 1/\mu$ implies that (5.13) is satisfied. By (5.10), $q_2 = \rho^2\mu > \rho$, since $\rho\mu > 1$. Thus, (5.12) is satisfied.

The above discussion demonstrates that if parameters ρ, μ and γ satisfy conditions (5.8) and (5.17) as well as (5.1) through (5.5), by Theorem 4.1, function h is the optimal transition function. Our computation indicates that up to $\rho \approx 0.366$, we may find values of μ and γ that satisfy those conditions. The next theorem gives typical parameter values.

Theorem 5.1: *Let* $\rho = 0.363$, $\beta = 3.1$ *and* $\alpha = 1$. *Moreover, let* μ *be that satisfying (5.8) for* $\gamma = \beta - 1$. *Then, function* h *is the optimal transition function, which is expansive and unimodal. Moreover,* $1/\beta$ *is a cyclical point of period 3 of transition function* h.

Proof: As is noted above, it suffices to demonstrate that these values of ρ, μ and γ satisfy (5.8) and (5.17) as well as (5.1) through (5.5). By construction, (5.8) is satisfied. Note $\mu \approx 2.840$. Thus, $\rho\mu \approx 1.031$. Thus, conditions (5.1) through (5.4) are satisfied. Moreover, $(\rho\mu)^3 \approx 1.095$ and $\gamma^2 = 4.41$; (5.5) is satisfied. Finally, $\frac{-1+\sqrt{1+4\gamma^2}}{2\mu} \approx 0.3638 > \rho$; (5.17) is satisfied. Q.E.D.

5.2 Case of Period 6

Recall $k_t = h^t(1/\beta)$. We consider a cyclical path of period 6 that starts from $1/\beta$ and satisfies the following.

$$k_t > 1/\beta \text{ for } t = 1, 3, 4, 5; \tag{5.18}$$

$$k_t < 1/\beta \text{ for } t = 2. \tag{5.19}$$

Then, by the definition of h,

$$h^6(1/\beta) - 1/\beta = -\frac{1/\alpha}{\gamma^4(1+\gamma)} \\ (\mu - 1)(\mu - \gamma)(\mu^4 - \gamma\mu^2 - \gamma^3). \tag{5.20}$$

By (5.20), it is easy to check that $1/\beta$ is a cyclical point of period 6 if, given (5.1), (5.2) and (5.3)

$$\mu^2 = \frac{\gamma + \sqrt{\gamma^2 + 4\gamma^3}}{2}. \tag{5.21}$$

In order to find a price sequence that satisfies conditions B of Theorem 4.1, as before, first take an arbitrary q_1. Let $q_6 = q_0$ so that condition B-i is satisfied. Determine $q_5, \ldots, q_1$ inductively by using the equations in condition B-iii; here, by (5.18), use the equation in condition B-iii-b for $t = 1, 3, 4, 5$ and, by (5.19), the equation in condition B-iii-a for $t = 2$. The $q_0, \ldots, q_6$ constructed in this way satisfies conditions B-i and B-iii of Theorem 1 if and only if they satisfy

$$q_t > 1/\beta \text{ for } t = 1, 3, 4, 5; \tag{5.22}$$

and

$$0 < q_t < 1/\beta \text{ for } t = 2. \tag{5.23}$$

Moreover, the construction of $q_0, \ldots, q_6$ results in

$$(\rho\mu)^5 q_0 - \gamma^4 q_1 = \rho(1+\gamma)(\rho\mu)^4 - \rho\gamma(1+\gamma)(\rho\mu)^3 \\ + \rho\gamma^2(1+\gamma)(\rho\mu)^2 - \rho\gamma^3(1+\gamma). \tag{5.24}$$

This is the counterpart of equation (5.15). Call the line defined by this equation line L_6.

A necessary condition for the price sequence to satisfy conditions B is the line L_6 intersects region Γ. Since (5.5) implies the slope of line L_6, $(\rho\mu)^6/\gamma^4$, is flatter than the slope of line X, $1/(\rho\mu)$, line

L_6 intersects region Γ if point A lies below line L_6. This condition is equivalent to

$$(\rho\mu - 1)(\rho\mu - \gamma)[(\rho\mu)^4 + \gamma^2(\rho\mu)^2 - \gamma^3] > 0. \tag{5.25}$$

As is noted above, given (5.5), $(\rho\mu)^2 < \gamma^2$. Thus, $\rho\mu < \gamma$. Therefore, condition (5.25) is equivalent to

$$(\rho\mu)^2 < \frac{-\gamma^2 + \sqrt{\gamma^4 + 4\gamma^3}}{2}. \tag{5.26}$$

Now, set $q_0 = q_6 = \rho$, as in the case of period 3. Determine, inductively, values of $q_5, ..., q_1$ by following the process indicated above. Because, as in the case of period 3, it is possible to demonstrate that the $q_5, ..., q_1$ satisfy conditions (5.21) and (5.26), the $q_0, ..., q_6$ satisfy condition B of Theorem 4.1. (In order to prove this, a rather long computation is required. One way to dispense with this proof is to compute actual values of $q_0, ..., q_6$ for the numerical values given in Theorem 5.1 and to check that, in fact, the $q_0, ..., q_6$ with those values actually satisfy conditions B.)

The above discussion demonstrates that if parameters ρ, μ and γ satisfy conditions (5.21) and (5.26) as well as (5.1) through (5.5), by Theorem 4.1, function h is the optimal transition function. Our computation indicates that up to $\rho \approx 0.614$, we may find a value of ρ with which μ and γ satisfy those conditions.

Theorem 5.2: *Let $\rho = 0.61$, $\beta = 1.5$ and $\alpha = 1$. Moreover, let μ be that satisfying (5.21) for $\gamma = \beta - 1$. Then, function h is the optimal transition function, which is expansive and unimodal. Moreover, $1/\beta$ is a cyclical point of period 6 of transition function h.*

Proof: As in the case of Theorem 5.1, it suffices to demonstrate that the parameter values chosen above satisfy (5.1) through (5.5) as well as (5.26). Note $\mu \approx 1.654$. Thus, $\rho\mu \approx 1.015$. Thus, conditions (5.1) through (5.4) are satisfied. Moreover, $(\rho\mu)^3 \approx 1.044$ and $\gamma^2 \approx 2.25$; (5.5) is satisfied. Since $\left(\frac{-\gamma^2+\sqrt{\gamma^4+4\gamma^3}}{2\mu}\right)^{1/2} \approx 0.614$, (5.26) is satisfied. Q.E.D.

5.3 Case of Period 12

Recall $k_t = h^t(1/\beta)$. Define the following set of time indices.

$$T_{12} = \{t|t = 1, 3, 4, 5, 7, 9, 11\} \tag{5.27}$$

and

$$T'_{12} = \{t|t = 2, 6, 8, 10\}. \tag{5.28}$$

We consider a cyclical path of period 12 that starts from $1/\beta$ and satisfies the following.

$$k_t > 1/\beta \text{ for } t \in T_{12}; \tag{5.29}$$

$$k_t < 1/\beta \text{ for } t \in T'_{12} \tag{5.30}$$

Then, by the definition of h,

$$h^{12}(1/\beta) - 1/\beta = -\frac{1/\alpha}{\gamma^7(1+\gamma)} (\mu - 1)(\mu - \gamma)(\mu^2 - \gamma)(\mu^8 - \gamma^3\mu^4 - \gamma^5). \tag{5.31}$$

By (5.31), it is easy to check that $1/\beta$ is a cyclical point of period 12 if

$$\mu^4 = \frac{\gamma^3 + \sqrt{\gamma^6 + 4\gamma^5}}{2}. \tag{5.32}$$

In order to find a price sequence that satisfies conditions B of Theorem 4.1, as before, first take an arbitrary q_1. Let $q_{12} = q_0$ so that condition B-i is satisfied. Determine $q_{11}, ..., q_1$ inductively by using the equations in condition B-iii; here, by (5.29), use the equation in condition B-iii-b for $t \in T_{12}$ and, by (5.30), the equation in condition B-iii-a for $t \in T'_{12}$. The $q_0, ..., q_{12}$ constructed in this way satisfies conditions B-i and B-iii of Theorem 4.1 if and only if they satisfy

$$q_t > 1/\beta \text{ for } t \in T_{12}; \tag{5.33}$$

and

$$0 < q_t < 1/\beta \text{ for } t \in T'_{12}. \tag{5.34}$$

Moreover, the construction of $q_0, ..., q_{12}$ results in

$$(\rho\mu)^{11} q_0 + \gamma^7 q_1 = C_{12}, \tag{5.35}$$

where

$$C_{12} = \rho(1+\gamma)[(\rho\mu)^{10} - \gamma(\rho\mu)^8 + \gamma^2(\rho\mu)^6 - \gamma^3(\rho\mu)^4 + \gamma^4(\rho\mu)^3 - \gamma^5(\rho\mu)^2 + \gamma^6]. \tag{5.36}$$

Call the line defined by equation (5.36) line L_{12}.

Given (5.5), the slope of line L_{12}, $-(\rho\mu)^{11}/\gamma^7$, is flatter than the slope of line Y, $-\gamma/(\rho\mu)$. Point A lies below line L_3 if and only if

$$(\rho\mu - 1)(\rho\mu - \gamma)[(\rho\mu)^2 - \gamma][(\rho\mu)^8 + \gamma^2(\rho\mu)^4 - \gamma^5] < 0. \tag{5.37}$$

Since $\rho\mu > 1$ by (5.2), by (5.5), it holds that $\rho\mu < \gamma$ and $(\rho\mu)^2 < \gamma$. Therefore, condition (5.37) is equivalent to

$$(\rho\mu)^4 < \frac{-\gamma^2 + \sqrt{\gamma^4 + 4\gamma^5}}{2}. \tag{5.38}$$

In a way similar to the case of period 6, it is possible to check that there are $q_0, ..., q_{12}$ that satisfy condition B of Theorem 4.1. Therefore, if parameters ρ, μ and γ satisfy conditions (5.32) and (5.38) as well as (5.1) through (5.5), by Theorem 1, function h is the optimal transition function. Our computation indicates that up to $\rho \approx 0.785$, we may find a value of ρ with which μ and γ satisfy those conditions.

Theorem 5.3: *Let $\rho = 0.78$, $\beta = 2.22$ and $\alpha = 1$. Moreover, let μ be that satisfying (5.32) for $\gamma = \beta - 1$. Then, function h is the optimal transition function, which is expansive and unimodal. Moreover, $1/\beta$ is a cyclical point of period 12 of transition function h.*

Proof: By construction (5.32) is satisfied. Note $\mu \approx 1.292$. Thus, $\rho\mu \approx 1.008$. Thus, conditions (5.1) through (5.4) are satisfied. Moreover, $(\rho\mu)^3 = 1.023$ and $\gamma^2 = 1.488$; (5.5) is satisfied.

Finally, $\left[\frac{-\gamma^2+\sqrt{\gamma^4+4\gamma^5}}{2}\right]^{1/4} \approx 0.785$; condition (5.38) is satisfied. Choose $q_0 = q_{12} = 0.78$ to construct prices $q_0, ..., q_{12}$ that satisfy condition B of Theorem 4.1. Q.E.D.

5.4 Case of Period 24

Recall $k_t = h^t(1/\beta)$. Define the following set of time indices.

$$T_{24} = \{t | t = 1, 3, 4, 5, 7, 9, 11, 12, \\ 13, 15, 16, 17, 19, 20, 21, 23\} \tag{5.39}$$

and

$$T'_{24} = \{t | t = 2, 6, 8, 10, 14, 18, 22\}. \tag{5.40}$$

We consider a cyclical path of period 24 that starts from $1/\beta$ and satisfies the following.

$$k_t > 1/\beta \text{ for } t \in T_{24};$$

$$k_t < 1/\beta \text{ for } t \in T'_{24}. \tag{5.42}$$

Then, by the definition of h,

$$h^{24}(1/\beta) - 1/\beta = \frac{1/\alpha}{\gamma^{16}(1+\gamma)}(\mu - 1)(\mu - \gamma) \\ (\mu^2 - \gamma)(\mu^4 - \gamma^3) \\ (\mu^{16} - \gamma^5\mu^8 - \gamma^{11}). \tag{5.43}$$

Therefore, $1/\beta$ is a cyclical point of period 12 if, by (5.43),

$$\mu^8 = \frac{\gamma^5 + \sqrt{\gamma^{10} + 4\gamma^{11}}}{2}. \tag{5.44}$$

In order to find a price sequence that satisfies conditions B of Theorem 4.1, as before, first take an arbitrary q_1. Let $q_{24} = q_0$ so that condition B-i is satisfied. Determine $q_{23}, ..., q_1$ inductively by using the equations in condition B-iii; here, by (5.41), use the equation in condition B-iii-b for $t \in T_{24}$ and, by (5.42), the equation in condition B-iii-a for $t \in T'_{24}$. The $q_0, \ldots, q_{24}$ constructed in this way satisfies conditions B-i and B-iii of Theorem 4.1 if and only if they satisfy

$$q_t > 1/\beta \text{ for } t \in T_{24}; \tag{5.45}$$

and

$$0 < q_t < 1/\beta \text{ for } t \in T'_{24}. \tag{5.46}$$

Moreover, the construction of $q_0, \ldots, q_{24}$ results in

$$(\rho\mu)^{23} q_0 - \gamma^{16} q_1 = C_{24}, \tag{5.47}$$

where

$$\begin{aligned} C_{24} = \rho(1+\gamma)[&(\rho\mu)^{22} - \gamma(\rho\mu)^{20} + \gamma^2(\rho\mu)^{19} - \\ &\gamma^3(\rho\mu)^{18} + \gamma^4(\rho\mu)^{16} - \gamma^5(\rho\mu)^{15} + \gamma^6(\rho\mu)^{14} - \\ &\gamma^7(\rho\mu)^{12} + \gamma^8(\rho\mu)^{11} - \gamma^9(\rho\mu)^{10} + \gamma^{10}(\rho\mu)^8 - \\ &\gamma^{11}(\rho\mu)^6 + \gamma^{12}(\rho\mu)^4 - \gamma^{13}(\rho\mu)^3 + \gamma^{14}(\rho\mu)^2 - \gamma^{15}] \end{aligned} \tag{5.48}$$

Call the line defined by equation (5.49) line L_{24}.

Given (5.5), the slope of line L_{24}, $(\rho\mu)^{23}/\gamma^{16}$, is flatter than the slope of line X, $1/(\rho\mu)$. Thus, line L_{24} intersects region Γ if and only if point A lies below line L_{24}. The latter condition is equivalent to

$$\begin{aligned} &(\rho\mu - 1)(\rho\mu - \gamma)[(\rho\mu)^2 - \gamma] \\ &[(\rho\mu)^4 - \gamma^3] \\ &[(\rho\mu)^{16} + \gamma^6(\rho\mu)^8 - \gamma^{11}] \\ &< 0. \end{aligned} \tag{5.49}$$

Since $\rho\mu > 1$ by (5.2), by (5.5), this implies $\rho\mu < \gamma$ and $(\rho\mu)^2 < \gamma$. Moreover, $(\rho\mu)^2 < \gamma$ implies $(\rho\mu)^4 < \gamma^2 < \gamma^3$ since $\gamma > \rho\mu > 1$. Therefore, condition (5.49) is equivalent to

$$(\rho\mu)^8 < \frac{-\gamma^6 + \sqrt{\gamma^{12} + 4\gamma^{11}}}{2}. \tag{5.50}$$

In a way similar to that of the case of period 6, we may find actual values of $q_0, ..., q_{24}$. Therefore, if parameters ρ, μ and γ satisfy conditions (5.44) and (5.50) as well as (5.1) through (5.5), by Theorem 4.1, function h is the optimal transition function. Our computation in indicates that up to $\rho \approx 0.886$, we may find a value of ρ with which γ and μ satisfy those conditions.

Theorem 5.4: *Let $\rho = 0.886$, $\beta = 2.1$, and $\alpha = 1$. Moreover, let μ be that satisfying (5.44) for $\gamma = \beta - 1$. Then, function h is the optimal transition function, which is expansive and unimodal. Moreover, $1/\beta$ is a cyclical point of period 24 of transition function h.*

Proof: By construction, (5.44) is satisfied. Note $\mu \approx= 1.131$. Thus, $\rho\mu \approx 1.002$. Thus, conditions (5.1) through (5.4) are satisfied. Moreover, $(\rho\mu)^3 = 1.006$ and $\gamma^2 = 1.21$; (5.5) is satisfied. Finally, $\left[\frac{-\gamma^6+\sqrt{\gamma^{12}+4\gamma^{11}}}{2}\right]^{1/8} \approx 0.887$; condition (5.5) is satisfied. Q.E.D.

6. Some Remarks

We have demonstrated that every time the order of periodicity is doubled (starting from 3 through 24), the discount factor with which chaotic optimal accumulation may appear is approximately "square-rooted". This suggests that for the case of period 48, chaotic optimal dynamics may appear for a discount factor around $\rho = \sqrt{0.886} \approx 0.94$, because $\rho \approx 0.886$ for the case of period 24. If, moreover, $1/\beta$ is a cyclical point of period $96 = 48 \cdot 2$, the corresponding discount factor may be around $\rho = \sqrt{0.94} \approx 0.97$. To find such examples would require, however, quite complicated computations and is left for future investigations. (In dealing with the case of period 24, we have found several cyclical paths of period 24, and only one of them, which we report here, have made a significant improvement upon the case of period 12 with respect to the value of ρ.)

The relationship between the order of periodicity and the discount factor agreeing with chaotic optimal accumulation is, at least through period 24, a natural consequence of the upper bounds of discount factors that the conditions of periodicity and optimality impose. Recall that in the case of period 3, by (5.8) and (5.17), ρ must satisfy

$$\rho < \rho' \tag{6.1}$$

where $\rho' = (-1+\sqrt{1+4\gamma})/(\gamma+\sqrt{\gamma^2+4\gamma})$. By (5.32) and (5.38), in the case of period 12, too, ρ must satisfy

$$\rho^4 < \rho'. \tag{6.2}$$

In contrast, in the case of period 6, by (5.21) and (5.26), ρ must satisfy

$$\rho^2 < \rho'', \tag{6.3}$$

where $\rho'' = (-\gamma+\sqrt{\gamma^2+4\gamma^1})/(1+\sqrt{1+4\gamma^2})$. In the case of period 24, by (5.44) and (5.50), ρ must satisfy, again,

$$\rho^8 < \rho''. \tag{6.4}$$

For $N = 0, 1$, this relationship can be summarized as follows. The supremum of discount factors with which transition function h may be optimal is in a neighborhood of $\overline{\rho} \approx \rho^{2^{2N}}$ if $1/\beta$ is a cyclical point of period $3 \cdot 2^{2N}$ and $\overline{\rho} \approx \rho^{2^{2N+1}}$ if $1/\beta$ is a cyclical point of period $3 \cdot 2^{2N+1}$. (Note that ρ' and ρ'' do not correspond to the supremum of discount factors with which the transition function h is optimal; the actual supremum must be defined in relation to the other restrictions on parameters such as $\rho\mu > 1$.) Whether or not this extends for $N > 1$ is an interesting question, to which Nishimura and Yano (1995b) provide a positive answer by demonstrating the existence of ergodically chaotic optimal paths for the case in which the discount factor is arbitrarily close to 1.

In the present model, the possibility of chaotic optimal accumulation depends on the values of discount factor ρ, the marginal produce of capital in the capital good sector, μ, and the degree of a factor intensity differential, γ. Once μ and γ are fixed, the existence of chaotic optimal accumulation for a particular value of ρ does not imply the existence for any value of ρ smaller than that particular value; condition (2.10), i.e., $\rho > 1/\mu$, provides a lower bound of ρ under which chaos may appear. The existence of a similar lower bound has been pointed out in Deneckere and Pelikan (1986) and Boldrin and Montrucchio (1986).

Leontief production functions do not allow for substitutability between inputs. However, our results may be extended to the substitutable cause by using CES production functions. As is well-known, a Leontief production function can be written as the limit of a CES production function. Moreover, the basic structure of optimality in the limit case is similar to that before the limit. It is of interest to investigate if the expansibility of the optimal transition function in the present case extends to the case of CES production functions.

In this chapter, we have focused on the case in which optimal accumulation can be expressed by a one-dimensional transition function. It may be possible, however, to extend part of our results in the case of a higher dimension. In the model with two capital good sectors, for example, the graph of an optimal transition function may consist of several flat faces in the four-dimensional space of beginning- and end-of-a-period stocks. Thus, it may be possible to characterize those faces of the graph in a way similar to that of this study. Such a characterization may reveal, in the multiple-dimensional case, the existence of optimal chaos with respect to a suitably extended concept of chaos; it is straight-forward to extend, for example, expansiveness and geometric sensitivity in the multiple-dimension case.

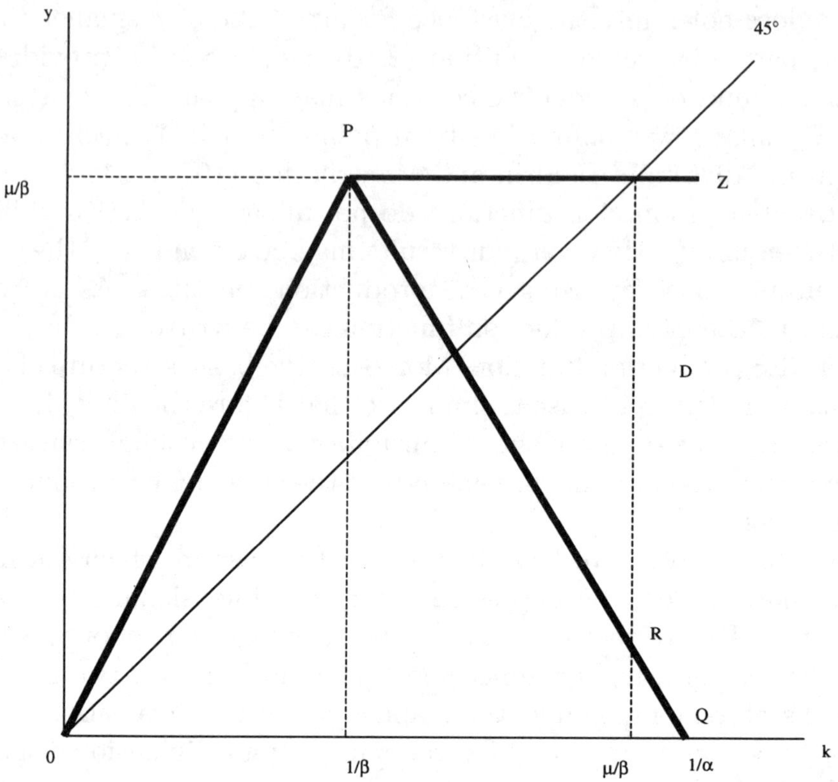

Figure 1

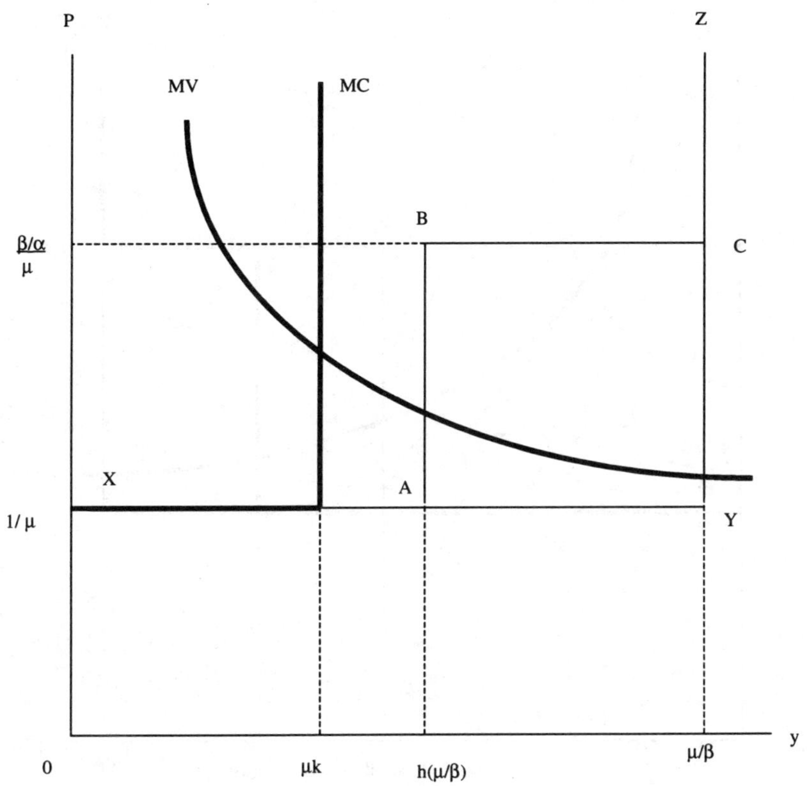

Figure 2

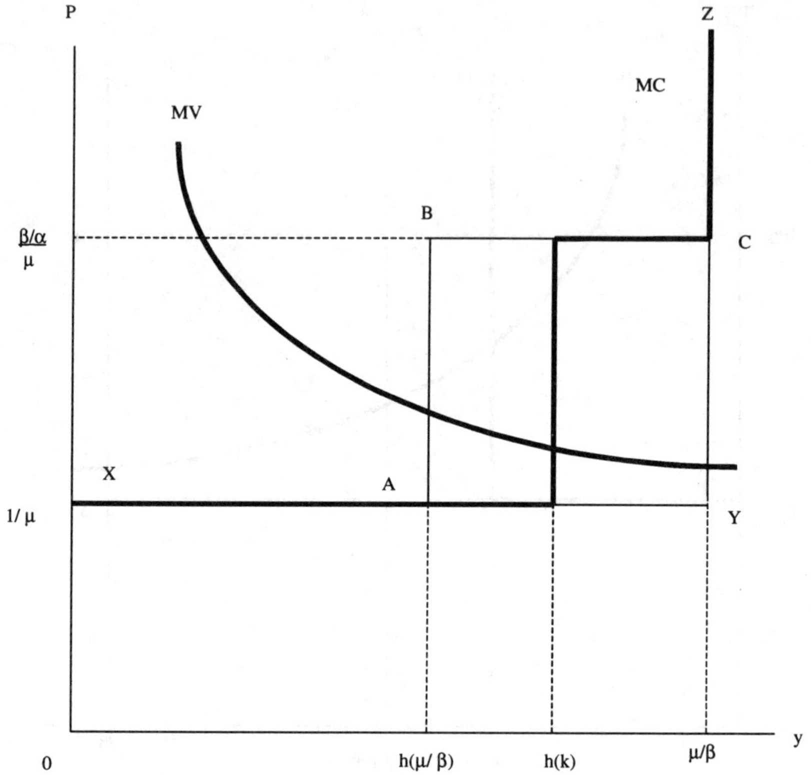

Figure 3

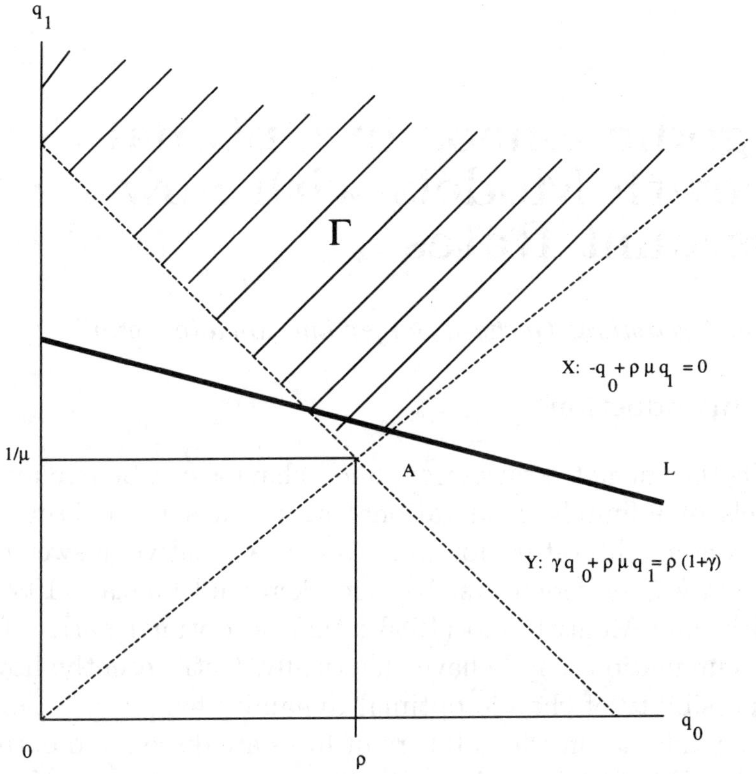

Figure 4

9

Ergodic Chaos in Optimal Growth Models with Low Discount Rates

Kazuo Nishimura, Gerhard Sorger and Makoto Yano

1. Introduction

Whether or not erratic economic behavior can be explained by models of infinitely-lived rational agents has been discussed in the economic literature for some years. A positive answer to this question was provided first by Deneckere and Pelikan (1986) and Boldrin and Montrucchio (1986); that is, optimal paths of capital accumulation may behave chaotically. Until recently, however, the possibility of chaotic optimal dynamics has been established only for the case in which future utilities are discounted extremely heavily. This fact is in line with the intuition obtained from the turnpike literature, namely, that weak myopia tends to simplify the dynamic behavior of optimal growth paths (see, e.g, Brock and Scheinkman (1976), Cass and Shell (1976), McKenzie (1976, 1983), Scheinkman (1976), Yano (1990)). In Sorger (1992a,b, 1995) this intuition is further strengthened by the minimum impatience theorems which demonstrate that for any given dynamical system generating complicated dynamics there is an upper bound for the set of those discount factors with which that system is the optimal policy function of an optimal growth model. These results, however, do not exclude the possibility of chaotic optimal dynamics for arbitrary low myopia. In other words, it remains an open question whether or not there are optimal growth models with arbitrary small discount rates which have chaotic optimal paths.

In a continuous-time framework, Montrucchio (1989) provided an affirmative answer to this question by using a time transformation; the orbits of any dynamical system defined by a sufficiently smooth differential equation can be the optimal paths of a strictly concave model with an arbitrary low discount rate. This sort of "slow chaos" is therefore not contradicting the rational behavior of infinitely-lived and patient economic agents. In discrete-time models, however, the "trick" with the time transformation does not work and only very recently a rigorous proof of the possibility of chaos for low time preference rates was given in a paper by Nishimura and Yano (Chapter 8). Since the present chapter contains results which are closely related to those in Nishimura and Yano (Chapter 8) we would like to point out the similarities as well as the differences between the two approaches.

In Nishimura and Yano (Chapter 8), a sufficient condition is derived for a simple two-sector optimal growth model with Leontief production functions to have an optimal policy function which is "tent-shaped" and which exhibits ergodic chaos. It is then shown that this sufficient condition for ergodic chaos can be met for a sequence of models with discount factors ρ_1, ρ_2, ρ_3, ... such that $\lim_{n\to\infty} \rho_n=1$. The strength of that paper is that the production functions and the (linear) utility function are taken as primitives and that the model is not considered in the so-called reduced form of optimal growth theory. The price one has to pay for this strong result is the analytical complexity of the argument which is based on the existence of periodic optimal growth paths of period 3×2^n for all $n \geq 0$. In the present chapter, we try to circumvent the analytical difficulties of Nishimura and Yano (Chapter 8) by resorting to reduced form optimal growth models. We construct a parametrized family of optimal growth models which have tent-shaped and chaotic optimal policy functions very similar to those in Nishimura and Yano (Chapter 8). However, due to the simpler framework of reduced form models, we are able to prove rigorously that the set $\Pi(\rho)$ of parameter values which give rise to these chaotic optimal policy functions is a non-empty open set for every discount factor ρ in the interval $(0, 1)$. This shows in a conclusive way that also in discrete-time optimal growth models chaos and complicated dynamics are perfectly compatible with rational behavior of patient agents. Moreover, the fact that $\Pi(\rho)$ is an open set in the parameter space implies that our result is robust with respect to small perturbations of the model. The issue of robustness of chaos in optimal growth models has also been addressed in Majumdar and Mitra (Chapter 7).

The plan of the chapter is as follows. In Section 2 we describe the standard reduced form model of optimal growth, present the basic assumptions, and state a few preliminary results from the literature. In Section 3 we construct parametrized examples of the reduced form model that satisfy all the assumptions. We then derive the optimal policy functions for these examples and prove that those functions generate chaotic optimal growth paths. In Section 4 we interpret our examples of the reduced form model in the two-sector setting. Similar to Majumdar and Mitra (Chapter 7) we use utility functions that depend both on the consumption level and the current capital stock, that is, they include a wealth effect. Section 5 presents a few concluding remarks.

2. Assumptions and Definitions

Let us denote by $\mathcal{M}$ the following one-dimensional infinite horizon optimization model:

$$\max \sum_{t=0}^{\infty} \rho^t V(x_t, x_{t+1}) \tag{2.1}$$

$$\text{subject to } x_t \in [0,1],\ t = 0,1,2,\ldots, \tag{2.2}$$

$$(x_t, x_{t+1}) \in T = \{(x,y) \,|\, 0 \le x \le 1\,,\, 0 \le y \le s(x)\}, \quad t = 0,1,2,\ldots. \tag{2.3}$$

This is an instance of the so-called reduced form optimal growth model which has been used extensively in the economic literature (see, e.g., Boldrin and Montrucchio (1986), Deneckere and Pelikan (1986), McKenzie (1976), Sorger (1992b), Stokey and Lucas (1989)). The set $[0,1]$ is the state space of the model (i.e., the set of feasible capital stocks), the set T is the production technology set, the function V is the short-run utility function, and the constant ρ is the discount factor.

A sequence $(x_0, x_1, x_2, \dots)$ of real numbers is called a feasible growth path if conditions (2.2) and (2.3) are satisfied. The set of all feasible growth paths will be denoted by F. Moreover, for every $x \in [0,1]$, we denote by $F(x)$ the set of those feasible growth paths $(x_0, x_1, x_2, \dots)$ which satisfy the initial condition $x_0 = x$.

In this chapter we impose the following assumptions on the model $\mathcal{M}$:

A1: *The production technology frontier* $s : [0,1] \mapsto [0,1]$ *is a continuous, concave, and non-decreasing function which satisfies* $s(0) = 0$ *and* $s(x) > 0$ *for all* $x \in (0,1]$.

A2: *The short-run utility function* $V : T \mapsto \Re$ *is continuous and strictly concave. Moreover, the mapping* $x \mapsto V(x,y)$ *is strictly increasing for every* y*, and the mapping* $y \mapsto V(x,y)$ *is strictly decreasing for every* x.

A3: *The discount factor* ρ *satisfies* $0 < \rho < 1$.

All of these assumptions are standard in the economic literature and require no explanation. Assumption A1, in particular, implies that the set $F(x)$ is non-empty and convex for every $x \in [0,1]$. We call a path $(\bar{x}_0, \bar{x}_1, \bar{x}_2, \dots) \in F$ an optimal path if the inequality

$$\sum_{t=0}^{\infty} \rho^t V(\bar{x}_t, \bar{x}_{t+1}) \geq \sum_{t=0}^{\infty} \rho^t V(x_t, x_{t+1})$$

holds for all feasible paths $(x_0, x_1, x_2, \dots) \in F(\bar{x}_0)$. Assumptions A2 and A3 imply that both infinite series in this inequality are absolutely convergent.

The optimal value function $W : [0,1] \mapsto \Re$ is well-defined by

$$W(x) = \sup \left\{ \sum_{t=0}^{\infty} \rho^t V(x_t, x_{t+1}) \ |(x_0, x_1, x_2, \dots) \in F(x) \right\}$$

and it is a continuous, strictly increasing, and strictly concave function. More specifically, it is the unique continuous function defined on the interval $[0,1]$ which satisfies the Bellman equation

$$W(x) = \sup \{ V(x,y) + \rho W(y) \,|\, 0 \leq y \leq s(x) \} \quad \text{for all } x \in [0,1]. \tag{2.4}$$

The supremum on the right hand side of (2.4) is attained at a unique point $y = h(x) \in [0, s(x)]$ and the function $h : X \mapsto [0,1]$ defined in this way is continuous on $[0,1]$. We call h the optimal policy function of the problem $\mathcal{M}$. It should be noted that the results mentioned so far imply that a unique optimal growth path exists for every initial state $x_0 \in [0,1]$.

We summarize those properties of the model $\mathcal{M}$ which are most important for our analysis in the following proposition (a proof can be found in Stokey and Lucas (1989, Sec. 4.1 and 4.2)).

Proposition 2.1: *Let $\mathcal{M}$ be an optimal growth model satisfying* Assumptions A1–A3. *For every initial state $x \in [0,1]$ there exists a unique optimal growth path $(x_0, x_1, x_2, \dots) \in F(x)$. A feasible growth path $(x_0, x_1, x_2, \dots)$ is an optimal growth path if and only if it satisfies*

$$W(x_t) = V(x_t, x_{t+1}) + \rho W(x_{t+1}) \textit{for all } t = 0, 1, 2, \dots .$$

Here, $W : [0,1] \mapsto \Re$ is the optimal value function of $\mathcal{M}$ and it is the only continuous and strictly concave function on $[0,1]$ which satisfies (2.4).

Proposition 2.1 together with the above remarks implies that the optimal growth paths $(x_0, x_1, x_2, \dots)$ of model (2.1)-(2.3) coincide with the trajectories of the dynamical system

$$x_{t+1} = h(x_t), \tag{2.5}$$

where $h : [0,1] \mapsto [0,1]$ is the optimal policy function of the model. We now introduce a few definitions concerning the qualitative behavior of this dynamical system. For a more elaborate discussion of these terms we refer the reader to Boldrin and Woodford (1990), Day and Pianigiani (1991), and Eckmann and Ruelle (1985).

First of all, let us introduce the notation $h^{(t)}$ for the t-th iterate of h, that is, $h^{(0)}(x) = x$ for all $x \in [0,1]$ and $h^{(t+1)}(x) = h(h^{(t)}(x))$ for all $t = 0, 1, 2, \dots$ and all $x \in [0,1]$. The two central notions that we are interested in are "ergodic chaos" and "geometric sensitivity". Let us first define these concepts in precise mathematical terms and then discuss some of their implications on a more intuitive level.

Definition 2.1: We say that the dynamical system (2.5) exhibits ergodic chaos if there exists an absolutely continuous probability measure μ on the interval $[0, 1]$ which is invariant and ergodic (i.e., indecomposable) under h. Here, a measure μ on the set $[0, 1]$ is said to be absolutely continuous if it has a Radon-Nikodym derivative with respect to the Lebesgue measure, and it is said to be invariant under h if $\mu(\{x \in [0, 1] \,|\, h(x) \in B\}) = \mu(B)$ for all measurable sets $B \subseteq [0, 1]$. The invariant measure μ is said to be ergodic if for every measurable set $B \subseteq [0, 1]$ which satisfies $\{x \in [0, 1] \,|\, h(x) \in B\} = B$ we have either $\mu(B) = 0$ or $\mu(B) = 1$.

Definition 2.2: We say that the dynamical system (2.5) exhibits geometric sensitivity if there is a real constant $\gamma > 1$ such that the following is true: for any $\tau = 0, 1, 2, \ldots$ there exists $\epsilon > 0$ such that for all $x, y \in [0, 1]$ with $|x - y| < \epsilon$ and for all $t \in \{0, 1, \ldots, \tau\}$ it holds that

$$|h^{(t)}(x) - h^{(t)}(y)| \geq \gamma^t |x - y|.$$

The most important property of a dynamical system which exhibits ergodic chaos is that the (asymptotic) statistical properties of the deterministic trajectories can be approximated by the invariant and ergodic measure μ. In other words, time averages along a trajectory $(x_0, x_1, x_2, \ldots)$ can be replaced by space averages in the following form:

$$\lim_{T \to \infty} \frac{1}{T} \sum_{t=0}^{T-1} \phi(h^{(t)}(x)) = \int_{[0,1]} \phi(z)\mu(dz) \tag{2.6}$$

This formula, which is known as the Ergodic Theorem, is valid for μ-almost all initial points $x \in [0, 1]$ and for all continuous functions $\phi : [0, 1] \mapsto \Re$. In the definition of ergodic chaos the absolute continuity of μ is important. The mere existence of an invariant and ergodic measure does not imply complicated dynamics. To see this, consider system (2.5) and assume that h admits a unique and globally asymptotically stable stationary point. It is straightforward to verify that (2.6) holds for all $x_0 \in [0,1]$ and for all continuous functions ϕ if μ denotes the Dirac measure concentrated at the stationary point.

Geometric sensitivity implies that small perturbations of the initial conditions are magnified at a geometric rate over arbitrary but finite time periods. Of course, the geometric magnification cannot last indefinitely because the state space $[0,1]$ of the dynamical system is bounded. Note also that geometric sensitivity implies that there is no stable periodic path of the dynamical system (2.5).

One of the simplest sufficient conditions for the existence of ergodic chaos and geometric sensitivity is stated in the following proposition.

Proposition 2.2: *Assume that there is a number $\hat{x} \in [0,1]$ and a constant $\gamma > 1$ such that $h : [0,1] \mapsto [0,1]$ has the following properties:*

i) h is continuous on $[0,1]$ and twice continuously differentiable on both $(0,\hat{x})$ and $(\hat{x},1)$.

ii) h is strictly increasing on $(0,\hat{x})$ and strictly decreasing on $(\hat{x},1)$, that is, h is unimodal.

iii) $|h'(x)| \geq \gamma$ for all $x \in (0,\hat{x}) \cup (\hat{x},1)$, that is, h is expansive.

Then it follows that the dynamical system (2.5) exhibits ergodic chaos as well as geometric sensitivity.

Proof: The statement concerning ergodic chaos follows from results by Lasota and Yorke (1973) and Li and Yorke (1978) (see also Boldrin and Woodford (1990) and Day and Pianigiani (1991)). The assertion concerning geometric sensitivity is an immediate consequence of assumption iii) of the proposition. We omit the formal proof which can be found in Nishimura and Yano (Chapter 8). Q.E.D.

It should be noted that there are other definitions of chaos available in the literature. One of the most commonly used notions is "topological chaos". It is very easily checked by the Li-Yorke condition (Li and Yorke (1975)) which requires that there exists a periodic orbit of period three. The set of initial conditions of aperiodic and complicated trajectories for a topologically chaotic dynamical system, however, can be small (of Lebesgue measure zero). This means that it could be extremely hard to detect a chaotic trajectory since for almost all initial conditions the system could be well behaved. In the case of ergodic chaos, on the other hand, one knows that the attractor of the dynamical system coincides with the support of the invariant measure μ which cannot be a set of Lebesgue measure zero because of the absolute continuity of μ. Consequently, ergodic chaos has often been called "observable" chaos since it is more easy to detect by numerical simulation than topological chaos.

3. The Main Result

We now construct a parametric family of optimal growth models. On the interior of the feasible set, the reduced form utility function of a model in this family is twice continuously differentiable (in fact, quadratic) except at a downward sloping straight line. We choose the parameter values in such a way (i) that the graph of the optimal policy function $h(x)$, wherever it is in the interior of the feasible set, lies on that downward sloping line, (ii) that the rest of the graph lies on that part of the boundary of the feasible set which starts from the origin and is upward sloping, and (iii) that the resulting optimal policy function is unimodal and expansive.

This method of deriving an ergodically chaotic optimal policy function is similar to that of Nishimura and Yano (Chapter 8) except that, in this study, we introduce an artificial boundary to a predefined reduced form utility function. The use of an artificial boundary as a part of the graph of the optimal policy function is adopted from an earlier work of Nishimura and Yano (Chapter 4). In Section 4 below, however, we shall show how the reduced form model can be interpreted as a two-sector optimal growth model and how the artificial boundary arises naturally from the assumption of Leontief production technologies in both sectors.

We denote by $H_{\leq}$ the half space of $\Re^2$ defined by the inequality $y \leq 2 - \gamma x$ and by $H_{\geq}$ the half space defined by $y \geq 2 - \gamma x$. The optimal growth models are parametrized by the constants α, β, γ, and δ as well as by the discount factor ρ. More specifically, the model $\mathcal{M}_{\alpha,\beta,\gamma,\delta,\rho}$ corresponding to these parameters consists of the artificial production technology frontier $s(x)$ and the reduced form utility function $V(x,y)$ defined as follows (compare with (2.1)-(2.3)):

$$s(x) = \min\{\gamma x, 1\} \tag{3.1}$$

$$V(x,y) = \begin{cases} (\beta - \frac{2}{\gamma})x - (\alpha - 1)x^2 - (\beta\rho - \frac{2}{\gamma^2})y \\ -(\frac{1}{\gamma^2} - \alpha\rho)y^2\,;\ \text{if } (x,y) \in H_{\leq} \\ (4\gamma\delta + \beta)x - (\alpha + \gamma^2\delta)x^2 - (\beta\rho - 4\delta)y - \\ (\delta - \alpha\rho)y^2 \\ -2\gamma\delta xy - 4\delta\,;\ \text{if } (x,y) \in H_{\geq} \end{cases} \tag{3.2}$$

It should be noted that the parameter ρ, which is the discount factor of the model, also appears in the short-run utility function V. First, we have to make sure that $\mathcal{M}_{\alpha,\beta,\gamma,\delta,\rho}$ satisfies Assumptions A1-A3.

Lemma 3.1: *Under the assumptions*

$$\gamma^2\rho < \frac{1}{\alpha} < 1\,,\ \beta > 2\max\left\{\frac{1}{\gamma^2\rho}, \alpha + \gamma\delta(\gamma - 1)\right\},$$
$$1 < \gamma < 2\,,\ \delta > \frac{\alpha\rho}{1-\gamma^2\rho}, 0 < \rho < 1 \tag{3.3}$$

it follows that the optimal growth model $\mathcal{M}_{\alpha,\beta,\gamma,\delta,\rho}$ satisfies Assumptions A1-A3.

Proof: Assumptions A1 and A3 are trivial to verify. To see that A2 is also satisfied, we first show that V is continuous. Since discontinuities can only occur along the line $y = 2 - \gamma x$, it is sufficient to prove that the expression on the first line of (3.2) coincides with the expression on the second line of (3.2) whenever $y = 2 - \gamma x$ holds. Substituting $y = 2 - \gamma x$ into (3.2) we obtain on both lines the same expression, namely,

$$V(x, 2-\gamma x) = -\alpha(1-\gamma^2\rho)x^2 + \\ (\beta + \beta\gamma\rho - 4\alpha\gamma\rho)x - 2(\beta - 2\alpha)\rho.$$

To prove strict concavity of V we proceed in three steps. First, we show that V is strictly concave on the half plane $H_{\leq}$, then we show that it is strictly concave on the half plane $H_{\geq}$, and finally we show that V is strictly concave along every straight line crossing the line $y = 2 - \gamma x$. If $(x, y) \in H_{\leq}$ then the Hessian of V is given by

$$D_2V = 2 \begin{pmatrix} 1-\alpha & 0 \\ 0 & \alpha\rho - \dfrac{1}{\gamma^2} \end{pmatrix}$$

which is negative definite if (3.3) is satisfied. This proves strict concavity on $H_{\leq}$. Similarly, if $(x, y) \in H_{\geq}$, then the Hessian of V is given by

$$D_2V = 2 \begin{pmatrix} -(\alpha + \gamma^2\delta) & -\gamma\delta \\ -\gamma\delta & -(\delta - \alpha\rho) \end{pmatrix}$$

which is also negative definite if (3.3) is satisfied. This shows that V is strictly concave on $H_{\geq}$. Finally, let us compute the directional derivative $V'(x, y \,|\, \xi, \eta)$ at a point $(x, y) = (x, 2 - \gamma x)$ into the direction (ξ, η). We obtain

$$V'(x, 2 - \gamma x \,|\, \xi, \eta) = \left[\beta - \frac{2}{\gamma} - 2(\alpha - 1)x\right] \xi + \left[4\alpha\rho - \beta\rho - \frac{2}{\gamma^2} + 2\left(\frac{1}{\gamma} - \alpha\gamma\rho\right) x\right] \eta$$

if $\gamma\xi + \eta \leq 0$ and

$$V'(x, 2 - \gamma x \,|\, \xi, \eta) = [\beta - 2\alpha x]\, \xi + [4\alpha\rho - \beta\rho - 2\alpha\gamma\rho x]\, \eta$$

if $\gamma\xi + \eta \geq 0$. To prove that V is strictly concave along every straight line crossing the line $y = 2 - \gamma x$ it is sufficient to show that for every vector $(\xi, \eta) \in \Re^2$ it holds that

$$V'(x, 2 - \gamma x \,|\, \xi, \eta) + V'(x, 2 - \gamma x \,|\, -\xi, -\eta) \leq 0. \tag{3.4}$$

Without loss of generality we may assume that $\gamma\xi + \eta \leq 0$ (otherwise we replace the vector (ξ, η) by $(-\xi, -\eta)$). Substituting the expressions for the directional derivatives from above into (3.4) we obtain

$$V'(x, 2 - \gamma x \mid \xi, \eta) + V'(x, 2 - \gamma x \mid -\xi, -\eta) = \frac{2}{\gamma^2}(\gamma \xi + \eta)(\gamma x - 1). \tag{3.5}$$

Now observe that for every point (x, y) in the production technology set T we must have $0 \leq y \leq s(x) \leq 1$. Consequently, if the point (x, y) lies in T and on the line $y = 2 - \gamma x$, then it must hold that $y = 2 - \gamma x \leq 1$ or, equivalently, $\gamma x - 1 \geq 0$. This shows that the expression in (3.5) is indeed non-positive and the proof of the strict concavity of V is completed.

Finally, to prove the monotonicity part of Assumption A2, we calculate the partial derivatives of V. This yields

$$\begin{aligned} V_x(x,y) &\geq \beta - \frac{2}{\gamma} - 2(\alpha - 1) && \text{for } (x,y) \in H_{\leq} \\ V_x(x,y) &\geq \beta - 2\alpha - 2\gamma\delta(\gamma - 1) && \text{for } (x,y) \in H_{\geq} \\ V_y(x,y) &\leq \frac{2}{\gamma^2} - \beta\rho && \text{for } (x,y) \in H_{\leq} \\ V_y(x,y) &\leq 2\alpha\rho - \beta\rho && \text{for } (x,y) \in H_{\geq} \end{aligned}$$

To derive the last line we again used that $y \geq 2 - \gamma x$ and $\gamma x \geq 1$ whenever $(x, y) \in T \cap H_{\geq}$. Now it is straightforward to verify that the assumptions in (3.3) imply that $V_x(x, y) > 0$ and $V_y(x, y) < 0$ for all $(x, y) \in T$. This completes the proof of the lemma. Q.E.D.

It will be shown shortly that the conditions of the above lemma can be satisfied. Before we do that, however, we derive the optimal policy function of $\mathcal{M}_{\alpha,\beta,\gamma,\delta,\rho}$.

Lemma 3.2: *The optimal value function of model $\mathcal{M}_{\alpha,\beta,\gamma,\delta,\rho}$ is given by $W(x) = \beta x - \alpha x^2$ and the optimal policy function is given by*

$$h(x) = \begin{cases} \gamma x & \text{for } x \in [0, 1/\gamma] \\ 2 - \gamma x & \text{for } x \in [1/\gamma, 1] \end{cases} \tag{3.6}$$

Proof: We use Proposition 2.1 to verify the claim. To this end we compute the right hand side of the Bellman equation (2.4). Consider first the case $(x, y) \in H_{\leq}$. In this case we have

$$\begin{aligned} V(x,y) + \rho W(y) &= (\beta - \frac{2}{\gamma})x - (\alpha - 1)x^2 + \frac{2}{\gamma^2}y - \frac{1}{\gamma^2}y^2 \\ &= W(x) - \frac{1}{\gamma^2}(y - \gamma x)[y - (2 - \gamma x)] \end{aligned}$$

It follows from this equation that $V(x, y) + \rho W(y) \leq W(x)$ for all $(x, y) \in T \cap H_{\leq}$ with equality holding if and only if $y = h(x)$. Now consider the case $(x, y) \in H_{\geq}$. Here, we have

$$\begin{aligned} V(x,y) + \rho W(y) = (4\gamma\delta + \beta)x - (\alpha + \gamma^2\delta)x^2 + \\ 4\delta y - \delta y^2 - 2\gamma\delta xy - 4\delta \\ = W(x) - \delta[y - (2 - \gamma x)]^2 \end{aligned}$$

As in the first case we see that $V(x, y) + \rho W(y) \leq W(x)$ for all $(x, y) \in T \cap H_{\geq}$ with equality holding if and only if $y = h(x)$. In conclusion, we have shown that the function W satisfies the Bellman equation (2.4) and that the supremum on the right hand side of (2.4) is obtained uniquely at $y = h(x)$ as defined in (3.6). These results imply that W is the optimal value function of $\mathcal{M}_{\alpha,\beta,\gamma,\delta,\rho}$ and, according to Proposition 2.1, that the function h from (3.6) is the optimal policy function of $\mathcal{M}_{\alpha,\beta,\gamma,\delta,\rho}$. This completes the proof. Q.E.D.

Our next step is to show that the optimal policy function of model $\mathcal{M}_{\alpha,\beta,\gamma,\delta,\rho}$ exhibits ergodic chaos. This is an easy application of the conditions stated in Proposition 2.2.

Lemma 3.3: *Under the parameter restrictions* (3.3) *it follows that the optimal policy function of $\mathcal{M}_{\alpha,\beta,\gamma,\delta,\rho}$ exhibits ergodic chaos.*

Proof: It follows from Lemma 3.2 that the optimal policy function is given by (3.6). Since $\gamma > 1$ was assumed in (3.3) it is easily seen that the conditions of Proposition 2.2 are satisfied with $\hat{x} = 1/\gamma$. Q.E.D.

We are now ready to prove the main result of this chapter. It states that we can find optimal growth models which generate chaotic optimal growth paths for discount factors arbitrary close to 1.

Theorem 3.1: *For every $\rho \in (0,1)$ there exists a non-empty open set $\Pi(\rho) \subset \Re^4$ such that for all $(\alpha, \beta, \gamma, \delta) \in \Pi(\rho)$ the following is true: the optimal growth model $\mathcal{M}_{\alpha,\beta,\gamma,\delta,\rho}$ has an optimal policy function which exhibits ergodic chaos.*

Proof: Let $\rho \in (0,1)$ be a given discount factor. To find parameter values α, β, γ, and δ which satisfy (3.3) we first choose $\gamma \in (1,2)$ such that $\gamma^2\rho < 1$. This is always possible since $1/\sqrt{\rho} > 1$ for all $\rho \in (0,1)$. Next we choose any α satisfying $1 < \alpha < 1/(\gamma^2\rho)$ which is also possible by our choice of γ. Now it is easy to see that we can satisfy (3.3) by choosing the remaining parameters δ and β sufficiently large. Therefore, the set $\Pi(\rho)$ defined by (3.3) is not the empty set. Moreover, since the conditions in (3.3) are strict inequalities this set $\Pi(\rho)$ must be open. The theorem follows now from Lemma 3.3. Q.E.D.

4. A Two-Sector Model

One way to interpret a reduced form model such as the one analyzed in the previous sections is to consider individual processes of production and consumption. In this section we provide such an interpretation by constructing a two-sector economy with a representative consumer which gives rise to the reduced form model discussed above.

To this end, consider a (pure) consumption good and a (pure) capital good sector each of which has a Leontief production technology. Denote their production functions as follows:

$$c_t = \min\{K_{C,t-1}, 2L_{C,t}/\gamma\} \geq 0, \tag{4.1}$$

$$y_t = \min\{\gamma K_{K,t-1}, L_{K,t}\} \geq 0. \tag{4.2}$$

Note that c_t is the amount of consumption in period t and that y_t is that of capital good output in period t. Moreover, $(K_{C,t-1}, L_{C,t})$ and $(K_{K,t-1}, L_{K,t})$ are, respectively, the vectors of capital and labor inputs that the two sectors use in order to produce outputs in period t. The labor endowment of the economy is constant over time and assumed, by suitable normalization, to be unity. Thus,

$$L_{C,t} + L_{K,t} \leq 1. \tag{4.3}$$

The aggregate level of capital good input cannot exceed the level of the capital good existing at the beginning of the period, say x_{t-1}, i.e.,

$$K_{C,t-1} + K_{K,t-1} \leq x_{t-1}. \tag{4.4}$$

Given the production processes described above it is possible to write down the maximal consumption good output as a function of input and output levels of the capital good, (x_{t-1}, y_t). That is,

$$g(x_{t-1}, y_t) = \max c_t \text{ subject to conditions (4.1) through (4.4).} \tag{4.5}$$

By actually solving (4.5) we obtain the explicit forms of the function $g : T \to \Re$ and its domain $T \subset \Re^2$ as follows:

$$g(x_{t-1}, y_t) = \begin{cases} x_{t-1} - \dfrac{1}{\gamma} y_t & \text{if } y_t \leq 2 - \gamma x_{t-1} \\ \dfrac{2}{\gamma} - \dfrac{2}{\gamma} y_t & \text{if } y_t \geq 2 - \gamma x_{t-1}, \end{cases} \tag{4.6}$$

and

$$T = \{(x_{t-1}, y_t) \in \Re^2_+ \mid y_t \leq s(x_{t-1})\},$$

where the function s is defined by (3.1).

In order to relate the output level of the consumption good, $c_t = g(x_{t-1}, y_t)$, to the utility of the representative consumer, u_t, we adopt a utility function considered by Majumdar and Mitra (Chapter 3). That is, we assume that utility u_t in period t depends both on the consumption c_t as well as on the stock of the capital good existing at the beginning of the period, x_{t-1}, in the form $u_t = u(c_t, x_{t-1})$. With this utility function consider the two-sector optimal growth model described by

$$\max \sum_{t=1}^{\infty} \rho^{t-1} u(c_t, x_{t-1}) \tag{4.7}$$

subject to conditions (4.1)-(4.4) and $x_t \leq y_t$.

With these preparations the next result follows immediately.

Theorem 4.1: *Let the functions s, V, and g be given by conditions* (3.1), (3.2), *and* (4.6) *and suppose that the utility function u is defined so as to satisfy $V(x, y) = u(g(x, y), x)$. Then it follows that the two-sector optimal growth model described by* (4.7) *and the reduced form optimal growth model described by* (2.1)-(2.3) *are equivalent. In other words, both models have the optimal policy function h specified in* (3.6) *and, consequently, both models have identical optimal paths.*

Our result can be compared with those of Nishimura and Yano (Chapter 8) and Majumdar and Mitra (Chapter 3). Similar as (Chapter 3), the present chapter is concerned with a non-standard utility function which depends not only on consumption levels but also on the level of the capital good. In the traditional literature on dynamic two-sector models it has been assumed that a utility function depends simply on consumption levels. For a study of that case we refer to Nishimura and Yano (Chapter 8) which demonstrates a result similar to that of the present chapter (i.e., the possibility of chaotic optimal accumulation for discount factors arbitrary close to 1).

Both this chapter and Nishimura and Yano (Chapter 8) are built on the ergodic chaos that appears from an expansive and unimodal optimal policy function. These studies, moreover, utilize the possibility that an optimal path hits the boundary of the feasible set, $s(x)$. In Majumdar and Mitra (Chapter 7), on the other hand, ergodic chaos is generated by the logistic function and the optimal paths lie in the interior of the feasible set.

5. Concluding Remarks

In this chapter we have shown that reduced form optimal growth models with strictly concave utility functions can have ergodically chaotic optimal paths even if the discount factor is arbitrarily close to 1. Furthermore, we have shown that the reduced form models used in our study can be interpreted as two-sector optimal growth models with a non-standard utility function that depends on both the consumption level and the capital stock. To summarize our results, we can say that there is no contradiction between chaotic behavior and rationality of infinitely lived economic agents regardless of their rate of time preference.

At first glance one might think that this result is not compatible with the turnpike theorems and the minimum impatience theorems (see the references given in Section 1). However, as we have already pointed out in the Introduction there is no contradiction since the turnpike theorems consider the optimal growth model to be fixed and the minimum impatience theorems consider the optimal policy function as fixed. In the examples of the present study, on the other hand, we consider a whole family of optimal growth models and every member of this family generates a different optimal policy function. Moreover, we will show in this section that the intuition obtained from turnpike theory and minimum impatience theorems remains valid in the present example since the optimal policy functions become less and less chaotic as the discount factor approaches one. To make this claim precise we have to be able to measure how chaotic a dynamical system is. In the literature on non-linear dynamical systems and chaos one can find quite a number of different indicators of chaos and we shall discuss four of them in the following paragraphs.

As a first example of a chaos indicator let us consider the Lyapunov exponent. It measures the average exponential divergence of initially nearby points under the iteration of the given mapping. For a one-dimensional map like the one in (3.6) the Lyapunov exponent can be defined as follows (see, e.g., Rasband (1990))

$$\ell(x) = \lim_{t\to\infty} \frac{1}{t} \ln \left| \frac{dh^{(t)}(x)}{dx} \right|,$$

where $h^{(0)}(x) = x$ and $h^{(t+1)}(x) = h(h^{(t)}(x))$ for all $t = 0, 1, 2, \ldots$. This formula is valid for (Lebesgue) almost all points $x \in [0,1]$, namely, for all those points which are not eventually mapped onto the critical point $x = 1/\gamma$ where h is not differentiable. Because the mapping (3.6) is piecewise linear with the slope satisfying $|h'(x)| = \gamma$ it is easy to see that we have $\ell(x) = \ln\gamma$ for almost all initial points $x \in [0,1]$. From (3.3) we obtain $1 < \gamma < \rho^{-1/2}$ so that γ approaches one as the discount factor ρ converges to one. This implies that $\lim_{\rho\to 1} \ell(x) = 0$ for almost all $x \in [0,1]$. In other words, the optimal policy function of the model $\mathcal{M}_{\alpha,\beta,\gamma,\delta,\rho}$ becomes less and less sensitive to perturbations of the initial values when the decision maker becomes infinitely patient.

Next consider the size of the attractor of the mapping (3.6). Since $h(x) > x$ for all $x \in (0, 1/\gamma)$ it must hold that trajectories are monotonically increasing as long as they stay in the interval $(0, 1/\gamma)$. They can only become chaotic because they are eventually mapped into the interval $[1/\gamma, 1]$. However, since $h([1/\gamma, 1]) = [2-\gamma, 1]$ it follows that no trajectory starting at a strictly positive initial point can have a limit point outside the interval $[2-\gamma, 1]$. This implies that the chaotic attractor of the mapping (3.6) must be contained in this interval and consequently, it must have a Lebesgue measure smaller or equal to $\gamma - 1$. Since we have already seen that the parameter restrictions in (3.3) imply that $\gamma \to 1$ as $\rho \to 1$ we conclude the Lebesgue measure of the chaotic attractor converges to zero as the discount factor approaches one. One can interpret this result also as an indication that the dynamics of the optimal growth paths of $\mathcal{M}_{\alpha,\beta,\gamma,\delta,\rho}$ become more regular when the rate of time preference is decreased.

Our third example of a chaos indicator is the entropy. Loosely speaking the entropy of a mapping measures the mean rate at which points which are so close to each other that they are practically indistinguishable become distinguishable as one follows the trajectories they generate. In other words, the entropy measures the rate at which information is created by successive application of the given mapping. For a formal definition of entropy we refer to Eckman and Ruelle (1985), Frank and Stengos (1988) and Rasband (1990). There are different versions of entropy used in the literature and we restrict ourselves to the Kolmogorov entropy K. Under quite general assumptions it can be shown (see Eckman and Ruelle (1985)) that it coincides with the sum of the positive Lyapunov exponents which in our case implies $K = \ln \gamma$. This identity proves that the Kolmogorov entropy converges to zero as the discount factor approaches one and we obtain the same result as in the previous cases, namely, that the mapping (3.6) becomes less chaotic the more patient the decision maker is. In this regard we would like to mention that a very general result relating the discount factor of an optimal growth model satisfying strong concavity assumptions to the Kolmogorov entropy of the corresponding optimal policy function can be found in Montrucchio (1994).

Finally, a fourth possibility to rank different mappings according to their complexity is motivated by Sarkovskii's theorem (see Sarkovskii (1964) or Devaney (1986)). This famous result shows that a mapping from an interval into itself that possesses a periodic point of minimal period p also possesses periodic points of all periods q which are greater than or equal to p with respect to the Sarkovskii order $\leq_S$. The smallest numbers in this complete ordering of the positive integers are the odd numbers greater than one starting with three and proceeding in ascending order, that is, $3 \leq_S 5 \leq_S 7 \leq_S \ldots$. Because of this result one could say that a mapping which admits a period-three-cycle is "more chaotic" than a mapping which does not. As for the example discussed in the present chapter it is easy to check that a the optimal policy function given in (3.6) admits a period-three-cycle if and only if $\gamma \geq (1+\sqrt{5})/2$. Because of the parameter restrictions in (3.3) this is possible if and only if the discount factor ρ is smaller than 0.382. This finding is in accordance with a much more general result concerning the optimality of period-three-cycles (see Sorger, 1994a). As in Nishimura and Yano (Chapter 8) one could try to find larger values for ρ and smaller values for γ such that (3.3) is still satisfied and such that the function h from (3.3) admits a period-p-cycle with $p \neq 3$. It can be conjectured that one can find a feasible value for ρ arbitrary close to 1 as one increases p in the sense of Sarkovskii's order.

The above discussion shows that, regardless of how we measure the complexity of a dynamical system, there is a close relation between the rate of time preference and the complexity of the optimal growth paths in our model. It is very tempting to conjecture that this relation is not only valid in the model at hand but in all strictly concave optimal growth models on arbitrary finite-dimensional state spaces. It would be a worthwhile project for future research to provide a formal proof of this conjecture.

10

On the Minimum Rate of Impatience for Complicated Optimal Growth Paths

Gerhard Sorger

1. Introduction

The occurrence of complex dynamics in economic models has received wide attention by the economic profession during the last decade (see Boldrin and Woodford (1990) for a recent survey). One of the most interesting and surprising results obtained so far is the indeterminacy theorem of Boldrin and Montrucchio (1986b), which implies that virtually every dynamical behavior is fully compatible with the standard assumptions of decreasing returns, competitive markets, and perfect foresight. Deneckere and Pelikan (1986) have used a related approach to derive similar results for the special case of one-dimensional dynamics. Whereas the analysis in Boldrin and Montrucchio (1986b) and Deneckere and Pelikan (1986) is restricted to optimal growth models formulated in discrete time, an analogous result holds also for the continuous time case (see Montrucchio (1988) and Sorger (1990)). All of these indeterminacy theorems have been proved by a constructive approach which requires a sufficiently high rate of impatience on the side of the decision maker. In particular, to construct optimal growth models exhibiting some well known chaotic maps (like the logistic map, or the Henon map) as optimal policy functions, one needs time preference rates of more than 400%. This raises the question of whether certain dynamically complicated growth paths can be ruled out a priori by a low rate of impatience or whether it is just a flaw of Boldrin and Montrucchio's method that it does not work for moderate discount rates.

One might suspect that an answer to this question can be found in the turnpike literature where it is shown that for any given convex technology and strictly concave utility function, there exists a value $\bar{\rho} > 0$ such that only "simple" dynamics can occur when the rate of impatience ρ satisfies $\rho < \bar{\rho}$ (see, e.g., McKenzie (1986) for discrete time models and Rockafellar (1976) for the continuous time case). In general, however, the bound $\bar{\rho}$ depends on the given technology and on the utility function. Therefore, the turnpike theorem does not rule out that one can construct, for any given rate of impatience $\rho > 0$, a standard economic growth model with an optimal solution that exhibits complicated dynamics. It is this fact which suggests that the above mentioned question may be viewed from a slightly different angle. More specifically, let us rephrase the problem as follows: For a given trajectory $x : I \mapsto \Re^n$, does there exist a value $\bar{\rho} > 0$ such that this trajectory is *definitely not* an optimal growth path in any standard problem of economic growth with a time preference rate $\rho \leqslant \bar{\rho}$? Here, $I = \{0, 1, 2, \dots\}$ in the discrete time case and $I = [0, \infty)$ for continuous time models. The class of standard economic growth problems under consideration consists of all reduced form models which have a convex production technology set and a continuous and strictly concave utility function. In this chapter we shall show that such a uniform bound on the minimum rate of impatience can in fact be computed if one only knows a finite truncation of the given trajectory x and if this trajectory exhibits sufficiently complicated dynamics.

To prove these results we first develop a new type of necessary optimality condition for strictly concave problems of economic growth in discrete time and continuous time formulation, respectively. Unlike the Euler equations or the Bellman equation, which depend on the particular problem under consideration, our condition is independent of the given technology and the utility function as long as they satisfy some standard assumptions of optimal growth theory. These assumptions are the convexity of the set of technologically feasible points and the strict concavity of the utility function. No differentiability assumptions will be needed. The interpretation of our optimality condition is as follows: If the growth path x is optimal in *any* standard optimal growth problem with discount rate ρ, then there is no representation of the initial state x_0 as a weighted average of all future capital stocks along this path with the weights decreasing at a rate greater or equal to ρ.

It is obvious how this necessary optimality condition can be used to derive a lower bound on the rate of impatience under which a given path can be an optimal path. We just have to find one representation of x_0 as a weighted average of future capital stocks with sufficiently fast decreasing weights. Moreover, we can also use the result to obtain lower bounds on the time preference rate under which a given policy function can be optimal. A policy function h determines for any given capital vector $x(t)$ the optimal next period's capital vector $x(t+1) = h(x(t))$ (discrete time formulation) or the optimal net investment rate $\dot{x}(t) = h(x(t))$ (continuous time formulation). To obtain the lower bound we can just pick a couple of initial capital stocks and apply the policy function to generate the trajectories corresponding to these initial states. Each of these trajectories can then be used to obtain a bound on the minimal time-preference rate in the aforementioned manner and it is only a matter of perseverance and luck to find an initial state which yields a good bound. We shall apply this approach to derive bounds on the minimum rate of impatience required for the optimality of some famous chaotic maps. In particular, we shall show that the logistic map $x \mapsto 4x(1-x)$, the tent map $x \mapsto 1-|2x-1|$, and the Henon map $(x,y) \mapsto (y, 1+0.3x-1.4y^2)$ cannot be optimal for discount rates smaller than 110%, 100%, and 80%, respectively. This precludes their optimality in optimal growth models with realistic parameter values. As an example for continuous time dynamics we shall consider a simple two-dimensional system of non-linear differential equations which has a unique limit cycle. Once again a lower bound on the discount rate of more than 90% will be obtained.

Let us conclude this introduction by a final word of caution. Although we can prove that certain chaotic functions like the logistic map cannot be optimal policy functions in any strictly concave optimal growth model with a reasonable rate of time preference, we do not claim that any other type of complicated dynamics is impossible in such models. As a matter of fact, it is known that periodic solutions (more specifically, stable limit cycles) can be optimal in strictly concave models at any given rate of impatience (see Boldrin and Deneckere (1990), p. 638 for the discrete time case and Benhabib and Rustichini (1990) for the continuous time case). The purpose of the present chapter is merely

(i) to show in a rigorous way that the optimal solutions of strictly concave optimal growth models can reveal a certain minimum degree of impatience of the decision maker and

(ii) to motivate by examples that this minimum degree is likely to be high if the optimal solutions exhibit dynamic complexity.

The paper is organized as follows. In Section 2 we derive a lower bound on the minimum rate of impatience in discrete time models and in Section 3 we formulate an analogous result for continuous time models. We illustrate our results by various examples and corollaries. Finally, Section 4 presents some concluding remarks.

2. Discrete Time Models

In this section, we consider the following model of optimal economic growth:

$$\text{Max} \sum_{t=0}^{\infty} \delta^t V(x_t, x_{t+1}) \tag{2.1}$$

$$\text{Subject to } (x_t, x_{t+1}) \in T. \tag{2.2}$$

Definition 2.1: We call problem (2.1)-(2.2) an n-dimensional discrete time standard economic growth problem if the following four conditions are satisfied:

A1: *X is a non-empty, closed and convex subset of the n-dimensional Euclidean space $\Re^n$. The set X represents the set of feasible capital stocks.*

A2: *The production technology set T is a closed and convex subset of $X \times X$ which has a non-empty and compact x-section $T_x = \{y \in X | (x, y) \in T\}$ for all $x \in X$.*

A3: $V : T \mapsto \Re$ *is a bounded, continuous, and strictly concave utility function.*

A4: *The discount factor* δ *satisfies* $\delta \in (0,1)$.

The above assumptions have been discussed in detail in the literature (see, e.g., Stokey and Lucas (1989). We just mention that problem (2.1)-(2.2) is a so-called reduced form model in which the utility function $V(x,y)$ represents the maximal utility that can be derived from consumption under the given technological constraints if the current capital stock is x and if the capital stock in the next period is required to be at least y. We do not assume that this utility function is differentiable nor do we assume that the set X is compact. The discount factor can also be written as $\delta = 1/(1+\rho)$ where $\rho > 0$ represents the time-preference rate or the rate of impatience of the decision maker. Deriving a lower bound $\bar{\rho}$ on the rate of impatience is obviously equivalent to deriving an upper bound $\bar{\delta}$ on the discount factor. Note also that several restrictions which are typical of optimal growth models have not been included in conditions A1-A4. This refers, for example, to certain monotonicity properties of the utility function $V(\cdot)$ or to the assumption of free disposal for the production technology set T. Omitting these assumptions from the model formulation enables one to apply the results of this chapter also to situations which are outside the realm of optimal economic growth theory. Despite this fact we shall continue to use the optimal growth terminology.

The model (2.1)-(2.2) is completely specified by the 5-tuple $(n, X, V(\cdot), T, \delta)$. Throughout this section we will refer to an n-dimensional standard economic growth problem in discrete time simply as a standard model.

Let us denote by $Z^+ = \{1, 2, \dots\}$ the set of positive integers and by $Z_0^+ = Z^+ \cup \{0\}$ the set of non-negative integers. We assume that a bounded sequence $\hat{x} = (x_t)_{t \in Z_0^+}$ is given, where $x_t \in \Re^n$ for all $t \in Z_0^+$, and we consider the following set of $n+1$ equations and countably many inequalities in the unknown variables $\delta \in (0,1)$

and $z_t \in [0, \infty)$, $t \in Z^+$:

$$\sum_{t=1}^{\infty} z_t \delta^t = 1 \tag{2.3}$$

$$\sum_{t=1}^{\infty} z_t \delta^t x_t = x_0 \tag{2.4}$$

$$z_1 \geqslant z_2 \geqslant \ldots \geqslant z_t \geqslant \ldots \geqslant 0 \tag{2.5}$$

For ease of notation we shall henceforth denote the set of feasible solutions to the system (2.3)-(2.5) as $S(\hat{x})$; that is, the pair $(\delta, \hat{z})$ is an element of $S(\hat{x})$ if and only if δ is a real number in the interval $(0, 1)$, $\hat{z} = (z_t)_{t \in Z^+}$ is a sequence of non-negative real numbers, and (2.3)-(2.5) hold. We can now state the main result of this section.

Theorem 2.1: *Let $\hat{x} = (x_t)_{t \in Z_0^+}$ be a given bounded sequence of vectors $x_t \in \Re^n$ with $x_0 \neq x_1$ and assume that there exists a feasible solution $(\overline{\delta}, \hat{z}) \in S(\hat{x})$. If the trajectory $\hat{x}$ is an optimal growth path in a standard problem $(n, X, T, V(\cdot), \delta)$, then it holds that $\delta < \overline{\delta}$.*

The proof of this theorem will be carried out in three steps which we formulate as independent lemmas.

Lemma 2.1: *Let $\hat{x} = (x_t)_{t \in Z_0^+}$ be a sequence of vectors $x_t \in \Re^n$. If $\hat{x}$ is an optimal growth path in the standard problem $(n, X, T, V(\cdot), \delta)$ then there exist continuous functions $W : X \mapsto \Re$ and $G : T \mapsto \Re$ such that the following statements are true:*

(a) $G(x, y)$ is non-positive for all $(x, y) \in T$, and the equality $G(x_t, y) = 0$ holds if and only if $y = x_{t+1}$.

(b) The functions $x \mapsto W(x)$ and $y \mapsto G(x, y)$ are strictly concave for all $x \in X$.

(c) For all $(x, y) \in T$ it holds that

$$V(x, y) = G(x, y) + W(x) - \delta W(y). \tag{2.6}$$

Proof: It is well known (see, e.g., Stokey and Lucas (1989)) that for any given initial capital stock x_0, there exists a unique optimal growth path for problem $(n, X, T, V(\cdot), \delta)$ and that the optimal value function $W : X \mapsto \Re$ is well defined and satisfies the Bellman equation

$$W(x) = \max\{V(x,y) + \delta W(y) \,|\, (x,y) \in T\}.$$

Moreover, the maximum on the right-hand side of this equation is attained by a unique capital vector $y = h(x)$. The function $h : X \mapsto X$ is referred to as the optimal policy function of $(n, X, T, V(\cdot), \delta)$ and it characterizes completely the dynamical behavior of any optimal growth path $(x_t)_{t \in Z_0^+}$. In particular, we have $x_{t+1} = h(x_t)$ for all $t \in Z_0^+$. It is also known that the optimal value function $W(\cdot)$ of a standard problem is continuous and strictly concave. Lemma 2.1 follows immediately from these properties and from the definition $G(x,y) = V(x,y) - W(x) + \delta W(y)$ for $(x,y) \in T$. Q.E.D.

Lemma 2.2: *If $(\bar{\delta}, \hat{z})$ is a point in $S(\hat{x})$ and if $\delta \in [\bar{\delta}, 1)$, then there exists a sequence $\hat{c} = (c_t)_{t \in Z^+}$ of real numbers such that $(\delta, \hat{c}) \in S(\hat{x})$.*

Proof: The result is easily verified by defining $c_t = z_t(\bar{\delta}/\delta)^t$ for all $t \in Z^+$. Q.E.D.

Lemma 2.2 states that for a high discount factor $\delta \geqslant \bar{\delta}$ one can always find a corresponding sequence $\hat{c}$ such that the pair $(\delta, \hat{c})$ is an element of $S(\hat{x})$. It is important to note that low discount factors $\delta < \bar{\delta}$ do not necessarily share this property. To put it in a different way, Lemma 2.2 shows that the projection of the set $S(\hat{x})$ onto the δ-space is a subinterval of $(0,1)$ with the right-hand endpoint being equal to one.

Lemma 2.3: *Let $\hat{x} = (x_t)_{t \in Z_0^+}$ be a given bounded sequence of vectors $x_t \in \Re^n$ with $x_0 \neq x_1$. If there exists a feasible solution $(\delta, \hat{c}) \in S(\hat{x})$ then the trajectory $\hat{x}$ cannot be an optimal growth path in any standard problem $(n, X, T, V(\cdot), \delta)$ with discount factor δ.*

Proof: Assume the contrary; i.e., there exists a standard problem $(n, X, T, V(\cdot), \delta)$ such that $\hat{x}$ is an optimal path for $(n, X, T, V(\cdot), \delta)$. Now choose an arbitrary real number $c_0 \geqslant c_1$ and define $\lambda_t = c_t\delta^t/(1+c_0)$, $t \in Z_0^+$. It follows from these specifications that $0 \leqslant \lambda_t < 1$ for all $t \in Z_0^+$, that $\sum_{t=0}^{\infty} \lambda_t = 1$, and that $\sum_{t=0}^{\infty} \lambda_t x_t = x_0$. Moreover, defining the vector $y_0 \in \Re^n$ by $y_0 = \sum_{t=0}^{\infty} \lambda_t x_{t+1}$ we obtain

$$\begin{aligned} y_0 &= \sum_{t=1}^{\infty} \lambda_{t-1} x_t = \frac{1}{\delta(1+c_0)} \sum_{t=1}^{\infty} c_{t-1}\delta^t x_t \\ &= \frac{1}{\delta(1+c_0)} \left[\sum_{t=1}^{\infty} c_t \delta^t x_t + \sum_{t=1}^{\infty} (c_{t-1} - c_t)\delta^t x_t \right]. \end{aligned}$$

Since $(\delta, \hat{c}) \in S(\hat{x})$, this implies

$$y_0 = \frac{1}{\delta(1+c_0)} x_0 + \sum_{t=1}^{\infty} \frac{c_{t-1} - c_t}{1+c_0} \delta^{t-1} x_t. \tag{2.7}$$

Because $(\delta, \hat{c}) \in S(\hat{x})$ and because of the definition of y_0 it follows that the vector (x_0, y_0) is a convex combination of the vectors (x_t, x_{t+1}), $t \in Z_0^+$. Since $\hat{x}$ is assumed to be an optimal path for $(n, X, T, V(\cdot), \delta)$, we have $(x_t, x_{t+1}) \in T$ for all $t \in Z_0^+$. Convexity of T implies that $(x_0, y_0) \in T$. Now observe that $(\delta, \hat{c}) \in S(\hat{x})$ implies $c_1 > 0$ and, hence, $\lambda_0 > \lambda_1 > 0$. Therefore, it follows from strict concavity of the utility function $V(\cdot)$ and the assumption that $x_0 \neq x_1$ that

$$V(x_0, y_0) > \sum_{t=0}^{\infty} \lambda_t V(x_t, x_{t+1}). \tag{2.8}$$

We are now in a position to use Lemma 2.1. Substituting the expression for $V(x, y)$ from (2.6) into (2.8) we obtain

$$\begin{aligned} &G(x_0, x_0) + W(x_0) - \delta W(y_0) \\ &> \sum_{t=0}^{\infty} \lambda_t [G(x_t, x_{t+1}) + W(x_t) - \delta W(x_{t+1})]. \end{aligned} \tag{2.9}$$

From Lemma 2.1 (a) it follows that the terms $G(x_t, x_{t+1})$, $t \in Z_0^+$, on the right-hand side of (2.9) vanish and that the term $G(x_0, y_0)$ on the left-hand side is non-positive. Therefore, we obtain

$$\begin{aligned} W(x_0) - \delta W(y_0) &> \sum_{t=0}^{\infty} \lambda_t [W(x_t) - \delta W(x_{t+1})] \\ &= \sum_{t=0}^{\infty} \lambda_t W(x_t) - \delta \sum_{t=1}^{\infty} \lambda_{t-1} W(x_t) \\ &= \lambda_0 W(x_0) + \sum_{t=1}^{\infty} (\lambda_t - \delta\lambda_{t-1}) W(x_t) \\ &= \frac{c_0}{1+c_0} W(x_0) + \sum_{t=1}^{\infty} \frac{c_t - c_{t-1}}{1+c_0} \delta^t W(x_t). \end{aligned}$$

Substituting the expression for y_0 from (2.7) and rearranging terms yields

$$\begin{aligned} &\frac{W(x_0)}{\delta(1+c_0)} + \sum_{t=1}^{\infty} \frac{c_{t-1} - c_t}{1+c_0} \delta^{t-1} W(x_t) \\ &> W\left(\frac{1}{\delta(1+c_0)} x_0 + \sum_{t=1}^{\infty} \frac{c_{t-1} - c_t}{1+c_0} \delta^{t-1} x_t \right). \end{aligned} \tag{2.10}$$

On the other hand, $(\delta, \hat{c}) \in S(\hat{x})$ implies that $1/[\delta(1+c_0)] \geqslant 0$ and $(c_{t-1} - c_t)\delta^{t-1}/(1+c_0) \geqslant 0$ for all $t \in Z^+$. Moreover, we have

$$\begin{aligned} \frac{1}{\delta(1+c_0)} + \sum_{t=1}^{\infty} \frac{c_{t-1} - c_t}{1+c_0} \delta^{t-1} &= \frac{1}{1+c_0} \left[\frac{1}{\delta} + \sum_{t=0}^{\infty} c_t \delta^t - \frac{1}{\delta} \sum_{t=1}^{\infty} c_t \delta^t \right] \\ &= \frac{1}{1+c_0} \left[\frac{1}{\delta} + c_0 + 1 - \frac{1}{\delta} \right] = 1. \end{aligned}$$

Since the numbers $1/[\delta(1+c_0)]$ and $(c_{t-1} - c_t)\delta^{t-1}/(1+c_0)$, $t \in Z^+$, are all non-negative and sum to one, we can use the strict concavity of the function $W(\cdot)$ to obtain

$$W\left(\frac{1}{\delta(1+c_0)}x_0+\sum_{t=1}^{\infty}\frac{c_{t-1}-c_t}{1+c_0}\delta^{t-1}x_t\right)$$
$$\geqslant \frac{W(x_0)}{\delta(1+c_0)}+\sum_{t=1}^{\infty}\frac{c_{t-1}-c_t}{1+c_0}\delta^{t-1}W(x_t).$$

Obviously, this is a contradiction to (2.10) and the lemma is proven. Q.E.D.

It is obvious that Lemma 2.2 and Lemma 2.3 together imply Theorem 2.1. In the rest of this section we shall discuss the implications of Theorem 2.1 and illustrate its application by various examples and corollaries. We begin with a brief technical remark.

Remark 2.1: We have assumed in Theorem 2.1 that $\hat{x}$ is a bounded sequence with $x_0 \neq x_1$. However, it is evident from the proof of Lemma 2.1 that, if $\hat{x}$ is an optimal path, then it has to be the trajectory of the dynamical system generated by the optimal policy function $h(\cdot)$. Therefore, sequences which are apparently not trajectories of any dynamical system (like the real sequence $\hat{x} = (0, 1, 0, 2, \dots))$ could have been excluded from consideration without altering the result. We have chosen not to do so, because the property of being a trajectory is not used anywhere in the proof of the theorem. We would also like to point out that the assumption $x_0 \neq x_1$ can be translated as "x_0 is not a fixed point of the optimal policy function $h(\cdot)$". This simply shows that we cannot use Theorem 1 to obtain a lower bound on the rate of impatience required for a stationary growth path to be optimal.

Let us now come to a theoretical point. Scrutinizing the system (2.3)-(2.5) reveals that Theorem 2.1 has a very simple verbal formulation which reads as follows: *The initial capital stock of any non-stationary optimal growth path in a standard optimal growth problem cannot be represented as a weighted average of all future capital stocks along this path when the weights are required to decrease at a rate exceeding the discount rate.* Formally, this means that for an optimal path $\hat{x}$ the set $S(\hat{x})$ must be empty. If this condition is not satisfied then it follows that the path $\hat{x}$ exhibits such high variations of the capital stocks that there exist arbitrage possibilities by which one can improve the path.

From a mathematical point of view, we would like to emphasize that Theorem 2.1 provides a necessary optimality condition not only for a single problem but for the entire class of all standard problems. It depends on the problem $(n, X, T, V(\cdot), \delta)$ only via the discount factor δ. Because of this reason Theorem 2.1 has to be distinguished from problem specific optimality conditions like the Euler equations or the Bellman equation. Whereas the latter conditions are used to characterize all optimal solutions of a given problem, our condition can be used to characterize (in terms of the discount factor) all problems which generate a given optimal solution. Strictly speaking, we should therefore regard Theorem 2.1 as a necessary condition for the inverse optimal problem (see, e.g., Kurz (1969) for an early reference concerning the inverse optimal problem).

Let us now discuss a more practical question, namely, the question of how to apply Theorem 2.1. In order to compute a good bound on the minimum rate of impatience required for the optimality of a given path $\hat{x}$, it is reasonable to try to solve the infinite-dimensional non-linear programming problem of minimizing δ over the feasible region $S(\hat{x})$. Let us call this the auxiliary problem. Even if we knew the complete trajectory $\hat{x}$, solving the auxiliary problem would be a formidable task. One possible way to circumvent this difficulty is to specify sufficiently many of the variables z_t in advance and to solve the remaining (smaller) auxiliary problem. Of course, this will lead to a suboptimal bound $\bar{\delta}$. To illustrate this further, let us briefly consider the one-dimensional case. In that case, we have two equality constraints, namely (2.3) and (2.4), which implies that we have to keep at least one of the z_t variables free (note that the variable δ has to remain free in any case as it represents the objective functional of the auxiliary problem). We can accomplish this, for example, by adding the constraint $z_t = z_s$ for all $t, s \in Z^+$. This means that all z_t variables are constrained to have the same value, say, z_0. From (2.3) it follows that $z_0 = (1-\delta)/\delta$. Substituting this into (2.4), we obtain the equation

$$\sum_{t=0}^{\infty} \delta^t (x_{t+1} - x_0) = 0. \tag{2.11}$$

Since condition (2.5) holds trivially, any solution $\bar{\delta}$ of (2.11) which satisfies $\bar{\delta} \in (0,1)$ qualifies as a lower bound on the time-preference rate required for the optimality of the path $\hat{x}$. As an example of the application of (2.11) let us discuss the logistic map $h(x) = 4x(1-x)$ which is probably the most prominent example of a chaotic dynamical system (see Devaney (1986)). Consider the trajectory emanating from the initial state $x_0 = [1+\cos(3\pi/8)]/2$. It is easy to verify that x_0 is eventually mapped on the fixed point 0 and that its orbit is given by $x_1 = [1+\cos(\pi/4)]/2$, $x_2 = \frac{1}{2}$, $x_3 = 1$, and $x_t = 0$ for all $t \geqslant 4$. Substituting these values into (2.11), we obtain the cubic equation

$$\delta^3 - \delta^2/2 + (\delta/2)\cos(\pi/4) - \sin(\pi/16)\sin(5\pi/16) = 0,$$

which has a solution at approximately 0.475. This shows that the logistic map cannot be the optimal policy function of a standard problem with a discount factor greater than or equal to 0.475 or, equivalently, a time-preference rate smaller than or equal to roughly 110%.

It shoudl be noted that the restriction to $n = 1$ is strictly speaking not necessary for the validity of the above argument. However, (2.11) will most likely not have any solution in the case $n > 1$. This follows simply from the fact that (2.11) represents n equations for the single variable δ.

A further result for one-dimensional maps which can easily be obtained from (2.11) is the following.

Corollary 2.1: *Let X denote the interval $[a,b]$ and assume that the policy function $h : X \mapsto X$ is surjective and satisfies $h(a) = h(b) = a$. Then the discount factor of any standard optimal growth problem which has $h(\cdot)$ as its optimal policy function is smaller than $\bar{\delta} = (b-c)/(b-a) < 1$, where $c = \max\{x \in [a,b] \mid h(x) = b\}$.*

Proof: Let c be the largest number in $[a, b]$ such that $h(c) = b$. After two iterations of $h(\cdot)$, the point c is mapped on the fixed point a. Therefore, the sequence $\hat{x} = (x_t)_{t \in Z_0^+}$ defined by $x_0 = c$, $x_1 = b$, and $x_t = a$ for all $t \geqslant 2$ is a trajectory generated by the policy function $h(\cdot)$. Substituting this trajectory into (2.11), we obtain the equation $b - c + (a - c)\delta/(1 - \delta) = 0$, which has the unique solution $\overline{\delta} = (b - c)/(b - a) \in (0, 1)$. Q.E.D.

The logistic map as well as the tent map $h(x) = 1 - |2x - 1|$ satisfies the above corollary with $a = 0$, $b = 1$, and $c = \frac{1}{2}$. Therefore, both of these maps cannot be optimal in a standard economic growth model with a discount factor greater than or equal to $\frac{1}{2}$ which corresponds to a discount rate of exactly 100%. In passing, we note that the assumptions of Corollary 2.1 are implied by those of Proposition 6 in Deneckere and Pelikan (1986). The latter result is a sufficient condition for chaotic dynamics in one-sector growth models.

In the above examples, we have used trajectories which are eventually mapped onto a fixed point of the system. The reason for this choice was that we know these trajectories on the whole planning interval Z_0^+ which is necessary to solve (2.11). By the same token, we can use initial points x_0 which are eventually mapped onto a periodic point of some period p. In both of these cases, (2.11) can be transformed into an equation $f(\delta) = 0$, where $f(\cdot)$ is a polynomial.

We now return to the general case $n \geqslant 1$. Since (2.3) and (2.4) constitute $n+1$ equality constraints for the set $S(\hat{x})$ we must have at least $n + 1$ free variables in the auxiliary problem to obtain a feasible point. One of these variables has to be δ, since it is the value to be maximized. Let us, therefore, consider now a sequence of capital vectors in n-dimensional space from which we know only a finite truncation $(x_0, x_1, \ldots, x_m)$. The only way to deal with the unknown values x_t, $t > m$ in system (2.3)-(2.5) is to set $z_t = 0$ for all $t > m$. What remains to be done is to find a feasible point

$(\delta, z_1, z_2, \ldots, z_m)$ to the following constraints:

$$\sum_{t=1}^{m} z_t \delta^t = 1$$

$$\sum_{t=1}^{m} z_t \delta^t x_t = x_0$$

$$z_1 \geqslant z_2 \geqslant \ldots \geqslant z_m \geqslant 0.$$

Let us denote the set of these feasible points by $S_m(\hat{x})$. As an example we consider the Henon map $(x, y) \mapsto (y, 1+0.3x-1.4y^2)$. It can easily be verified that the point $(\bar{\delta}, c_1, c_2, \ldots, c_7)$ defined by $\bar{\delta} = 0.555$, $c_1 = c_2 = c_3 = 0.81535$, $c_4 = c_5 = c_6 = 0.81363$, and $c_7 = 0.80922$ (values rounded to five significant decimal places) is an element of $S_7(\hat{x})$, where $\hat{x}$ is the trajectory emanating from the initial state $(x_0, y_0) = (0.59, 0.58)$. Therefore, the Henon map cannot be optimal for a discount factor greater than or equal to 0.555 or, equivalently, a discount rate smaller than or equal to 80%.

3. Continuous Time Models

In this section, we consider the continuous time analogue to problem (2.1)-(2.2) and show that Theorem 2.1 carries over to this case with some obvious modifications. Although the basic idea of the proof remains the same, the technical details become a little bit more subtle. In particular, the additional smoothness hypotheses required for the validity of the Bellman equation force us to impose an interiority assumption which was not necessary in the discrete time case. It will be demonstrated, however, that this assumption does not substantially impede the applicability of our main result. Since the interpretations of the various assumptions and results are completely analogous to the discrete time case, we shall restrict ourselves for the most part to the mathematical analysis.

The problem under consideration is to

$$\text{Max} \int_0^\infty e^{-\rho t} V(x(t), \dot{x}(t))\, dt \tag{3.1}$$

$$\text{Subject to } (x(t), \dot{x}(t)) \in T. \tag{3.2}$$

Definition 3.1: We call problem (3.1)-(3.2) an n-dimensional continuous time standard economic growth problem, if the following four conditions are satisfied:

1. X is a non-empty, closed, convex subset of the n-dimensional Euclidean space $\Re^n$ representing the set of feasible capital stocks.

2. The production technology set T is a closed and convex subset of $X \times \Re^n$ which has a non-empty and compact x-section $T_x = \{y \in \Re^n | (x, y) \in T\}$ for all $x \in X$.

3. $V : T \mapsto \Re$ is a bounded, continuous, and strictly concave utility function..

4. The discount rate ρ is strictly positive.

Throughout this section we refer to an n-dimensional standard economic growth problem in continuous time simply as a standard model and we denote it by the 5-tuple
$(n, X, T, V(\cdot), \rho)$.

Remark 3.1: Solutions of problem (3.1)-(3.2) are sought in the set of absolutely continuous functions from $[0, \infty)$ to $\Re^n$. On the other hand, we are mainly interested in deriving bounds on the mimum rate of impatience required for the optimality of a given optimal policy function $h(\cdot)$ relating the current capital stock to the optimal net investment rate. If $x : [0, \infty) \mapsto \Re^n$ is the trajectory of the dynamical system $\dot{x}(t) = h(x(t))$, where $h(\cdot)$ is a continuous function, then $x(\cdot)$ is by necessity not only absolutely continuous but also continuously differentiable. Therefore, it would be no restriction to assume $x(\cdot)$ to be continuously differentiable. For the sake of generality, however, we state the result for absolutely continuous growth paths. The derivative of such a path is determined only up to modifications on sets of Lebesgue measure zero. Consequently, properties of derivatives are usually only stated for almost every $t \in [0, \infty)$.

In contrast to the discrete time case, the existence of feasible (let alone optimal) solutions is not guaranteed by the assumptions 1-4 in Defintion 3.1. Let us therefore denote by X_0 the set of all initial states $x \in X$ for which a feasible growth path with $x(0) = x$ exists. It is easy to see that X_0 is a convex subset of X and that $x(t) \in X_0$ must hold for all $t \in [0, \infty)$ and for all feasible paths $x(\cdot)$. Furthermore, we shall call a feasible path $x(\cdot)$ an almost interior path if the condition $x(t) \in \text{int } X_0$ holds for almost every $t \in [0, \infty)$. Here int X_0 denotes the topological interior of X_0. Note that an almost interior path can have contact points with the boundary of X_0 but that there cannot be any non-trivial boundary segments.

We assume that a bounded and absolutely continuous function $x : [0, \infty) \mapsto \Re^n$ is given and we consider the following set of conditions:

$$\int_0^\infty z(t) e^{-\rho t} \, dt = 1 \tag{3.3}$$

$$\int_0^\infty z(t) e^{-\rho t} x(t) \, dt = x_0 \tag{3.4}$$

$$z : [0, \infty) \mapsto \Re \text{ is non-negative and non-increasing.} \tag{3.5}$$

The function $z(\cdot)$ is required to be piecewise absolutely continuous, that is, we assume that there exists a strictly increasing sequence $(t_i)_{i \in Z^+}$ of positive numbers with no accumulation points such that $z(\cdot)$ is absolutely continuous on each of the intervals $(0, t_1)$ and (t_i, t_{i+1}) for $i \in Z^+$ and that the possible discontinuities at the points t_i are jumps of finite height. Furthermore, ρ is assumed to be a strictly positive real number. The set of all such pairs $(\rho, z(\cdot))$ satisfying the system (3.3)-(3.5) will be denoted as $S(x(\cdot))$. We can now state the continuous time analogue to Theorem 2.1.

Theorem 3.1: *Let* $x : [0, \infty) \mapsto \Re^n$ *be a given bounded and absolutely continuous function such that* $x(\cdot)$ *is not constant on any interval of the form* $[0, \varepsilon)$ *with* $\varepsilon > 0$. *Moreover, assume that there exists a feasible solution* $(\bar{\rho}, z(\cdot)) \in S(x(\cdot))$. *If the trajectory* $x(\cdot)$ *is an almost interior optimal growth path in a standard problem* $(n, X, T, V(\cdot), \rho)$ *satisfying* $x(0) \in \text{int } X_0$, *then it holds that* $\rho > \bar{\rho}$.

The proof of this theorem will be subdivided into the same three steps as in the discrete time case. In what follows, $\partial W(x)$ denotes the subdifferential at x of an extended-real-valued and concave function $W : \Re^n \mapsto \Re$, i.e.,

$$\partial W(x) = \{x^* \in \Re^n \mid W(y) \leqslant W(x) + x^* \cdot (y - x) \text{ for all } y \in \Re^n\}.$$

Note that $\partial W(x)$ is a non-empty, compact, and convex set for all x in the interior of the effective domain of $W(\cdot)$ (see Rockafellar (1970)). Moreover, we denote by $W'(x; y)$ the directional derivative in direction y of the function $W(\cdot)$ at the point x. This derivative is well defined (possibly $+\infty$ or $-\infty$) for all x in the effective domain of $W(\cdot)$. It is also known that for all x in the interior of the effective domain of $W(\cdot)$ it holds that

$$W'(x; y) = \inf\{x^* \cdot y | x^* \in \partial W(x)\}. \tag{3.6}$$

For a proof of this and other properties of concave functions we refer to Rockafellar (1970).

Lemma 3.1: *Let* $x : [0, \infty) \mapsto \Re^n$ *be a continuously differentiable function and assume that* $x(\cdot)$ *is an almost interior optimal growth path in the standard problem* $(n, X, T, V(\cdot), \rho)$. *Moreover, define the set* T_0 by $T_0 = \{(x, y) \in T | x \in X_0\}$. *Then there exist functions* $W : X_0 \mapsto \Re$ *and* $G : T_0 \mapsto \Re$ *such that the following statements are true:*

(a) $G(x, y) \leqslant 0$ *for all* $x \in \text{int } X_0$ *and for all* $y \in T_x$. *For almost all* $t \in [0, \infty)$, *the equality* $G(x(t), y) = 0$ *holds if and only if* $y = \dot{x}(t)$.

(b) The functions $x \mapsto W(x)$ *and* $y \mapsto G(x, y)$ *are strictly concave for all* $x \in X_0$.

(c) For all $(x, y) \in T_0$ it holds that

$$V(x, y) = G(x, y) + \rho W(x) - W'(x; y). \tag{3.7}$$

Proof: The proof is based on the Bellman equation

$$\rho W(x) = \sup\{V(x, y) + W'(x; y)\},$$

which holds for all x in the interior of X_0. Here, $W(\cdot)$ is the optimal value function of problem (3.1)-(3.2) extended to $\Re^n$ by the specification $W(x) = -\infty$ for all x outside X_0. We omit the details which are similar to those in the proof of Lemma 2.1.[1] Q.E.D.

Lemma 3.2: *If $(\bar{\rho}, z(\cdot)) \in S(x(\cdot))$ and if $0 < \rho \leqslant \bar{\rho}$, then there exists a piecewise absolutely continuous function $c : [0, \infty) \mapsto \Re$ such that $(\rho, c(\cdot)) \in S(x(\cdot))$.*

Proof: The result is easily verified by defining $c(t) = z(t)e^{(\rho - \bar{\rho})t}$ for all $t \in [0, \infty)$. Q.E.D.

Lemma 3.3: *Let $x : [0, \infty) \mapsto \Re^n$ be a given bounded and absolutely continuous function such that $x(\cdot)$ is not constant on any interval of the form $[0, \varepsilon)$ with $\varepsilon > 0$. If there exists a feasible solution $(\rho, c(\cdot)) \in S(x(\cdot))$, then the trajectory $x(\cdot)$ cannot be an almost interior optimal growth path in any standard problem $(n, X, T, V(\cdot), \rho)$ with discount rate ρ and $x(0) \in \text{int } X_0$.*

Proof: Assume the contrary; i.e., there exists a standard problem $(n, X, T, V(\cdot), \rho)$ such that $x(\cdot)$ is an almost interior optimal path for $(n, X, T, V(\cdot), \rho)$. Defining $\lambda(t) = c(t)e^{-\rho t}$, $t \in [0, \infty)$, it follows that $\lambda(t) \geqslant 0$ for all $t \in [0, \infty)$, that $\int_0^\infty \lambda(t)\, dt = 1$, and that $\int_0^\infty \lambda(t)x(t)\, dt = x(0)$. Now denote by $Z = \{t_1, t_2, \ldots\}$ the points of discontinuity of the function $c(\cdot)$ and define $t_0 = 0$.

---

[1]The author is not aware of any published proof of the Bellman equation under the specific assumptions made in this chapter, that is, infinite horizon, no differentiability assumptions, convexity of T, and concavity of $V(\cdot)$, etc. Related results can be found, for example, in Clarke (1983), Chapter 3 or in Vinter and Lewis (1980). A proof under the present assumptions can be obtained from the author upon request.

Moreover, denote by $c(t_i^-)$ and $c(t_i^+)$ the left-hand limit, respectively, the right-hand limit of the function $c(\cdot)$ at t_i. Using these notations and defining the vector $y \in \Re^n$ by $y = \int_0^\infty \lambda(t)\dot{x}(t)\,dt$ we obtain

$$\begin{aligned} y &= \sum_{i=0}^{\infty} \int_{t_i}^{t_{i+1}} c(t)e^{-\rho t}\dot{x}(t)\,dt \\ &= \sum_{i=0}^{\infty} \left\{ c(t)e^{-\rho t}x(t) \Big|_{t_i}^{t_{i+1}} - \int_{t_i}^{t_{i+1}} \left[\dot{c}(t) - \rho c(t)\right] e^{-\rho t}x(t)\,dt \right\} \\ &= \sum_{i=0}^{\infty} \left[c(t_{i+1}^-)e^{-\rho t_{i+1}}x(t_{i+1}) - c(t_i^+)e^{-\rho t_i}x(t_i)\right] \\ &\quad - \sum_{i=0}^{\infty} \int_{t_i}^{t_{i+1}} \dot{c}(t)e^{-\rho t}x(t)\,dt + \rho x(0). \end{aligned}$$

Rearranging the terms in the first sum on the right-hand side of this equation we obtain

$$\begin{aligned} y = (\rho - c(0))x(0) &+ \sum_{i=1}^{\infty} \left[c(t_i^-) - c(t_i^+)\right] e^{-\rho t_i}x(t_i) \\ &- \sum_{i=0}^{\infty} \int_{t_i}^{t_{i+1}} \dot{c}(t)e^{-\rho t}x(t)\,dt. \end{aligned} \tag{3.8}$$

Because $(\rho, c(\cdot)) \in S(x(\cdot))$ and because of the definition of y it follows that the vector $(x(0), y)$ is a convex combination of the vectors $(x(t), \dot{x}(t))$, $t \in [0, \infty)$. Since $x(\cdot)$ is assumed to be an optimal path for $(n, X, T, V(\cdot), \rho)$, we have $(x(t), \dot{x}(t)) \in T$ for almost every $t \in [0, \infty)$. Convexity of T implies that $(x(0), y) \in T$. Moreover, strict concavity of the utility function $V(\cdot)$ together with the assumption that $x(\cdot)$ is not constant on any initial interval implies that

$$\begin{aligned} V(x(0), y) &> \int_0^\infty \lambda(t)V(x(t), \dot{x}(t))dt = \\ &= \sum_{i=0}^{\infty} \int_{t_i}^{t_{i+1}} \lambda(t)V(x(t), \dot{x}(t))dt. \end{aligned} \tag{3.9}$$

We are now in a position to use Lemma 3.1. Substituting the expression for $V(x,y)$ from (3.7) into (3.9) we obtain

$$G(x(0),y) + \rho W(x(0)) - W'(x(0);y)$$

$$> \sum_{i=0}^{\infty} \int_{t_i}^{t_{i+1}} \lambda(t)\left[G(x(t),\dot{x}(t)) + \rho W(x(t)) - W'(x(t);\dot{x}(t))\right] dt.$$

From Lemma 3.1 (a) it follows that the terms $G(x(t),\dot{x}(t))$, $t \in [0,\infty)$, on the right-hand side of this inequality vanish except on a set of Lebesgue measure zero and that the term $G(x(0),y)$ on the left-hand side is non-positive. Therefore, we obtain

$$\rho W(x(0)) - W'(x(0);y) > \sum_{i=0}^{\infty} \int_{t_i}^{t_{i+1}} c(t)e^{-\rho t}[\rho W(x(t)) - W'(x(t);\dot{x}(t))]\, dt.$$

It is shown in Clarke (1983, p. 158) that the function $t \mapsto W(x(t))$ is absolutely continuous and that $(d/dt)W(x(t)) \in \{x^* \cdot \dot{x}(t) \mid x^* \in \partial W(x(t))\}$. Therefore, we obtain from the above inequality and from (3.6) that for some $x^* \in \partial W(x(0))$ the following is true:

$$\rho W(x(0)) - x^* \cdot y > \sum_{i=0}^{\infty} \int_{t_i}^{t_{i+1}} c(t)e^{-\rho t} \left[\rho W(x(t)) - \frac{d}{dt} W(x(t))\right] dt$$

$$= \rho \int_0^{\infty} c(t)e^{-\rho t} W(x(t))\, dt - \sum_{i=0}^{\infty} \Big\{ c(t)e^{-\rho t} W(x(t))\Big|_{t_i}^{t_{i+1}} - \int_{t_i}^{t_{i+1}} [\dot{c}(t) - \rho c(t)]\, e^{-\rho t} W(x(t))\, dt \Big\}$$

$$= -\sum_{i=0}^{\infty} \left[c(t_{i+1}^-)e^{-\rho t_{i+1}} W(x(t_{i+1})) - c(t_i^+)e^{-\rho t_i} W(x(t_i))\right] + \sum_{i=0}^{\infty} \int_{t_i}^{t_{i+1}} \dot{c}(t)e^{-\rho t} W(x(t))\, dt$$

$$= c(0)W(x(0)) - \sum_{i=1}^{\infty} \left[c(t_i^-) - c(t_i^+)\right] e^{-\rho t_i} W(x(t_i))$$

$$+ \sum_{i=0}^{\infty} \int_{t_i}^{t_{i+1}} \dot{c}(t) e^{-\rho t} W(x(t))\, dt.$$

Substituting the expression for y from (3.8) and rearranging terms yields

$$(\rho - c(0)) \left[W(x(0)) - x^* \cdot x(0)\right]$$

$$> \sum_{i=1}^{\infty} \left[c(t_i^+) - c(t_i^-)\right] e^{-\rho t_i} \left[W(x(t_i)) - x^* \cdot x(t_i)\right]$$

$$+ \sum_{i=0}^{\infty} \int_{t_i}^{t_{i+1}} \dot{c}(t) e^{-\rho t} \left[W(x(t)) - x^* \cdot x(t)\right] dt. \tag{3.10}$$

On the other hand, $(\rho, c(\cdot)) \in S(x(\cdot))$ implies that $c(t_i^+) - c(t_i^-) \leqslant 0$ for all $i \in Z^+$ and that $\dot{c}(t) \leq 0$ for almost every $t \in (t_i, t_{i+1})$ and for all $i \in Z_0^+$. Moreover, we have

$$\sum_{i=1}^{\infty} [c(t_i^+) - c(t_i^-)] e^{-\rho t_i} + \sum_{i=0}^{\infty} \int_{t_i}^{t_{i+1}} \dot{c}(t) e^{-\rho t} dt$$

$$= \rho \sum_{i=0}^{\infty} \int_{t_i}^{t_{i+1}} c(t) e^{-\rho t}\, dt - c(0) = \rho - c(0).$$

By strict concavity of $W(\cdot)$ it follows that for all $x^* \in \partial W(x(0))$ and for all $t \in [0, \infty)$ the following is true:

$$W(x(t)) - x^* \cdot x(t) \leqslant W(x(0)) - x^* \cdot x(0).$$

From these properties we obtain

$$(\rho - c(0)) \left[W(x(0)) - x^* \cdot x(0)\right]$$

$$= \sum_{i=1}^{\infty} \left[c(t_i^+) - c(t_i^-)\right] e^{-\rho t_i} \left[W(x(0)) - x^* \cdot x(0)\right]$$

$$+ \sum_{i=0}^{\infty} \int_{t_i}^{t_{i+1}} \dot{c}(t) e^{-\rho t} \left[W(x(0)) - x^* \cdot x(0)\right] dt$$

$$\leqslant \sum_{i=1}^{\infty} \left[c(t_i^+) - c(t_i^-)\right] e^{-\rho t_i} \left[W(x(t_i)) - x^* \cdot x(t_i)\right]$$
$$+ \sum_{i=0}^{\infty} \int_{t_i}^{t_{i+1}} \dot{c}(t) e^{-\rho t} \left[W(x(t)) - x^* \cdot x(t)\right] dt$$

which is obviously a contradiction to (3.10). This completes the proof. Q.E.D.

It is obvious that Lemma 3.2 and Lemma 3.3 together imply Theorem 3.1. The verbal formulation of Theorem 2.1 remains valid also for Theorem 3.1. We just have to include the interiority condition which was not present in Theorem 2.1 and which guarantees that the Bellman equation holds. The other remarks concerning the economic and the mathematical content, respectively, of Theorem 2.1 apply also to Theorem 3.1 without any change.

To derive lower bounds on the minimum rate of impatience which is necessary for the optimality of a given path $x(\cdot)$, we can proceed in a similar manner as in the discrete time case. To obtain a tractable auxiliary problem we reduce the dimensionality of system (3.3)-(3.5) by introducing additional constraints. One reasonable way to do this is to require that the function $z(\cdot)$ is piecewise constant. Moreover, if we just know the finite truncation of the trajectory $x(\cdot)$ on the interval $[0, s)$, then we have to set $z(t) = 0$ for all $t \geqslant s$. We illustrate these ideas by the following example.

Consider the system of two non-linear differential equations

$$\dot{x}(t) = x(t) + y(t) - x(t)\left(x(t)^2 + y(t)^2\right)$$
$$\dot{y}(t) = -x(t) + y(t) - y(t)\left(x(t)^2 + y(t)^2\right).$$

By transforming this system into polar coordinates one can easily show that there exists a unique stable limit cycle $x(t) = \cos t$, $y(t) = -\sin t$. More generally, the solution starting in $x(0) = x_0$, $y(0) = 0$ is given by (see Boyce and DiPrima (1977, p. 444))

$$x(t) = \cos t \left[1 + ((1/x_0)^2 - 1)e^{-2t}\right]^{-1/2}$$
$$\dot{y}(t) = -\sin t \left[1 + ((1/x_0)^2 - 1)e^{-2t}\right]^{-1/2}$$

For our calculations we have chosen the trajectory $(x(\cdot), y(\cdot))$ starting in $x(0) = [3e^{4\pi} + 1]^{-1/2}$ and $y(0) = 0$. This trajectory passes through the point $(0.5, 0)$ at time $t = 2\pi$. Specifying the function $z(\cdot)$ by

$$z(t) = \begin{cases} 2.16795 & \text{for } 0 \le t < 0.3 \\ 0.52345 & \text{for } 0.3 \le t < 4.1 \\ 0.52321 & \text{for } 4.1 \le t < 7.2 \\ 0 & \text{otherwise} \end{cases}$$

(all figures rounded to 5 decimal places) one can check that $(0.919, z(\cdot)) \in S((x(\cdot), y(\cdot)))$. Therefore, we can conclude that the given trajectory cannot be an optimal growth path in any standard problem with a discount rate smaller than roughly 92%. Note that we do not have to add the qualification "almost interior" since the trajectory can only be an interior path if it is feasible at all. This follows simply from the convexity of X_0 and from $(x(t), y(t)) \in X_0$ for all $t \in [0, \infty)$. As a matter of fact, we can conclude from the feasibility of the trajectory under consideration that the open unit disc has to be a subset of X_0. A similar argument can be helpful in many cases where we start from a given policy function h. Consequently, the interiority condition for the path $x(\cdot)$ is not a severe restriction of Theorem 3.1.

4. Concluding Remarks

In this chapter, we have developed a new type of necessary optimality condition for strictly concave discounted dynamic optimization problems. This condition is not problem specific as the usual first order conditions or the Bellman equation but it is valid for all optimal solutions to all problems satisfying the general convexity assumptions. Using this property of our condition we were able to derive lower bounds on the time-preference rates which are required for the optimality of given policy functions. We have illustrated our results by deriving such bounds for various examples.

Let us briefly summarize what we consider as the strong points of our results. First of all, we impose only minimal assumptions on the class of standard economic growth problems. Except for the usual convexity hypotheses almost no restrictions are made. In particular, we do not assume differentiability of the utility function. A second and probably more important point is that our results do not rely on very accurate knowledge of the given policy function. All we need to know is a finite truncation of one single trajectory generated by the policy function.

Our primary goal in this study was to derive necessary conditions for the occurrence of complicated or chaotic optimal growth paths. We have derived such a condition in the form of a minimum rate of impatience. Complicated dynamics or chaos can be defined in a number of ways. Therefore, it might well be possible to derive better bounds than ours by exploiting more information on the policy function (for example, the largest Ljapunov exponent or an invariant ergodic measure). Research in this direction is an ongoing project.

Finally, we would like to note that we have no clue as to how sharp our bound is. The known examples of optimal growth problems which are solved by the logistic map require discount factors much smaller than our bound of 0.475 (see Boldrin and Montrucchio (1986b), Deneckere and Pelikan (1986)). These examples, however, are based on very smooth, in fact, analytic utility functions and it is not clear if a more sophisticated construction would make the logistic map optimal also for higher discount factors. As far as the Henon map is concerned, we are not so far off the best known example. In fact, it is reported in Boldrin and Woodford (1990, p. 201) that this map can be optimal for a discount factor of 0.2, whereas our bound is 0.555. In this regard we have to admit that we did not devote an extremely large amount of time and effort to the search for the "best" bounds, but that we stopped the search once a reasonably good bound was achieved, which allowed us to exclude the given policy function from being optimal in a realistic growth model.

11

An Exact Discount Factor Restriction for Period-Three Cycles in Dynamic Optimization Models

Tapan Mitra

1. Introduction

Consider a standard aggregative dynamic optimization framework (Ω, u, δ), where Ω is the transition possibility (technology) set, u is a (reduced form) utility function defined on this set, and $0 < \delta < 1$ a discount factor. Can an optimal program in this framework exhibit a period-three cycle?

I will show, in this chapter, that a complete answer to this question can be given as follows. There is a "universal constant"[1] defined by

$$M \equiv [(\sqrt{5} - 1)/2]^2 \approx 0.3819 \tag{1.1}$$

such that (i) if an optimal program of any dynamic optimization model (Ω, u, δ) exhibits a period-three cycle, then $\delta < M$, and (ii) if $\delta < M$, then it is possible to construct a transition possibility set,

[1]The constant, M, is closely related to the well-known "golden ratio", Φ. We recall, briefly, how the golden ratio is defined. If we divide a line into two parts (A and B) such that the ratio of A to B is the same as the ratio of the line itself to A, then the ratio of A to B is called the *golden ratio*. The relevant equation to solve is: $(A/B) = (A+B)/A$. Denoting (A/B) by x, we get the equation: $x = 1 + (1/x)$, which is a quadratic: $x^2 - x - 1 = 0$. The positive solution (denoted by Φ) is given by $\Phi = [1+\sqrt{5}]/2 \approx 1.618$. [The number, Φ, figures prominently in Fibonacci series: 1, 1, 2, 3, 5, 8, 13, 21, 34, ..., where each entry is obtained by adding the previous two entries, beginning traditionally with 1 and 1. The successive ratios: 2/1, 3/2, 5/3, 8/5, 13/8,..., converge to the golden ratio Φ.] It is clear from the way in wihch we solved for Φ that the reciprocal of Φ is $(1/\Phi) = \Phi - 1 = (\sqrt{5} - 1)/2 \approx 0.618$. Thus $(1/\Phi) = \sqrt{M}$, and so $(1/M) = \Phi^2 = \Phi + 1$. This means that since M is an exact *discount factor* restriction for period-three cycles, $(1/M) - 1 = \Phi$ is an exact *discount rate* restriction for period-three cycles.

Ω, and a reduced-form utility function, u, such that the dynamic optimization model (Ω, u, δ) has an optimal program exhibiting a period-three cycle.[2]

The question stated above has a history which is useful to review in order to place this chapter in proper perspective.. It stems from the issue of whether chaotic economic behavior can be explained by models of infinitely lived rational agents. Boldrin and Montrucchio (1986b) took a significant step in addressing this issue when they showed that any twice continuously differentiable function can be a policy function of an appropriately defined dynamic optimization model. Thus, in particular, if we take the C^2 function to be the "logistic map" ($h(x) = 4x(1-x)$ for $x \in [0,1]$), we can show that optimal trajectories can be chaotic for some suitably constructed dynamic optimization model. (This was also independently demonstrated by Deneckere and Pelikan (1986).)

It turned out that the dynamic optimization model (Ω, u, δ) constructed to yield the logistic map as its policy function had an extremely small discount factor (about 0.01). This raised the question of whether chaotic trajectories can be ruled out when more "reasonable" discount factors prevailed, or whether this feature of the dynamic optimization model was simply a shortcoming of the particular method used in its construction.

Sorger (1992a) has constructed examples of dynamic optimization models (using somewhat different methods from those in Boldrin and Montrucchio (1986b)), for which the logistic map is the policy function when the discount factor is about 0.04. While this is four times larger than the previous discount factors used, it is still very small, entailing a discount rate of over 2000%.

A major step in understanding the relation between discounting and the existence of chaotic optimal trajectories was taken by Sorger (Chapter 10) when he showed (among other things) that if the policy function of *any* dynamic optimization model is the logistic map, then its associated discount factor must be smaller

---

[2] After completing this chapter, I learned that essentially the same result has been independently obtained by Nishimura and Yano (Chapter 12). Their maintained assumptions on the dynamic optimization model differ somewhat from those used in this chapter. Their approach in obtaining the basic result is quite different from mine.

than 0.5. The remarkable aspect of this discount factor restriction is that it is completely independent of the particular nature of the transition possibility set or of the utility function [beyond the standard properties imposed in such models].

While the result of Sorger (Chapter 10) settled one issue, it naturally raised another. It was clearly not necessary to generate a logistic map as the policy function of a dynamic optimization model in order to show that optimal trajectories can be chaotic. Any continuous policy function which generates a period three cycle is chaotic, and the question arises whether the existence of a period-three optimal cycle by itself (that is, without knowing other aspects of the continuous policy function) leads to suitable restrictions on the discount factor. Using the theory of stochastic dominance, Sorger (1992b) showed that if any dynamic optimization model (Ω, u, δ) generates a periodic optimal program of period three, then the discount factor, δ, must satisfy

$$\delta < (\sqrt{5} - 1)/2 \approx 0.618 \tag{1.2}$$

This bound has been later refined to $\delta < 0.5479$ in Sorger (1994a).

Montrucchio (1994) independently arrived at some discount rate restrictions for dynamic complexity of optimal programs by exploiting the theory of strongly concave functions. He established (under a strong concavity assumption on the utility function) a relation between the discount factor of a dynamic optimization model and the topological entropy of its policy function. I have shown elsewhere (see Mitra (Chapter 13)) that this result can be used to establish the discount factor restriction (1.2) for any dynamic optimization model (Ω, u, δ) which generates a period-three optimal cycle.

The present chapter differs from the literature in two significant dimensions. First, we provide an *exact* discount factor restriction, as indicated in (1.1); it also happens to be a *better* restriction than any proposed in the literature so far for period-three cycles.

Second, the method used in this chapter is completely distinct from those employed in this literature. It has often been noted that when there is chaotic dynamic behavior, there must be intertemporal arbitrage opportunities for high discount factors, and the discount factor restrictions for optimal chaotic paths is simply a reflection of this fact. However, the methods that have been used in the literature in obtaining these discount factor restrictions do not illustrate this intuition. In contrast, in this chapter, we rely heavily on the theory of dual variables or shadow prices associated with optimal programs, and all our discount factor restrictions are obtained from the simple observation that at these "supporting" prices, no activity yields a higher (generalized) profit at any date than the activity chosen along the optimal program at that date. In other words, there are no intertemporal arbitrage opportunities at the prevailing "supporting" prices. The technical advantage of this method for the problem studied in this chapter is that it makes all the proofs entirely elementary.

This method, which I refer to as the "value-loss approach" is, of course, very familiar to optimal growth theorists. It is most closely associated with McKenzie's contributions to "turnpike theory" (see McKenzie (1986) for a survey). It has also figured prominently in (i) the literature on the stability of Hamiltonian dynamical systems (see especially Cass and Shell (1976)) and in (ii) the literature on the intertemporal decentralization of the transversality condition (see especially Brock and Majumdar (1988) and Dasgupta and Mitra (1988)).

2. Chaos

2.1 Periodic Orbits

Let I be an interval in $\Re$, the set of reals. Let $f : I \to I$ be a continuous map of the interval I into itself. The pair (I, f) is called a *dynamical system*; I is called the *state space* and f the *law of motion* of the dynamical system.

We write $f^0(x) = x$ and for any integer $n \geqslant 1$, $f^n(x) = f[f^{n-1}(x)]$. If $x \in I$, the sequence $\tau(x) = \{f^n(x)\}_0^\infty$ is called the *trajectory* from (the initial condition) x. The *orbit* from x is the set $\gamma(x) = \{y : y = f^n(x) \text{ for some } n \geqslant 0\}$.

A point $x \in I$ is a *fixed point* of f if $f(x) = x$. A point $x \in I$ is called a *periodic point* of f if there is $k \geqslant 1$ such that $f^k(x) = x$. The smallest such k is called the *period* of x. [In particular, if

$x \in I$ is a fixed point of f, it is periodic with period 1]. If $x \in I$ is a periodic point with period k, we also say that the orbit of x (or trajectory from x) is periodic with period k.

The following striking result, due to Sarkovskii (1964), is a fundamental result on the existence of periodic orbits. A good discussion of this result is contained in Block and Coppel (1992).

Proposition 2.1: *Let the positive integers be totally ordered in the following way:*

$$3 \prec 5 \prec 7 \prec 9 \prec ... \prec 2.3 \prec 2.5 \prec ... \prec 2^2.3 \prec 2^2.5 \prec ... \prec$$
$$2^3 \prec 2^2 \prec 2 \prec 1.$$

If f has a periodic orbit of period n and if $n \prec m$, then f also has a periodic orbit of period m.

2.2 Aperiodic Orbits

In order to study the nature of trajectories which are not periodic, it is useful to defined a "scrambled" set. A set $S \subset I$ is called a *scrambled set* if it possesses the following two properties:

(i) If x, $y \in S$ with $x \neq y$, then

$$\lim_{n \to \infty} \sup |f^n(x) - f^n(y)| > 0$$

and

$$\lim_{n \to \infty} \inf |f^n(x) - f^n(y)| = 0$$

(ii) If $x \in S$ and y is any periodic point of f,

$$\lim_{n \to \infty} \sup |f^n(x) - f^n(y)| > 0.$$

Thus trajectories starting from points in a scrambled set are not even "asymptotically periodic". Furthermore, for any pair of initial states in the scrambled set, the trajectories move apart and return close to each other infinitely often.

The following theorem, due to Li and Yorke (1975), is fundamental in establishing a connection between the existence of period-three cycles and the existence of an uncountable scrambled set.

Proposition 2.2: *Assume that there is some point x^* in I such that*

$$f^3(x^*) \leqslant x^* < f(x^*) < f^2(x^*)$$
$$(\text{or } f^3(x^*) \geqslant x^* > f(x^*) > f^2(x^*)). \tag{2.1}$$

Then

(i) for every positive integer $k = 1, 2, ...,$ there is a periodic point of period k.

(ii) there is an uncountable scrambled set $S \subset I$.

We will say that the dynamical system (I, f) is *chaotic* if conditions (i) and (ii) of Proposition 2.2 are satisfied.[3] Note that if (I, f) has a periodic point of period three, we can order the three values so that $a < c < b$. Then either (i) $f(c) = b$, in which case $f(b) = a$, and $f(a) = c$, or (ii) $f(c) = a$, in which case $f(a) = b$ and $f(b) = c$. In case (i), choosing $x^* = a$, we have $f(x^*) = f(a) = c > a = x^*$; $f^2(x^*) = f(c) = b > c = f(x^*)$; $f^3(x^*) = f(b) = a = x^*$. In case (ii), choosing $x^* = b$, we have $f(x^*) = f(b) = c < b = x^*$; $f^2(x^*) = f(c) = a < c = f(x^*)$; $f^3(x^*) = f(a) = b = x^*$. Thus, in either case, condition (2.1) of Proposition 2.2 is satisfied, and (I, f) is chaotic, On the other hand, if (I, f) is chaotic, then by condition (i), it has a periodic point of period three. Thus, (I, f) is chaotic if and only if (I, f) has a periodic point of period three.

---

[3]This is the notion of chaos, which is referred to in the literature as "Li-Yorke chaos". An alternate definition, proposed in the work of Li *et al.* (1982) involves the existence of (i) infinitely many periodic points of different periods and (ii) an uncountable scrambled set. They have shown that if f has a periodic point of period which is not a power of 2, then f is chaotic in this sense. We focus, in this chapter, on Li-Yorke chaos but our discussion in Section 7 relates to the alternate definition. *Both* definitions involve *topological chaos*, which is quite distinct from the notion of *ergodic chaos*, an important concept in the study of complex dynamic behavior. We do not discuss the concept of ergodic chaos in this chapter.

3. Dynamic Optimization

3.1 The Model

The framework is described by a triplet (Ω, u, δ), where Ω, a subset of $\Re_+ \times \Re_+$, is a *transition possibility set*, $u : \Omega \to \Re$ is a *utility function* defined on this set, and δ is the *discount factor* satisfying $0 < \delta < 1$.

The transition possibility set describes the states $z \in \Re_+$ to which it is possible to go tomorrow, if one is in the state $x \in \Re_+$ today. We define a correspondence $\phi : \Re_+ \to \Re_+$ by $\phi(x) = \{y \in \Re_+ : (x, y) \in \Omega\}$ for each $x \in \Re_+$.

A program $\{x_t\}_0^\infty$ from $\mathbf{x} \in \Re_+$ is a sequence satisfying

$$x_0 = \mathbf{x} \text{ and } (x_t, x_{t+1}) \in \Omega \text{ for } t \geqslant 0.$$

If one is in state x today and one moves to state z tomorrow (with $(x, z) \in \Omega$) then there is an immediate utility (or "reward" or "return") generated, measured by the utility function, u.

The discount factor, δ, is the weight assigned to tomorrow's utility (compared to today's) in the objective function. The *discount rate* (associated with the discount factor, δ) is given by $\rho = (1/\delta) - 1$.

The following assumptions[4] are imposed on the transition possibility set, Ω, and the utility function, u:

A1: *(i)* $(0,0) \in \Omega$, *(ii)* $(0, z) \in \Omega$ *implies* $z = 0$.

A2: Ω *is (i) closed, and (ii) convex.*

A3: *There is* $\xi > 0$ *such that* $(x, z) \in \Omega$ *and* $x \geqslant \xi$ *implies* $z < x$.

[4]An observation about the set of maintained assumptions (A.1)-(A.7) is in order at this point. We are concerned with providing an exact discount factor restriction for period-three cycles as indicated in (1.1) in the introduction. This involves two parts, the "necessity" part (i) and the "sufficiency" part (ii). As a general principle, with more assumptions, the sufficiency part becomes harder to establish, the necessity part may become easier. Thus, the choice of the set of maintained assumptions can become an important one. I have chosen to keep the assumptions as close as possible to "standard" ones used in the literature on intertemporal allocation theory. The reader will note that this has made some assumptions (like the "free-disposal" assumption (A.4) and the "monotonicity" assumption (A.7)) superfluous in providing the "necessity" results (Theorems 5.1 and 6.1). They have, however, made the "sufficiency" part somewhat harder to establish.

A4: *If* $(x, z) \in \Omega$ *and* $x' \geqslant x$, $0 \leqslant z' \leqslant z$ *then* $(x', z') \in \Omega$.

Clearly, we can pick $0 < \zeta < \xi$, such that if $x > \zeta$ and $(x, z) \in \Omega$, then $z < x$.[5] It is straightforward to verify that if $(x, z) \in \Omega$, then $z \leqslant \max(\zeta, x)$. It follows from this that if $\{x_t\}_0^\infty$ is a program from $\mathbf{x} \in \Re_+$, then $x_t \leqslant \max(\zeta, \mathbf{x})$ for $t \geqslant 0$. In particular, if $\mathbf{x} \leqslant \zeta$, then $x_t \leqslant \zeta$ for $t \geqslant 0$. This leads us to choose the closed interval, $[0, \zeta]$ as the natural *state space* of our model, which we will denote by X. We denote the interval $[0, \xi]$ by Y.

The following assumptions are imposed on the utility function, u:

A5: u *is concave on* Ω; *further, if* (x, z) *and* (x', z') *are in* Ω, *and* $x \neq x'$, *then for every* $0 < \lambda < 1$, $u(\lambda(x, z) + (1 - \lambda)(x', z')) > \lambda u(x, z) + (1 - \lambda) u(x', z')$.

A6: u *is upper semi-continuous on* Ω.

A7: *If* $x, x' \in Y$, $(x, z) \in \Omega$, $x' \geqslant x$ *and* $0 \leqslant z' \leqslant z$, *then* $u(x', z') \geqslant u(x, z)$.

A program $\{\hat{x}_t\}_0^\infty$ from $\mathbf{x} \geqslant 0$ is an *optimal program* if

$$\sum_0^\infty \delta^t u(x_t, x_{t+1}) \leqslant \sum_0^\infty \delta^t u(\hat{x}_t, \hat{x}_{t+1})$$

for every program $\{x_t\}_0^\infty$ from $\mathbf{x}$.

Under (A1)-(A7), there is a unique optimal program from every $\mathbf{x} \in \Re_+$.

---

[5] The fuss about making a distinction between ξ and ζ should perhaps be explained. Clearly, ζ can be a "maximum sustainable stock", while ξ cannot. Thus, $Y = [0, \xi]$ is a somewhat larger closed interval than the state space $X = [0, \zeta]$, where the important dynamics will take place. We wanted to make assumption (A.7) on the monotone nature of the utility function in "standard" form; that is, without restricting x, x' in any way. But this created problems in the construction of the example in Section 6.2. So we settled for (A.7) in its present form, noting thereby that the monotone restriction on u can be maintained on a larger closed interval than the state space X, while preserving the example in Section 6.2. In fact, the reader will note that the method followed in constructing the example allows us to maintain the monotone restrictions on u on any large finite interval $[0, \xi']$ containing the state space, by suitable modification in the definition of u in the example.

3.2 Value and Policy Functions

The *value function* $V : \Re_+ \to \Re$ is defined by

$$V(x) = \sum_0^\infty \delta^t u(\hat{x}_t, \hat{x}_{t+1}),$$

where $\{\hat{x}_t\}_0^\infty$ is the optimal program from $x \in \Re_+$.

The *policy function* $h : \Re_+ \to \Re_+$ is defined by

$$h(x) = \hat{x}_1,$$

where $\{\hat{x}_t\}_0^\infty$ is the optimal program from $x \in \Re_+$.

The properties of the value and policy functions can be summarized in the following result. This is based on Dutta and Mitra (1989) and Stokey *et al.* (1989).

Proposition 3.1: *(i) The value function V is strictly concave*[6] *and continuous on $\Re_+$ and non-decreasing on Y. Further, V is the unique continuous function on $Y \equiv [0, \xi]$ which satisfies the functional equation of dynamic programming:*

$$V(x) = \max_{y \in \phi(x)} [u(x, y) + \delta V(y)].$$

(ii) The policy function h satisfies the following property: for each $x \in \Re_+$, $h(x)$ is the unique solution to the constrained maximization problem:

$$\begin{array}{ll} \text{maximize} & u(x,y) + \delta V(y) \\ \text{subject to} & y \in \phi(x) \end{array}$$

Further, h is continuous on $\Re_+$.

Remark 3.1: (i) In view of the definition of the policy function h, the optimal program from $x \in X$ is the trajectory $\{h^t(x)\}_0^\infty$

[6] The strict concavity of the value function is essential to this study (as it is in the literature discussed in Section 1). Assumption (A.5) is sufficient for this, but is not necessary (see Montrucchio and Sorger (1994) for a useful discussion). We have still maintained (A.5), since we view, for this model, the utility function, u, as a "primitive," and the value function, V, as a "derived" concept.

generated by the policy function. Thus, an optimal program from $x \in X$ can be called periodic (with period k) if x is a periodic point of h (with period k).

(ii) Since V is concave on $\Re_+$, it has well-defined left-hand and right-hand derivatives for all $x > 0$, which we denote by $\mu(x)$ and $v(x)$ respectively. For $x = 0$, the right-hand derivative $v(0)$ can be defined, but it may be infinite. If $x > y > 0$, then we have $\mu(y) \geqslant v(y) > \mu(x) \geqslant v(x)$, the strict inequality following from the strict concavity of V.

3.3 Price Characterization of Optimality

Optimality can be conveniently characterized in terms of dual variables or shadow prices. The theory, describing this characterization, is well known and a cornerstone of the theory of optimal economic growth. We state the result which we will use in the following section; a full discussion can be found in Weitzman (1973) and McKenzie (1986).

Proposition 3.2: *(a) If* $\{x_t\}_0^\infty$ *is an optimal program from* $\mathbf{x} \in X$ *and* $\mathbf{x} > 0$, *and there is some* $(\bar{x}, \bar{y}) \in \Omega$ *with* $\bar{y} > 0$ *then there is a sequence* $\{p_t\}_0^\infty$ *of non-negative prices such that for* $t \geqslant 0$,

(i) $\delta^t V(x_t) - p_t x_t \geqslant \delta^t V(x) - p_t x$ *for all* $x \geqslant 0$

(ii) $\delta^t u(x_t, x_{t+1}) + p_{t+1}x_{t+1} - p_t x_t \geqslant \delta^t u(x, y) + p_{t+1}y - p_t x$ *for all* $(x, y) \in \Omega$

(iii) $\lim_{t\to\infty} p_t x_t = 0$.

(b) If $\{x_t\}_0^\infty$ *is a program from* $\mathbf{x} \geqslant 0$, *and there is a sequence* $\{p_t\}_0^\infty$ *of non-negative prices such that for* $t \geqslant 0$, *(ii) and (iii) above are satisfied, then* $\{x_t\}_0^\infty$ *is an optimal program from* $\mathbf{x}$.

We should add a remark here regarding the proof of part (a) of Proposition 3.2. If we follow McKenzie (1986, Lemma 4.1), we can obtain a sequence $\{p_t\}_0^\infty$, with $p_t \in \Re$ for $t \geqslant 0$, such that (i) and (ii) of Proposition 3.2(a) are satisfied.[7] Since $\mathbf{x} \in X$ we

---

[7] In fact this is all that is needed in proving Theorems 5.1 and 6.1. The non-negativity of prices and the necessity of the transversality condition for optimal programs are not needed

have $x_t \leqslant \zeta$ for $t \geqslant 0$. By choosing $\zeta < x' \leqslant \xi$, and noting that V is non-decreasing on Y, $V(x') \geqslant V(x_t)$ for $t \geqslant 0$. Now, using (i), $p_t(x' - x_t) \geqslant \delta^t[V(x') - V(x_t)] \geqslant 0$, so that $p_t \geqslant 0$ for $t \geqslant 0$. The "transversality condition" (iii) can be obtained from (i) by choosing $x = (x_t/2)$, so that $0 \leqslant (1/2)p_t x_t \leqslant \delta^t[V(x_t) - V(x_t/2)]$. Since V is continuous on X, there is $m > 0$ such that $|V(x)| \leqslant m$ for all $x \in X$. Since $x_t \in X$ and $(x_t/2) \in X$, $\delta^t[V(x_t) - V(x_t/2)] \leqslant 2\delta^t m$. Since $2\delta^t m \to 0$ as $t \to \infty$, we can conclude that (iii) holds.

If $\{x_t\}_0^\infty$ is a program from $\mathbf{x} \geqslant 0$, and $\{p_t\}_0^\infty$ is non-negative sequence of prices satisfying (i), (ii), and (iii) of Proposition 3.2(a), we will say that the program $\{x_t\}_0^\infty$ is *price supported* by $\{p_t\}_0^\infty$.

Corollary 3.1: *If $\{x_t\}_0^\infty$ is a periodic optimal program of period $k \geqslant 2$, then there is a sequence $\{q_t\}_0^\infty$ which price supports $\{x_t\}_0^\infty$ such that $(q_t/\delta^t) = (q_{t-k}/\delta^{t-k})$ for all $t \geqslant k$.*

Proof: Let $\{x_t\}_0^\infty$ be a periodic optimal program of period $k \geqslant 2$. Then $x_t > 0$, and $x_t \in X$ for $t \geqslant 0$, so Proposition 3.2 is applicable. Let $\{p_t\}_0^\infty$ be a price support of $\{x_t\}_0^\infty$. Define the "current value" prices $P_t = (p_t/\delta^t)$ for $t \geqslant 0$.

One can show that a price support $\{q_t\}_0^\infty$ of $\{x_t\}_0^\infty$ can be found such that its "current value" prices, $Q_t = (q_t/\delta^t)$ for $t \geqslant 0$, satisfy the condition

$$Q_t = Q_{t-k} \text{ for } t \geqslant k$$

This can be done, by following the method of Sutherland (1967) and McKenzie (1986). Since V is continuous on X, there is $m > 0$ such that $|V(x)| \leqslant m$ for all $x \in X$. Denote $\min[x_0, \ldots, x_{k-1}]$ by ε. Then $\varepsilon > 0$, and by choosing $x = (\varepsilon/2)$ in Proposition 3.2(i),

$$2m \geqslant V(x_t) - V(\varepsilon/2) \geqslant P_t(\varepsilon/2)$$

so that $0 \leqslant P_t \leqslant (4m/\varepsilon) = m'$ for all $t \geqslant 0$.

---

for that. However, given our maintained set of assumptions (see footnote 4), these properties do follow easily and are noted to keep our exposition closely related to the literature on "price characterizations" of optimal programs.

Given $j \in [0, 1, 2, ..., k]$, denote for $N \geqslant 0$,

$$Q_j(N) \equiv (P_j + P_{j+k} + P_{j+2k} + ... + P_{j+Nk})/(N+1).$$

Note that $0 \leqslant Q_j(N) \leqslant m'$ for all j and N. Thus, there is a subsequence $\{N_i\}$, $i = 1, 2, ...$ such that for $j \in [0, 1, ..., k]$, $Q_j(N_i) \to Q_j$ as $i \to \infty$.

Given $j \in [0, 1, 2, ..., k-1]$, we can use Proposition 3.2 to get

$$V(x_j) - P_j x_j \geqslant V(x) - P_j x \text{ for all } x \geqslant 0$$
$$u(x_j, x_{j+1}) + \delta P_{j+1} x_{j+1} - P_j x_j \geqslant u(x, z) + \delta P_{j+1} z - P_j x$$
$$\text{for all } (x, z) \in \Omega$$

The same inequality holds if (P_{j+1}, P_j) are replaced by (P_{j+1+Nk}, P_{j+Nk}) for all $N \geqslant 0$ since $(x_j, x_{j+1}) = (x_{j+Nk}, x_{j+1+Nk})$. Thus, we get, by averaging over the first $(N+1)$ inequalities,

$$V(x_j) - Q_j(N) x_j \geqslant V(x) - Q_j(N) x$$
$$\text{for all } x \geqslant 0$$
$$u(x_j, x_{j+1}) + \delta Q_{j+1}(N) x_{j+1} - Q_j(N) x_j \geqslant u(x, z) +$$
$$\delta Q_{j+1}(N) z - Q_j(N) x \text{ for all } (x, z) \in \Omega.$$

Letting $N_i \to \infty$, we get

$$V(x_j) - Q_j x_j \geqslant V(x) - Q_j x \text{ for all } x \geqslant 0$$
$$u(x_j, x_{j+1}) + \delta Q_{j+1} x_{j+1} - Q_j x_j \geqslant u(x, z) + \delta Q_{j+1} z - Q_j x$$
$$\text{for all } (x, z) \in \Omega.$$

Now, note that for $N \geqslant 0$,

$$Q_k(N) = [P_k + P_{2k} + ... + P_{(N+1)k}]/(N+1)$$
$$= Q_0(N) + [P_{(N+1)k} - P_0)/(N+1)].$$

Thus we have

$$Q_k = \lim_{i \to \infty} Q_k(N_i) = \lim_{i \to \infty} Q_0(N_i) = Q_0.$$

Now, defining $Q_t = Q_{t-k}$ for $t > k$, and $q_t = Q_t \delta^t$ for $t \geqslant 0$, it is straightforward to check that $\{q_t\}_0^\infty$ is a price support of $\{x_t\}_0^\infty$. Q.E.D.

4. The Value-Loss Approach to Minimum Impatience Results

In this section, we describe our approach to obtaining minimum impatience results. It is based on the observation that at the prices supporting an optimal program, there is no activity which yields a higher "generalized profit" at any date (value of utility plus value of terminal stocks minus value of initial stocks at that date) than the activity chosen along the optimal program at that date. In other words, there are no arbitrage possibilities available at the supporting prices.

We now proceed to explain this more formally. Let $\{x_t\}_0^\infty$ be an optimal program with supporting prices $\{p_t\}_0^\infty$. Then, the *value loss* at date t if the economy chooses $(x, z) \in \Omega$ (instead of $(x_t, x_{t+1}) \in \Omega$) is

$$\begin{aligned} L_t(x, z; \{x_t, p_t\}_0^\infty) = \delta^t u(x_t, x_{t+1}) + p_{t+1} x_{t+1} - p_t x_t \\ - [\delta^t u(x, z) + p_{t+1} z - p_t x]. \end{aligned} \tag{4.1}$$

Since $\{p_t\}_0^\infty$ supports $\{x_t\}_0^\infty$, we have

$$L_t(x, z; \{x_t, p_t\}_0^\infty) \geqslant 0 \text{ for all } (x, z) \in \Omega. \tag{4.2}$$

Now let $\{y_t\}_0^\infty$ be another optimal program with supporting prices $\{q_t\}_0^\infty$. Then the value loss at date t if the economy chooses $(x, z) \in \Omega$ (instead of $(y_t, y_{t+1}) \in \Omega$) is

$$\begin{aligned} L_t(x, z; \{y_t, q_t\}_0^\infty) = \delta^t u(y_t, y_{t+1}) + q_{t+1} y_{t+1} - q_t y_t \\ - [\delta^t u(x, z) + q_{t+1} z - q_t x]. \end{aligned} \tag{4.3}$$

Since $\{q_t\}_0^\infty$ supports $\{y_t\}_0^\infty$, we have

$$L_t(x, z; \{y_t, q_t\}_0^\infty) \geqslant 0 \text{ for all } (x, z) \in \Omega. \tag{4.4}$$

If we apply (4.1) to $(y_t, y_{t+1}) \in \Omega$, we get

$$\delta^t u(x_t, x_{t+1}) + p_{t+1}x_{t+1} - p_t x_t = \delta^t u(y_t, y_{t+1}) + p_{t+1}y_{t+1} - p_t y_t + L_t(y_t, y_{t+1}, \{x_t, p_t\}_0^\infty). \tag{4.5}$$

Similarly, if we apply (4.3) to $(x_t, x_{t+1}) \in \Omega$, we get

$$\delta^t u(y_t, y_{t+1}) + q_{t+1}y_{t+1} - q_t y_t = \delta^t u(x_t, x_{t+1}) + q_{t+1}x_{t+1} - q_t x_t + L_t(x_t, x_{t+1}; \{y_t, q_t\}_0^\infty). \tag{4.6}$$

Combining (4.5) and (4.6), we get

$$L_t(x_t, x_{t+1}; \{y_t, q_t\}_0^\infty) + L_t(y_t, y_{t+1}; \{x_t p_t\}_0^\infty) = -(p_{t+1} - q_{t+1})(y_{t+1} - x_{t+1}) + (p_t - q_t)(y_t - x_t). \tag{4.7}$$

Using (4.2) and (4.4) in (4.7), we get

$$(p_{t+1} - q_{t+1})(y_{t+1} - x_{t+1}) \leqslant (p_t - q_t)(y_t - x_t). \tag{4.8}$$

Denoting (p_t/δ^t) by P_t and (q_t/δ^t) by Q_t for $t \geqslant 0$ (these are, of course, the "current value prices"), we get

$$\delta(P_{t+1} - Q_{t+1})(y_{t+1} - x_{t+1}) \leqslant (P_t - Q_t)(y_t - x_t). \tag{4.9}$$

By (A5), the inequality in (4.9) is strict if $x_t \neq y_t$.

This fundamental inequality yields *all* the discount factor restrictions that we will derive in this chapter. But, before we do, we need to know the sign of the typical term $(P_t - Q_t)(y_t - x_t)$. This is fairly easy. Since $\{p_t\}_0^\infty$ is a price support of $\{x_t\}_0^\infty$,

$$\delta^t V(x_t) - p_t x_t \geqslant \delta^t V(y_t) - p_t y_t. \tag{4.10}$$

Since $\{q_t\}_0^\infty$ is a price support of $\{y_t\}_0^\infty$,

$$\delta^t V(y_t) - q_t y_t \geqslant \delta^t V(x_t) - q_t x_t. \tag{4.11}$$

Combining (4.10) and (4.11) ,

$$(p_t - q_t)(y_t - x_t) \geqslant 0. \tag{4.12}$$

Dividing through in (4.12) by δ^t,

$$(P_t - Q_t)(y_t - x_t) \geqslant 0 \tag{4.13}$$

By the strict concavity of V, the inequality in (4.13) is strict if $y_t \neq x_t$. We summarize the above discussion in the following proposition.

Proposition 4.1: *Let (Ω, u, δ) be a dynamic optimization model. Suppose $\{x_t\}_0^\infty$ is an optimal program with price support $\{p_t\}_0^\infty$, and $\{y_t\}_0^\infty$ is an optimal program with price support $\{q_t\}_0^\infty$. Denoting (p_t/δ^t) by P_t and (q_t/δ^t) by Q_t for $t \geqslant 0$, we have*

(i) $\delta(P_{t+1} - Q_{t+1})(y_{t+1} - x_{t+1}) \leqslant (P_t - Q_t)(y_t - x_t)$ for $t \geqslant 0$

(ii) $(P_t - Q_t)(y_t - x_t) \geqslant 0$ for $t \geqslant 0$.

Furthermore, if $y_t \neq x_t$ for some t, then the inequalities in (i) and (ii) are strict for that t.

Let us demonstrate how the basic inequalities of Proposition 4.1 can be used to obtain minimum impatience theorems. The typical situation that we will be applying it to will involve an optimal program $\{x_t\}_0^\infty$ exhibiting the pattern

$$x_2 < x_0 < x_1.$$

That is, first the state variable increases, $(x_1 > x_0)$ and then it falls below the *original level* $(x_2 < x_0)$. We will show[8] that this immediately leads to a discount factor restriction

$$\delta < (x_1 - x_0)/(x_1 - x_2). \tag{4.14}$$

Proposition 4.2: *Let (Ω, u, δ) be a dynamic optimization model. Suppose $\{x_t\}_0^\infty$ is an optimal program.*

(i) If $x_2 < x_0 < x_1$, then $\delta < (x_1 - x_0)/(x_1 - x_2)$.

(ii) If $x_1 < x_0 < x_2$, then $\delta < (x_0 - x_1)(x_2 - x_1)$.

---

[8] Although Proposition 4.2 has not been explicitly noted in the literature, Gerhard Sorger has used it in his oral presentations on the subject, to show how just three observations on an (assumed) optimal program might yield a rather good upper bound on the discount factor.

Proof: We will only establish (i), since (ii) follows by similar arguments.

If $x_2 < x_0 < x_1$, then $x_0 > 0$, and $(x_0, x_1) \in \Omega$ with $x_1 > 0$. Thus, Proposition 3.2 is applicable, and let $\{p_t\}_0^\infty$ be a price support of $\{x_t\}_0^\infty$. The sequence $\{y_t\}_0^\infty$ defined by $y_t = x_{t+1}$ for $t \geqslant 0$ is a program from $y_0 \equiv x_1$, and it is clearly an optimal program from y_0. Defining a sequence $\{q_t\}_0^\infty$ by $q_t = (p_{t+1}/\delta)$ for $t \geqslant 0$, it is straightforward to check that $\{q_t\}_0^\infty$ is a price support of $\{y_t\}_0^\infty$. Thus, Proposition 4.1 is applicable.

Applying Proposition 4.1 for $t = 0$, and noting that $x_0 \neq x_1 = y_0$, $x_1 \neq x_2 = y_1$, we get

$$0 < \delta(P_1 - Q_1)(y_1 - x_1) < (P_0 - Q_0)(y_0 - x_0). \tag{4.15}$$

This yields the discount factor restriction

$$\delta < (P_0 - Q_0)(y_0 - x_0)/(Q_1 - P_1)(x_1 - y_1). \tag{4.16}$$

It follows from Proposition 3.2(i) that $v(x_t) \leqslant P_t$, $v(y_t) \leqslant Q_t$; furthermore, $P_t \leqslant \mu(x_t)$, $Q_t \leqslant \mu(y_t)$ whenever $x_t > 0$, $y_t > 0$. Thus, we get

$$P_0 \geqslant v(x_0) > \mu(x_1) = \mu(y_0) \geqslant Q_0 \tag{4.17}$$

since $x_0 < x_1$ and V is strictly concave. Similarly, we get

$$Q_1 \geqslant v(y_1) = v(x_2) > \mu(x_1) \geqslant P_1 \tag{4.18}$$

since $x_2 < x_1$ and V is strictly concave. So $(P_0 - Q_0) > 0$, $(Q_1 - P_1) > 0$, and

$$\begin{aligned}(Q_1 - P_1) &\geqslant [v(x_2) - P_1] > [\mu(x_0) - P_1] \\ &\geqslant [P_0 - P_1] = (P_0 - Q_0).\end{aligned} \tag{4.19}$$

Also $(y_0 - x_0) = (x_1 - x_0) > 0$, and $(x_1 - y_1) = (x_1 - x_2) > 0$. Thus, (4.16) yields

$$\delta < \frac{(y_0 - x_0)}{(x_1 - y_1)} = \frac{(x_1 - x_0)}{(x_1 - x_2)},$$

which establishes the result. Q.E.D.

Remark 4.1: Proposition 4.2 imposes strong discount factor restrictions on dynamic optimization models yielding the logistic map or the tent map as their policy functions.

If a dynamic optimization model (Ω, u, δ) has the tent-map ($h(x) = 2x$ for $0 \leqslant x \leqslant 0.5$, $h(x) = 2 - 2x$ for $0.5 \leqslant x \leqslant 1$) as its policy function, then $(x_0, x_1, x_2 ...) = (1/2, 1, 0, 0 ...)$ is an optimal program with $x_2 < x_0 < x_1$, so that δ must be less than $(1/2)$ by Proposition 4.2.

If (Ω, u, δ) has the logistic map ($h(x) = 4x(1-x)$ for $0 \leqslant x \leqslant 1$) as its policy function, then $(x_0, x_1, x_2, ...) = (0.625, 0.9375, 0.234375, ...)$ is an optimal program with $x_2 < x_0 < x_1$, so that $\delta < 0.4$ by Proposition 4.2. More generally, if (Ω, u, δ) has the quadratic map ($h(x) = \mu x(1-x)$ for $0 \leqslant x \leqslant 1$, with $3 < \mu \leqslant 4$) as its policy function, then choosing $x_0 = 0.5(1+\mu)/\mu$, $x_1 = h(x_0)$, $x_2 = h(x_1)$, one can show that $\delta < 4/(\mu - 1)^2$ by applying Proposition 4.2.

5. A Discount Factor Restriction for Period-Three Cycles

While Proposition 4.2 is useful in obtaining some discount factor restrictions (as noted in the previous section) its shortcoming is that it may not be very useful when the only information we have is that there is a period three cycle (i.e., when the nature of the policy function on the rest of its domain is not known).

Proceeding more formally, suppose $\{x_t\}_0^\infty$ is an optimal program of a dynamic optimization model (Ω, u, δ) exhibiting a period-three cycle. Then we can order the set $\{x_0, x_1, x_2\}$ on the real line. Let the highest value be b, the lowest value be a, and the middle value be c. Then, denoting the policy function by h, we have two possibilities (i) $h(c) = a$, (ii) $h(c) = b$. In (i), we must also have $h(a) = b$ and $h(b) = c$; similarly in case (ii), we must have $h(b) = a$ and $h(a) = c$. Thus, denoting the optimal program from c by $\{y_t\}_0^\infty$, we have two possibilities: (i) $y_2 < y_0 < y_1$, or (ii)

$y_1 < y_0 < y_2$. Then Proposition 4.2 is applicable yielding

$$\delta < (y_1 - y_0)/(y_1 - y_2) \tag{5.1}$$

But, not knowing exactly where y_0 is between y_1 and y_2, the discount factor restriction obtained from (5.1) can be quite weak (when y_0 is close to y_2).

The way to strengthen the restriction imposed by (5.1) is suggested by the observation that Proposition 4.2 does not fully exploit the fact that there is a period-three cycle. In particular, $y_3 = h(y_2) = y_0$ is nowhere used, and this information should yield a restriction in addition to (5.1). This additional restriction turns out to be

$$\delta^2 < (y_0 - y_2)/(y_1 - y_2). \tag{5.2}$$

Now the two restrictions can be combined to yield

$$\delta < [\sqrt{5} - 1]/2, \tag{5.3}$$

which is completely independent of the actual value of y_0 (or, for that matter, y_1 and y_2). This is formally demonstrated in the following theorem.

Theorem 5.1: *Let (Ω, u, δ) be a dynamic optimization model, with a policy function h. If h exhibits a period-three cycle, then*

$$\delta < [\sqrt{5} - 1]/2. \tag{5.4}$$

Proof: Let the three values of the period-three cycle be ordered so that $a < c < b$. There are then exactly two possibilities: (i) $h(c) = b$[in which case $h(b) = a$ and $h(a) = c$], or (ii) $h(c) = a$ [in which case $h(a) = b$ and $h(b) = c$]. We concentrate on the first possibility, since the second one can be analyzed similarly.

In case (i), let $\{x_t\}_0^\infty$ be the optimal program from c. Then $x_0 = c$, $x_1 = b$, $x_2 = a$, $x_3 = c$, and

$$x_2 < x_0 < x_1.$$

Applying Proposition 4.2, we have

$$\delta < (x_1 - x_0)/(x_1 - x_2). \tag{5.5}$$

Note that $x_0 > 0$, and $(x_0, x_1) \in \Omega$ with $x_1 > 0$, so Proposition 3.2 is applicable, and there is a sequence $\{p_t\}_0^\infty$ which provides a price-support to $\{x_t\}_0^\infty$.

The sequence $\{y_t\}_0^\infty$ defined by $y_t = x_{t+2}$ for $t \geqslant 0$ is a program from $y_0 = x_2$, and it is clearly an optimal program from y_0. Defining $\{q_t\}_0^\infty$ by $q_t = (p_{t+2}/\delta^2)$ for $t \geqslant 0$, it is easy to check that $\{q_t\}_0^\infty$ is a price support of $\{y_t\}_0^\infty$. Applying Proposition 4.1 for $t = 0$, and noting that $x_0 \neq x_2 = y_0$, $x_1 \neq x_0 = x_3 = y_1$, we get

$$0 < \delta(P_1 - Q_1)(y_1 - x_1) < (P_0 - Q_0)(y_0 - x_0). \tag{5.6}$$

Applying Proposition 4.1 for $t = 1$, and noting that $x_1 \neq x_0 = y_1$, $x_2 \neq x_1 = x_4 = y_2$, we get

$$0 < \delta(P_2 - Q_2)(y_2 - x_2) < (P_1 - Q_1)(y_1 - x_1). \tag{5.7}$$

Combining (5.6) and (5.7), we get

$$\delta^2 < \frac{(P_0 - Q_0)(y_0 - x_0)}{(P_2 - Q_2)(y_2 - x_2)} = \frac{(Q_0 - P_0)(x_0 - y_0)}{(P_2 - Q_2)(y_2 - x_2)} = \frac{(P_2 - P_0)(x_0 - y_0)}{(P_2 - Q_2)(y_2 - x_2)} \tag{5.8}$$

It follows from Proposition 3.2(i) that

$$P_0 \leqslant \mu(x_0) < v(x_2) \leqslant P_2 \tag{5.9}$$

since $x_0 > x_2$ and V is strictly concave. Similarly, we get

$$P_2 \geqslant v(x_2) > \mu(x_1) = \mu(y_2) \geqslant Q_2 \tag{5.10}$$

since $x_2 < x_1$ and V is strictly concave. Thus $(P_2 - P_0) > 0$ and $(P_2 - Q_2) > 0$, and

$$(P_2 - Q_2) \geqslant [P_2 - \mu(y_2)] = [P_2 - \mu(x_1)] > [P_2 - v(x_0)] \geqslant (P_2 - P_0). \tag{5.11}$$

Thus (5.8) yields the restriction

$$\delta^2 < (x_0 - y_0)/(y_2 - x_2) = (x_0 - x_2)/(x_1 - x_2). \qquad (5.12)$$

Denote $(x_1 - x_0)/(x_1 - x_2)$ by α. Then $0 < \alpha < 1$, and $(1 - \alpha) = (x_0 - x_2)/(x_1 - x_2)$. Thus using (5.5) and (5.12),

$$\delta < \min[\alpha, \sqrt{1 - \alpha}]. \qquad (5.13)$$

Denote $(\sqrt{5} - 1)/2$ by β, so that $\beta^2 = (1 - \beta)$. Thus if $\alpha \geqslant \beta$, then $\min[\alpha, \sqrt{1 - \alpha}] \leqslant \sqrt{1 - \alpha} \leqslant \sqrt{1 - \beta} = \beta$. And if $0 < \alpha < \beta$, then $\min[\alpha, \sqrt{1 - \alpha}] < \beta$. Thus in either case,

$$\min[\alpha, \sqrt{1 - \alpha}] \leqslant \beta = (\sqrt{5} - 1)/2 \qquad (5.14)$$

Combining (5.13) and (5.14), we get (5.4).

Remark 5.1: Sorger (1992b) obtained the discount factor restriction (5.4) of Theorems 5.1, by applying the theory of stochastic dominance. The restriction has been refined to $\delta < 0.5479$ in Sorger (1994), using the same methods.

Montrucchio (1994) does not address directly the problem of finding a discount factor restriction for period three cycles. However, he establishes (under strong concavity assumptions on the utility function) that if (Ω, u, δ) is a dynamic optimization model with policy function h, and the topological entropy of h is $\Psi(h)$, then the discount factor, δ, is related to the topological entropy by the inequality

$$\delta \leqslant e^{-\Psi(h)}. \qquad (5.15)$$

It has been shown in Mitra (Chapter 13) that if h exhibits a period-three cycle then (5.15) can be used to obtain the discount factor restriction (5.4) of Theorem 5.1. The inequality (5.15) actually holds under the assumptions used in this chapter: this follows from the recent work of Montrucchio and Sorger (1994).

6. An Exact Discount Factor Restriction for Period-Three Cycles

6.1 The Main Result

Theorem 5.1 provides a strong restriction on the discount factor for period-three cycles, but as we mentioned in the previous section, stronger restrictions have already been obtained. This suggests that, in establishing the Theorem, we have not exploited the full implication of the existence of a period-three cycle; in other words, we have thrown out useful information somewhere along the line of our argument.

In order to see most clearly what we have left out, assume that the value function is differentiable at the points of the period-three cycle. Then the "current value prices" are simply the derivatives of the value function at the respective points.

Now, if we look at the proof of Theorem 5.1 carefully, we notice that the restriction (5.5) is obtained from the inequality (4.16) in Proposition 4.2 which now reads

$$\delta < \frac{[V'(x_0) - V'(x_1)](x_1 - x_0)}{[V'(x_2) - V'(x_1)](x_1 - x_2)}. \tag{6.1}$$

Similarly, the restriction (5.12) is obtained from the inequality (5.8), which reads

$$\delta^2 < \frac{[V'(x_2) - V'(x_0)](x_0 - x_2)}{[V'(x_2) - V'(x_1)](x_1 - x_2)}. \tag{6.2}$$

We know that $[V'(x_0)-V'(x_1)]/[V'(x_2)-V'(x_1)]$ is between 0 and 1 and thus (6.1) yields (5.5). But nothing more is known about the ratio, and it could well be very close to 1. The same could be said about the ratio $[V'(x_2)-V'(x_0)]/[V'(x_2)-V'(x_1)]$ and the relation of (6.2) to (5.12). But *both* ratios cannot be close to 1: in fact, if one is close to 1, the other is close to zero independent of how concave V is. In this regard, these ratios play exactly the same role in the expressions (6.1) and (6.2), as the ratios involving the primal variables $(x_1-x_0)/(x_1-x_2)$ and $(x_0-x_2)/(x_1-x_2)$. Since we were able to exploit the inequalities (5.5) and (5.12), involving only the primal variables, to get a discount factor restriction of $(\sqrt{5}-1)/2$, we should be able to exploit (6.1) and (6.2) to yield a restriction which is exactly the *square* of this number. We now proceed to prove this formally. [No assumption regarding the differentiability of V will be made in the formal proof].

Theorem 6.1: *Let* (Ω, u, δ) *be a dynamic optimization model, with a policy function* h. *If* h *exhibits a period-three cycle, then*

$$\delta < [(\sqrt{5}-1)/2]^2. \tag{6.3}$$

Proof: As in the proof of Theorem 5.1, let the three values of the period-three cycle be ordered so that $a < c < b$. There are two possibilities to consider: (i) $h(c) = b$ [in which case $h(b) = a$ and $h(a) = c$], or (ii) $h(c) = a$ [in which case $h(a) = b$ and $h(b) = c$]. We concentrate on possibility (i); (ii) can be analyzed similarly.

In case (i), let $\{x_t\}_0^\infty$ be the optimal program from c. Then $x_0 = c$, $x_1 = b$, $x_2 = a$, $x_3 = c$, and $x_2 < x_0 < x_1$. Let $\{p_t\}_0^\infty$ be a price support of $\{x_t\}_0^\infty$.

The sequence $\{z_t\}_0^\infty$ defined by $z_t = x_{t+1}$ for $t \geqslant 0$ is a program from $z_0 = x_1$, and it is the optimal program from z_0. Defining a sequence $\{r_t\}_0^\infty$ by $r_t = (p_{t+1}/\delta)$ for $t \geqslant 0$, it is easy to check that $\{r_t\}_0^\infty$ is a price support of $\{z_t\}_0^\infty$. Denote (p_t/δ^t) by P_t and (r_t/δ^t) by R_t for $t \geqslant 0$.

Now, following the proof of Proposition 4.2, we get

$$\delta < \frac{(P_0 - R_0)(z_0 - x_0)}{(R_1 - P_1)(x_1 - z_1)}, \tag{6.4}$$

where $0 < (P_0 - R_0) < (R_1 - P_1)$, and $0 < (z_0 - x_0) < (x_1 - z_1)$. Now $R_0 = r_0 = (p_1/\delta) = P_1$, and $R_1 = (r_1/\delta) = (p_2/\delta^2) = P_2$. Also $z_0 = x_1$ and $z_1 = x_2$. Thus (6.4) can be rewritten as

$$\delta < \frac{(P_0 - P_1)(x_1 - x_0)}{(P_2 - P_1)(x_1 - x_2)}. \tag{6.5}$$

The sequence $\{y_t\}_0^\infty$ defined by $y_t = x_{t+2}$ for $t \geqslant 0$ is a program from $y_0 = x_2$, and it is the optimal program from y_0. Defining a sequence $\{q_t\}_0^\infty$ by $q_t = (p_{t+2}/\delta^2)$ for $t \geqslant 0$, it is easy to check that $\{q_t\}_0^\infty$ is a price-support of $\{y_t\}_0^\infty$. Denote (q_t/δ^t) by Q_t for $t \geqslant 0$.

Then, following the proof of Theorem 5.1, we get

$$\delta^2 < \frac{(Q_0 - P_0)(x_0 - y_0)}{(P_2 - Q_2)(y_2 - x_2)}, \tag{6.6}$$

where $0 < (Q_0 - P_0) < (P_2 - Q_2)$, and $0 < (x_0 - y_0) < (y_2 - x_2)$. Now, $Q_0 = q_0 = (p_2/\delta^2) = P_2$ and $Q_2 = (q_2/\delta^2) = (p_4/\delta^4) = P_4 = P_1$, the last equality following from Corollary 3.1. Also $y_0 = x_2$ and $y_2 = x_1$. Thus (6.6) can be rewritten as

$$\delta^2 < \frac{(P_2 - P_0)(x_0 - x_2)}{(P_2 - P_1)(x_1 - x_2)}. \tag{6.7}$$

Denote $(P_0 - P_1)/(P_2 - P_1)$ by α; then $0 < \alpha < 1$, and $(1 - \alpha) = (P_2 - P_0)/(P_2 - P_1)$. Denote $(x_1 - x_0)/(x_1 - x_2)$ by γ; then $0 < \gamma < 1$ and $(1 - \gamma) = (x_0 - x_2)/(x_1 - x_2)$. Thus (6.5) and (6.7) yield the restriction

$$\delta < \min[\alpha\gamma, \sqrt{(1 - \alpha)(1 - \gamma)}]. \tag{6.8}$$

Denoting $(\alpha + \gamma)/2$ by η, we have

$$\min[\alpha\gamma, \sqrt{(1 - \alpha)(1 - \gamma)}] \leqslant \min[\eta^2, \sqrt{(1 - \eta)^2}]. \tag{6.9}$$

Using (6.9) in (6.8), we get

$$\delta < \min[\eta^2, (1 - \eta)]. \tag{6.10}$$

As in the proof of Theorem 6.1, denote $(\sqrt{5} - 1)/2$ by β, so that $\beta^2 = 1 - \beta$. If $\eta \geqslant \beta$, then $(1 - \eta) \leqslant (1 - \beta) = \beta^2$, so that $\min[\eta^2, (1 - \eta)] \leqslant \beta^2$. If $0 < \eta < \beta$, then $\eta^2 < \beta^2$, so that $\min[\eta^2, (1 - \eta)] < \beta^2$. Thus, in either case,

$$\min[\eta^2, (1 - \eta)] \leqslant \beta^2. \tag{6.11}$$

Using (6.11) in (6.10), we get (6.3).

6.2 The Example

We now proceed to show that the discount factor restriction obtained in Theorem 6.1 for period three optimal cycles is the best possible. We do this by constructing an example, when $\delta < [\sqrt{5}-1)/2]^2$, of a transition possibility set, Ω, and a utility function, u, such that (Ω, u, δ) exhibits a period-three optimal cycle.

The constructed example is a simple adaptation, for our purpose, of the example given in Nishimura *et al.* (Chapter 9). Denote $[(\sqrt{5}-1)/2]^2$ by M, and assume that $0 < \delta < M$. Define, $\gamma = 1/\sqrt{M}$ so that $1 < \gamma < 2$ and $\gamma^2 M = 1$. Then $\gamma^2\delta < 1$, and it is possible to find $\alpha > 1$, such that $\gamma^2\delta < (1/\alpha)$. Choose $\theta > \alpha\delta/(1-\gamma^2\delta)$ and $\beta > 2\max\{(1/\gamma^2\delta), 2\alpha + \gamma\theta(2\gamma - 1)\}$.

Define the transition possibility set, Ω, by

$$\Omega = \{(x, z) \in \Re^2_+ : z \leqslant \min(\gamma x, 1)\}.$$

Then (A1)-(A4) are satisfied.

Let $H_{\leqslant} = \{(x, z) \in \Re^2 : z \leqslant 2 - \gamma x\}$ and $H_{\geqslant} = \{(x, z) \in \Re^2 : z \geqslant 2 - \gamma x\}$. Define the reduced form utility function $u : \Omega \to \Re$ by

$$\begin{aligned} u(x, z) = {} & [\beta - (2/\gamma)]x - (\alpha - 1)x^2 - \\ & [\beta\delta - (2/\gamma^2)]z - [(1/\gamma^2) - \alpha\delta]z^2 \\ & \text{for } (x, z) \in H_{\leqslant} \cap \Omega \end{aligned}$$

$$\begin{aligned} u(x, z) = {} & (4\gamma\theta + \beta)x - (\alpha + \gamma^2\theta)x^2 - (\beta\delta - 4\theta)z - \\ & (\theta - \alpha\delta)z^2 - 2\gamma\theta xz - 4\theta \\ & \text{for } (x, z) \in H_{\geqslant} \cap \Omega. \end{aligned}$$

Then, by using the proof of Lemma 3.1 in Nishimura *et al.* (Chapter 9) , (A5)-(A7) are satisfied.

By using the proof of Lemma 3.2 of Nishimura *et al.* (Chapter 9), the policy function of the constructed model (Ω, u, δ) is given by

$$h(x) = \begin{cases} \gamma x & \text{for } x \in [0, 1/\gamma] \\ 2 - \gamma x & \text{for } x \in [1/\gamma, 1] \end{cases}$$

Note that $h(1) = (2-\gamma)$, and $(2-\gamma) = 2-(1/\sqrt{M}) = 1-\sqrt{M} = M$, so that $0 < (2-\gamma) < \sqrt{M} = (1/\gamma)$. Thus $h(2-\gamma) = \gamma(2-\gamma) = (M/\sqrt{M}) = \sqrt{M}$, and $h(\sqrt{M}) = 1$ Thus $(1, M, \sqrt{M}, 1, M, \sqrt{M}, ...)$ is the optimal program starting from 1. This optimal program clearly exhibits a period-three cycle.

7. Discount Factor Restrictions for Periodic Optimal Programs

In this final section, we will find discount factor restrictions for any dynamic optimization model exhibiting a periodic optimal program of period not equal to a power of 2. To illustrate our method of deriving these restrictions most transparently, we first obtain a restriction for an important special case.

Specifically, we obtain an upper bound on the discount factor, δ, that must be satisfied in order that a dynamic optimization model (Ω, u, δ) yield a periodic optimal program of odd period greater than one. It is important to note that this bound holds *uniformly* for all odd periods greater than one.

In order to establish these results, we will find it convenient to present a preliminary result which compares the value and policy functions of two dynamic optimization models. To proceed more formally, let us refer to a triplet (Ω, u, δ) satisfying (A1)-(A7) as a *dynamic optimization model.* Now, let $\Im = (\Omega, u, \delta)$ be a dynamic optimization model, with value function V and policy function h. We can then construct[9] another dynamic optimization model $\Im^* = (\Omega^*, u^*, \delta^*)$ with $\delta^* = \delta^2$, such that the policy function, h^*, of $\Im^*$ is h^2 and the value function, V^*, of $\Im^*$ is V.

Proposition 7.1: *Let $\Im = (\Omega, u, \delta)$ be a dynamic optimization model, with value function V and policy function h. Then, there exists Ω^* and u^* satisfying (A1)-(A7) such that with $\delta^* = \delta^2$,* (i) *the value function V^*, of the dynamic optimization model $\Im^* =$*

[9]Sorger (1992a) proves a result similar to Proposition 7.1. However, his set of maintained assumptions on a dynamic optimization model is different from ours. It is not clear to us whether Proposition 7.1 is valid if the upper-semicontinuity of the utility function in (A.6) is replaced by continuity.

$(\Omega^*, u^*, \delta^*)$ *is given by* V, *and* (ii) *the policy function,* h^*, *of* $\Im^*$ *is* given by h^2.

Proof: Define $\Omega = \{(x, z) :$ there is $y \in \Re_+$ such that $(x, y) \in \Omega$ and $(y, z) \in \Omega\}$. Define, next, $u^* : \Omega^* \to \Re$ by

$$\left.\begin{array}{c} u^*(x, z) = \max[u(x, y) + \delta u(y, z)] \\ \text{subject to } (x, y) \in \Omega \\ \text{and } (y, z) \in \Omega. \end{array}\right\} \text{(P)},$$

It is straightforward to check that Ω^* and u^* satisfy assumptions (A.1)-(A.5) and (A.7).

To check (A6), one essentially applies the method used to obtain a version of a maximum theorem in Berge (1963; Theorem 2, p. 116).

Let (x^s, z^s), $s = 1, 2, ...,$ be a sequence of elements in Ω^* converging to (x^0, z^0). For (x^s, z^s), $s = 1, 2, ...,$ let y^s solve problem (P). Then since $x^s \to x^0$ as $s \to \infty$, and $(x^s, y^s) \in \Omega$, y^s is in a bounded set (by (A1) and (A2)). Let $(y^{s'})$ be an arbitrary convergent subsequence of (y^s), converging to y^0.

Since $(x^{s'}, y^{s'}) \to (x^0, y^0)$, and $(y^{s'}, z^{s'}) \to (y^0, z^0)$ as $s' \to \infty$, and $(x^{s'}, y^{s'}) \in \Omega$ and $(y^{s'}, z^{s'}) \in \Omega$ for all s', we have $(x^0, y^0) \in \Omega$ and $(y^0, z^0) \in \Omega$. Now, we have

$$\begin{aligned} \overline{\lim_{s' \to \infty}}\, u^*(x^{s'}, z^{s'}) &= \overline{\lim_{s' \to \infty}}\, [u(x^{s'}, y^{s'}) + \delta u(y^{s'}, z^{s'})] \\ &\leqslant \overline{\lim_{s' \to \infty}}\, u(x^{s'}, y^{s'}) + \delta\, \overline{\lim_{s' \to \infty}}\, u(y^{s'}, z^{s'}) \\ &\leqslant u(x^0, y^0) + \delta u(y^0, z^0), \end{aligned}$$

the last inequality following from the upper semicontinuity of u. By definition of u^*, we also have

$$u(x^0, y^0) + \delta u(y^0, z^0) \leqslant u^*(x^0, z^0).$$

Thus, we obtain

$$\overline{\lim_{s' \to \infty}}\, u^*(x^{s'}, z^{s'}) \leqslant u^*(x^0, z^0).$$

Since $(y^{s'})$ was an arbitrary convergent subsequence of (y^s), this establishes property (A6) of u^*.

Denote the value and policy functions of the dynamic optimization model $(\Omega^*, u^*, \delta^*)$, where $\delta^* = \delta^2$, by V^* and h^*.

To compare the value functions of the two dynamic optimization models, $\Im = (\Omega, u, \delta)$ and $\Im^* = (\Omega^*, u^*, \delta^*)$, we will find it convenient to refer to programs and optimal programs in the former by the prefix $\Im$ and in the latter by $\Im^*$.

Let $\mathbf{x} \geqslant 0$ be arbitrarily given, and let $\{x_t\}$ be $\Im^*$-optimal from $\mathbf{x}$. Then there is $\{y_t\}$ such that y_t solves (P) for $(x, z) = (x_t, x_{t+1})$. Define $k_t = x_{t/2}$ for $t = 0, 2, 4, \ldots$, and $k_t = y_{(t-1)/2}$ for $t = 1, 3, 5, \ldots$. Then $\{k_t\}$ is a $\Im$-program. Further,

$$
\begin{aligned}
V^*(\mathbf{x}) &= \sum_0^\infty (\delta^2)^t u^*(x_t, x_{t+1}) \\
&= \sum_0^\infty (\delta^2)^t [u(x_t, y_t) + \delta u(y_t, x_{t+1})] \\
&= \sum_0^\infty \delta^t u(k_t, k_{t+1}) \leqslant V(\mathbf{x}).
\end{aligned}
$$

On the other hand, let $\{k'_t\}$ be $\Im$-optimal from $\mathbf{x}$. Then, defining $\{x_t\}$ by $x_t = k'_{2t}$ for $t \geqslant 0$, $\{x_t\}$ is a $\Im^*$-program from $\mathbf{x}$, and

$$
\begin{aligned}
V(\mathbf{x}) &= \sum_0^\infty \delta^t u(k'_t, k'_{t+1}) \\
&= \sum_0^\infty (\delta^2)^t [u(k'_{2t}, k'_{2t+1}) + \delta u(k'_{2t+1}, k'_{2t+2})] \\
&\leqslant \sum_0^\infty (\delta^2)^t u^*(k'_{2t}, k'_{2t+2}) \leqslant V^*(\mathbf{x}).
\end{aligned}
$$

Thus, $V(\mathbf{x}) = V^*(\mathbf{x})$, which shows that $V = V^*$, since $\mathbf{x} \geqslant 0$ was arbitrary.

To compare the policy functions, let $\mathbf{x} \geqslant 0$ be arbitrary, and let $\{k_t\}$ be $\Im$-optimal from $\mathbf{x}$. Then defining $x_t = k_{2t}$ for $t \geqslant 0$, $\{x_t\}$

is a $\Im$-program from $\mathbf{x}$. Now,

$$\begin{aligned} V^*(\mathbf{x}) &\geqslant \sum_0^\infty (\delta^2)^t u^*(x_t, x_{t+1}) \\ &\geqslant \sum_0^\infty (\delta^2)^t [u(k_{2t}, k_{2t+1}) + \delta u(k_{2t+1}, k_{2t+2})] \\ &= \sum_0^\infty \delta^t u(k_t, k_{t+1}) = V(\mathbf{x}) = V^*(\mathbf{x}). \end{aligned}$$

Thus we must have equalities in all the above inequalities, so that $\{x_t\}$ is $\Im$-optimal from $\mathbf{x}$. This establishes that $h^*(\mathbf{x}) = h^2(\mathbf{x})$, and since $\mathbf{x} \geqslant 0$ was arbitrary, $h^* = h^2$.

Theorem 7.1: *Suppose (Ω, u, δ) is a dynamic optimization model with policy function h. Let $k > 1$ be any odd integer. If h has a periodic orbit of period k, then*

$$\delta < (\sqrt{5} - 1)/2. \tag{7.1}$$

Proof: If h has a periodic orbit of period k, where $k > 1$ is an odd integer, then by Sarkovskii's result (Proposition 2.1), h has a periodic orbit of period 6. This implies that h^2 has a periodic orbit of period 3.

By Proposition 7.1, there exists a dynamic optimization model, $\Im^* = (\Omega^*, u^*, \delta^*)$, with $\delta^* = \delta^2$, such that the policy function, h^*, of $\Im^*$ is given by h^2. By using Theorem 6.1, we can conclude that

$$\delta^* < [(\sqrt{5} - 1)/2]^2.$$

Since $\delta^* = \delta^2$, (7.1) follows immediately.

Remark 7.1: (i) Sorger (1992b) conjectured that one can find for every $\delta \in (0, 1)$ and odd number $p \geqslant 3$, a continuous function h from $[0, 1]$ to $[0, 1]$, such that h is the policy function of a dynamic optimization model (Ω, u, δ), and h has a periodic point of period p. Theorem 3 indicates that this conjecture is false for $1 > \delta \geqslant (\sqrt{5} - 1)/2$.

(ii) The restriction in (7.1) on the discount factor translates to the discount rate restriction

$$\rho > (\sqrt{5}-1)/2 \approx 0.618,$$

which still involves substantial discounting.

(iii) There is no claim in our result that the discount factor restriction (7.1) is the "best possible". In fact, applying the theory of turbulence to this problem (see Mitra (Chapter 13)) one can show that the bound in (7.1) can be improved.

The Sarkovskii ordering and the above method of proof can be used to obtain discount factor restrictions that must hold in order that optimal programs be periodic with period $p = 2^q.k$ where $q \geqslant 0$ and $k > 1$ is an odd integer.

Corollary 7.1: *Suppose $\Im = (\Omega, u, \delta)$ is a dynamic optimization model with policy function h. Let $k > 1$ be an odd integer, q be a non-negative integer, and $p = 2^q.k$. If h has a periodic orbit with period p, then*

$$\delta < [(\sqrt{5}-1)/2]^{1/2^q}. \tag{7.2}$$

Remark 7.2: Nishimura and Yano (1995b) have constructed a sequence of dynamic optimization models $\Im_n \equiv (\Omega_n, u_n, \delta_n)$ for $n = 0, 1, 2, \ldots$, such that $\Im_n$ exhibits a periodic point of period 3.2^n and $\delta_n \to 1$ as $n \to \infty$. This appears to be in accord with the discount rate restrictions we obtain in our result not being uniformly below 1. For $n = 0, 1, 2, 3$, they also noted that in obtaining each successive model exhibiting a period twice that in the previous model (6 compared to 3, 12 compared to 6, 24 compared to 12), the (minimum) discount factor needed was approximately "square rooted". This feature also appears to be in accord with the formula we obtain for the discount rate restrictions.

12

On the Least Upper Bound of Discount Factors that are Compatible with Optimal Period-Three Cycles

Kazuo Nishimura and Makoto Yano

1. Introduction

In this chapter, we derive, in the standard class of optimal growth models, the least upper bound of discount factors of future utilities for which a cyclical optimal path of period 3 may emerges.[1] On the one hand, Nishimura and Yano (1992) and Nishimura, Sorger and Yano (Chapter 9) construct examples in which a cyclical optimal path of period 3 emerges for discount factors around 0.36. On the other hand, Sorger (1992, 1994) demonstrates that if such a path emerges in an optimal growth model of the standard class, the model's discount factor cannot exceed 0.55. These results imply that the least upper bound of discount factors that can give rise to cyclical optimal paths of period 3 must lie between 0.36 and 0.55.[2] We demonstrate that the least upper bound

[1]The focus in the early literature on optimal growth is on the convergence of an optimal path to a stationary state. In that literature, it has been proved that an optimal path converges to a stationary state if the utility function is fixed and if the discounting of future utilities is sufficiently weak (see Brock and Scheinkman, 1976, Cass and Shell, 1976, Scheinkman, 1976, and McKenzie, 1983). In this chapter, we are concerned with the existence of a utility function that gives rise to an optimal cyclical path of period 3 for a given discount factor of future utilities.

[2]In relation to these results, Nishimura and Yano (Chapter 8) prove that for values of the discount factor ρ arbitrarily close to 1, ergodically chaotic optimal paths can emerge that are generated by a unimodal, expansive dynamical system. In that result, it is shown that part of the graph of the optimal dynamical system lies on a von Neumann facet containing the stationary state and that any optimal path is confined in a small neighborhood of the facet. In this respect, the result is closely related to the neighborhood turnpike theorem of McKenzie (1983), which implies that any optimal paths converge into a neighborhood of the von Neumann facet. See also the result of Nishimura, Sorger, and Yano (Chapter 9),

is $\hat{\rho} = \left(3 - \sqrt{5}\right)/2$.

The importance of finding this least upper bound stems from the following two facts: (i) given the long-run real interest rate *per annum,* the discount factor is positively correlated to the length of an individual period of the model;[3] (ii) the existence of a cyclical path of period 3 is a fundamental criterion for the emergence of complex non-linear dynamics, in particular of cyclical paths of any other periodicity (Sarkovskii, (1964), and Li and Yorke, (1975)). With these considerations, in the existing literature, a number of studies have been concerned with the existence of a cyclical optimal path of period 3 and the corresponding discount factor.[4]

The rest of this chapter is organized as follows. In Section 2, we will introduce our main result. In Section 3, we will discuss the basic idea of a proof. In Section 4, we will provide preliminary results. In Section 5, we will prove the main result. Section 6 is for the concluding remarks.

Note: Recently, Tapan Mitra's article (Chapter 11) has come to our attention, in which he obtains the same least upper bound by a method different from ours.

2. The Model

Think of a standard optimal growth model $M = \{u, \rho\}$ in which path $\{k_t\}$ is chosen so as to maximize $\sum_{t=1}^{\infty} \rho^{t-1} u(k_{t-1}, k_t)$ subject to the initial condition $k_0 = k$. If $\{k_t\}$ solves this optimization problem, it is called an *optimal path* from initial stock k. Function

---

which extends Nishimura and Yano (Chapter 8) for the case in which the von Neumann facet is trivial.

[3]In a standard optimal growth model, the discounted sum of utilities u_t, $\sum_{t=1}^{\infty} \rho^{t-1} u_t$, is maximized. It is known that, in such a model, the long-run real interest rate per period is $r = \rho^{-1} - 1$. If, therefore, the long-run real interest rate *per annum* is given by i, the length of an individual period of the model, ℓ, is given by

$$r = (1 + i)^{\ell} - 1$$

In other words, it holds that $\rho = 1/(1 + i)^{\ell}$.

[4]See Deneckere and Pelikan (1986), Boldrin and Montrucchio (1986), Boldrin and Deneckere (1990), Nishimura and Yano (1994) and Majumdar and Mitra (1994). These studes are concerned with deterministic economic flutuations. For broader issues on deterministic and indeterministic fluctuations, see, for example Shell (1977) and Cass and Shell (1983).

$u(x, y)$, which is called a reduced-form utility function, captures the maximum utility level that can be achieved in an individual period when the economy has stock x at the beginning of that period and is to leave stock y at the end of the period. Discount factor ρ satisfies

$$0 < \rho < 1 \tag{2.1}$$

[The model is standard; for a more detailed explanation, for example, see McKenzie, (1986).]

In this study, we take the case of a one-dimensional capital stock; i.e., the domain of function u is a subset of $R_+^2 = \{(x, y) \mid x \geq 0, y \geq 0\}$.

Definition 2.1: The single-capital optimal growth model is a combination of u and ρ, $M = \{u, \rho\}$, satisfying the following.

A1: *$u : D \to R$ is a continuous and concave function defined on a closed and convex subset $D \subset R_+^2$.*

A2: *$u(k, h)$ is strictly increasing in k and strictly decreasing in h.*

A3: $0 < \rho < 1$.

If, in particular, u is strictly concave, model M is called a strictly concave optimal growth model.

We are concerned with cyclical optimal paths of period 3. A path $\{z_t\}$ is a *cyclial path of period 3* if there are ζ_1, ζ_2 and ζ_3, $\zeta_1 < \zeta_2 < \zeta_3$, such that $z_{3\tau} = \zeta_1$, $z_{3\tau+1} = \zeta_2$ and $z_{3\tau+2} = \zeta_3$ for $\tau = 0, 1, \cdots$. If a dynamical system $z_{t+1} = \zeta(z_t)$, $t = 0, 1, 2, \cdots$, generates a cyclical path of period 3, it is often said to be *chaotic in the sense of Li and Yorke.*

In general, as is well-known, the optimal paths can be characterized by a set valued function that relates each stock level at a beginning of a period to the set of optimal stock levels at the end of the period. To state it more precisely, define $V(k)$ as follow.

$$V(k) = \max_{(k_1, k_1, \cdots)} \sum_{t=1}^{\infty} \rho^{t-1} u(k_{t-1}, k_t) \ s.t. k_0 = k, \tag{2.2}$$

Given this definition, it holds that

$$V(k) = \max_{y} \left[v(k,y) + \rho V(y) \right] \tag{2.3}$$

Denote by $F(k)$ the set of y that solves the maximization problem on the right-hand side of (2.3). If and only if $\{k_t\}$ is an optimal path solving maximization problem (2.2), it holds that

$$k_t \in F\left(k_{t-1}\right) \tag{2.4}$$

for $t = 1, 2, ...$ with $k_0 = k$. Because $F(k)$ describes the behavior of optimal paths of problem (2.2) and because it is generally a set-valued function, we call system (2.4) a *generalized optimal dynamical system.* If, in particular, $F(k)$ is a singleton for any k (at which $V(k)$ is well defined), we write $F(k) = \{f(k)\}$ and call $k_t = f(k_{t-1})$ an *optimal dynamical system.*

Our main result is as follows:

Proposition 2.1: *$\hat{\rho} = (3 - \sqrt{5})/2$ is the least upper bound of discount factors with which it is possible to construct an optimal growth model $M = \{u, \rho\}$ in which the optimal dynamical system is chaotic in the sense of Li and Yorke.*

3. The Basic Idea of a Proof

In the section, we will explain the basic idea of our proof. Think of activities $(x, y) \in D$, $(y, z) \in D$ and $(z, x) \in D$ such that

$$0 \leq x < y < z \tag{3.1}$$

and think of three numbers, a, b and c. In a three dimensional space, called the $k - h - v$ space, take the three points A, B and C as follows:

$$A = (x, y, a),\ B = (y, z, b) \text{ and } C = (z, x, c)\,. \tag{3.2}$$

By using these points, A, B and C, and their projections to the $k-h$, $h-v$, $v-k$ planes, construct the polyhedral P, as illustrated in Figure 1. Denote by $v(k,h)$ the function the graph of which coincides with polyhedral P. Given (2.1) and (3.1), $M' = \{v, \rho\}$ is an optimal growth model, i.e., satisfies conditions A-1, A-2 and A-3 of Definition 2.1, if and only if a, b and c satisfy

$$a < c \text{ and } b < c \tag{3.3}$$

For the moment, we will deal with model M', which we call a polyhedral (optimal growth) model. Denote by $G(k)$ the generalized optimal dynamical system of model $M' = \{v, \rho\}$; because the graph of v consists of flats (is not strictly concave), in general, $G(k)$ is not a dynamical system (cannot be described by a function from R_+ to R_+). We say that the generalized optimal dynamical system $G(k)$ of polyhedral model $M' = \{v, \rho\}$ generates an isolated period-3 path if there is (x, y, z) satisfying

$$G(x) = \{y\},\ G(y) = \{z\} \text{ and } G(z) = \{x\}. \tag{3.4}$$

We will focus on polyhedral models M' satisfying the following condition.

Condition 3.1 The generalized optimal dynamical system of polyhedral model M' generates an isolated period-3 path.

We will first characterize polyhedral P by using x, y, z, a, b and c. Half space S_1 is the region below the plane that goes through points A, B and C (i.e., contains face I of the Figure 1). That is,

$$S_1 = \left\{(k,h,v) \in R^3 \mid d_1 \geq v + H_1 h - K_1 k\right\}, \tag{3.5}$$

where d_1, K_1 and H_1 are positive and given by the following system.

$$\begin{cases} a + H_1 y - K_1 x = d_1 \\ b + H_1 z - K_1 y = d_1 \\ c + H_1 x - K_1 z = d_1 \end{cases} \tag{3.6}$$

This implies

$$\begin{aligned} K_1 &= \left((a-b)\,x + (b-c)\,y + (c-a)\,z\right)/\Delta_1 \\ H_1 &= \left((b-c)\,x + (c-a)\,y + (a-b)\,z\right)/\Delta_1 \\ \Delta_1 &= \tfrac{1}{2}(y-x)^2 + \tfrac{1}{2}(z-y)^2 + \tfrac{1}{2}(z-x)^2 > 0 \end{aligned} \tag{3.7}$$

Half space S_2 is the region below the plane that goes through points B and C and is perpendicular to the h-v plane (i.e., contains face II of the Figure 1). That is,

$$S_2 = \left\{(k, h, v) \in R^3 \mid d_2 \geq v + H_2 h\right\}, \tag{3.8}$$

where d_2 and H_2 are positive and given by following system.

$$\begin{cases} b + H_2 z = d_2 \\ c + H_2 x = d_2 \end{cases} . \tag{3.9}$$

This implies the following.

$$H_2 = (c - b)/\Delta_2, \text{ where } \Delta_2 = z - x > 0 \tag{3.10}$$

Half space S_3 is the region below the plane that goes through point C and is perpendicular to the h-v plane and the k-v plane (i.e., contains face III of the Figure 1). That is,

$$S_3 = \left\{(k, h, v) \in R^3 \mid c \geq v\right\}. \tag{3.11}$$

Half space S_4 is the region below the plane that goes through points A and C and is perpendicular to the k-v plane (i.e., contains face IV of the Figure 1). This is,

$$S_4 = \left\{(k, h, v) \in R^3 \mid d_4 \geq v - K_4 k\right\}, \tag{3.12}$$

where d_4 and K_4 are positive and given by the following system.

$$\begin{cases} a - K_4 x = d_4 \\ c - K_4 z = d_4 \end{cases} . \tag{3.13}$$

This implies

$$K_4 = (c - a)/\Delta_4, \text{ where } \Delta_4 = z - x \tag{3.14}$$

Half space S_5 is the region to the right of the plane that goes through point A and is perpendicular to the k-h plane and the k-v plane (i.e., contains face V of the Figure 1). That is,

$$S_5 = \left\{(k, h, v) \in R^3 \mid -x \geq -k\right\}, \tag{3.15}$$

where d_5 and K_5 satisfy

$$-x = d_5/K_5.$$

Half space S_6 is the region to the northeast of the plane that goes through points A and B and is perpendicular to the k-h plane (i.e., contains face VI of the Figure 1). That is,

$$S_6 = \left\{(k,h,v) \in R^3 \mid d_6 \geq h - K_6 k\right\}, \tag{3.16}$$

where $d_6,$ K_6 and H_6 are defined by the following system.

$$\begin{cases} y - K_6 x = d_6 \\ z - K_6 y = d_6 \end{cases}. \tag{3.17}$$

This is implies $(y-z) - K_6(x-y) = 0$.

$$K_6 = (z-y)/(y-x). \tag{3.18}$$

Half space S_7 is the region to the north of the plane that goes through point B and is perpendicular to the k-h plane and h-v plane (i.e., contains face VII of the Figure 1). This is,

$$S_7 = \left\{(k,h,v) \in R^3 \mid z \geq h\right\}. \tag{3.19}$$

Polyhedral P coincides with the intersection of the above half spaces, $S_{1,\cdots,}S_{7,}$ in the non-negative orthant of the $k-h-v$ space. It is said that the price vector $(p,q,1)$ *supports* polyhedral P (or utility function $v(k,h)$) at $(\hat{k},\hat{h},v(\hat{k},\hat{h}))$ if for all $(k,h,v) \in P$,

$$v(\hat{k},\hat{h}) + q\hat{h} - p\hat{k} \geq v(k,h) + qh - pk \tag{3.20}$$

Moreover the price vector $(p,q,1)$ *strongly supports* polyhedral P if (3.20) holds with strict inequality at any point $(k,h,v) \in P$ except $(\hat{k},\hat{h},v(\hat{k},\hat{h}))$.

Recall points A, B and C are defined by (3.2). We may prove the following lemmas (Lemmas 3.1 through 3.3 are more or less obvious; for a proof of Lemma 3.4, see McKenzie (1986).

Lemma 3.1: *Price vector* $(p_1, q_{1,}1)$ *strongly supports polyhedral* P *at point* A *if and only if there are* $t_1 > 0$, $t_4 > 0$, $t_5 > 0$, *and* $t_6 > 0$ *such that* $t_1 + t_4 = 1$ *and that*

$$(p_1, q_1, 1) = t_1(K_1, H_1, 1) + t_4(K_4, 0, 1) + t_5(1, 0, 0,) + t_6(K_6, 1, 0) \tag{3.21}$$

Lemma 3.2: *Price vector* $(p_2, q_{2,}1)$ *strongly supports polyhedral* P *at point* B *if there are* $n_1 > 0$, $n_2 > 0$, $n_6 > 0$, *and* $n_7 > 0$ *such that* $n_1 + n_2 = 1$ *and that*

$$(p_2, q_2, 1) = n_1(K_1, H_1, 1) + n_2(0, H_2, 1) + n_6(K_6, 1, 0) + n_7(0, 1, 0) \tag{3.22}$$

Lemma 3.3: *Price vector* $(p_3, q_{3,}1)$ *strongly supports polyhedral* P *at point* C *if and only if there are* $m_1 > 0$, $m_2 > 0$, $m_3 > 0$, *and* $m_4 > 0$ *such that* $m_1 + m_2 + m_3 + m_4 = 1$ *and that*

$$(p_3, q_3, 1) = m_1(K_1, H_1, 1) + m_2(0, H_2, 1) + m_3(0, 0, 1) + m_4(K_4, 0, 1) \tag{3.23}$$

Lemma 3.4: *Condition 1 holds if and only if the following condition is satisfied.*

C.o: *There are price vectors* $(p_i, q_i, 1)$, $i = 1, 2, 3$, *such that the following conditions are satisfied:*

(i): $(p_1, q_1, 1)$, $(p_2, q_2, 1)$ *and* $(p_3, q_3, 1)$, *respectively, strongly support polyhedral* P *at points* A, B *and* C, *respectively.*

(ii): $-q_1 + \rho p_2 = 0$, $-q_2 + \rho p_3 = 0$ and $-q_3 + \rho p_1 = 0$.

This lemma implies that a cyclical optimal path $\{k_t\}$ of period 3 is supported by a cyclical price path $\{p_t\}$ of period 3; that is, there are p_1, p_2 and p_3 such that for any $(k, h) \in D$

$$v(x, y) + p_2 y - \rho^{-1} p_1 x \geq v(k, h) + p_2 y - \rho^{-1} p_1 k \tag{3.24}$$

$$v(y, z) + p_3 z - \rho^{-1} p_2 y \geq v(k, h) + p_3 h - \rho^{-1} p_2 k \tag{3.25}$$

$$v(z,x) + p_1 x - \rho^{-1} p_3 z \geq v(k,h) + p_1 h - \rho^{-1} p_3 k \tag{3.26}$$

The next lemma follows from Lemmas 3.1 through 3.4.

Theorem 3.1: *The generalized optimal dynamical system of polyhedral model M' generates an isolated period-3 path (i.e., Condition 1 is satisfied) if and only if the following condition holds:*

C.i: *There is* $\pi = (m_1, m_2, m_3, m_4, t_1, t_4, t_5, t_6, n_1, n_2, n_6, n_7)$ $>> 70$ *that satisfies the following system:*[5]

$$\begin{cases} -(H_1 t_1 + t_6) + \rho (K_1 n_1 + K_6 n_6) = 0; \\ -(H_1 n_1 + H_2 n_2 + n_6 + n_7) + \rho (K_1 m_1 + K_4 m_4) = 0; \\ -(H_1 m_1 + H_2 m_2) + \rho (K_1 t_1 + K_4 t_4 + t_5 + K_6 t_6) = 0; \\ m_1 + m_2 + m_3 + m_4 = 1; \\ t_1 + t_4 = 1; \\ n_1 + n_2 = 1. \end{cases} \tag{3.27}$$

Proof: By Lemmas 3.1 through 3.4, C.i is equivalent to C.o of Lemma 3.4. Thus, the theorem follows from Lemma 3.4 and Condition 1. Q.E.D.

Our main theorem is as follows (a proof will be given in the next section).

Theorem 3.2: *If and only if* $0 < \rho < \hat{\rho} = \left(3 - \sqrt{5}\right)/2$, *system* (3.27) *has a strictly positive solution* $\pi >> 0$.

Given Theorem 2, we may demonstrate the following result.

Proposition 3.1: $\hat{\rho} = \left(3 - \sqrt{5}\right)/2$ *is the least upper bound of discount factors with which it is possible to construct a strictly concave optimal growth model* $M = \{u, \rho\}$ *in which the optimal dynamical system is chaotic in the sense of Li and Yorke.*

Proof: Given Theorem 3.2, it suffices to demonstrate that if and only if it is possible to construct a polyhedral model $M' =$

[5] $(x_1, ..., x_n) >> 0$ means that $x \in \Re^n$ and $x_1 > 0, ..., x_n > 0$.

$\{v, \rho\}$ satisfying Condition 1, it is possible to construct a strictly concave model $M = \{u, \rho\}$ in which the optimal dynamical system is chaotic in the sense of Li and Yorke. Suppose that model M' satisfies Condition 1. Then, it is possible to construct a strictly concave utility function $u = u(k, h)$ satisfying

$$a = u(x, y) \ , b = u(y, z) \ , c = u(z, x) \tag{3.28}$$

and that if function u is substituted for v, conditions (3.24), (3.25) and (3.27) holds for any $(k, h) \in D$. This implies that the period-3 path $x, y, z, x, y, z, x, \cdots$ is optimal. Since the solutions to a strictly concave model can be expressed by an optimal dynamical system, the optimal dynamical system of $M = \{u, \rho\}$ is chaotic in the sense of Li and Yorke.

Conversely, suppose that the optimal dynamical system of a strictly concave model M is chaotic in the sense of Li and Yorke. Then, it may be proved that there are prices p_1, p_2 and p_3 such that $(\rho^{-1}p_1, p_2)$, $(\rho^{-1}p_2, p_3)$ and $(\rho^{-1}p_3, p_1)$, respectively, strictly support $u(k, h)$ at (x, y), (y, z) and (z, x), respectively; see McKenzie (1986) for a proof. By using (3.28) define a, b and c, and construct polyhedral P and utility function v. Then, $(\rho^{-1}p_1, p_2)$, $(\rho^{-1}p_2, p_3)$ and $(\rho^{-1}p_3, p_1)$, respectively, strictly support $v(k, h)$ at (x, y), (y, z) and (z, x), respectively. Thus, by Lemma 4, Condition 1 hold. Q.E.D.

Proof of Proposition 2.1: In order to prove Proposition 2.1, we need to extend Proposition 3.1 to the case including general concave utility functions. As the proof below demonstrate that the possibility of such an extension is straightforward due to the continuity of the model.

4. Proof of Theorem 3.2: Preliminary Steps

We will establish our main results (Propositions 2.1 and 3.1) by proving Theorem 3.2. For this purpose, we use Theorem 3.1. System (3.27) of Theorem 3.1, however, involves too may variables. In this section, we will reduce C.i to a more manageable form that does not involve variables π.

Step 1: At the outset, we will erase t_4, n_2 and m_3 from system (3.27) and deal with

$$\begin{cases} -(H_1 t_1 + t_6) + \rho(K_1 n_1 + K_6 n_6) = 0; \\ -((H_1 - H_2) n_1 + H_2 + n_6 + n_7) + \rho(K_1 m_1 + K_4 m_4) = 0; \\ -(H_1 m_1 + H_2 m_2) + \rho((K_1 - K_4) t_1 + K_4 + t_5 + K_6 t_6) = 0; \\ m_1 + m_2 + m_4 < 1; \\ t_1 < 1; \\ n_1 < 1. \end{cases} \tag{4.1}$$

Sublemma 4.1: *Condition C.i of Theorem 3.1 is satisfied if and only if the following condition is satisfied:*

C.ii: *There is* $\pi' = (m_1, t_1, n_1, t_5, n_7, m_2, m_4, , t_6, n_6) >> 0$ *satisfying* (4.1).

Before moving to the next step, define $\theta_1 = \rho^3 K_1^2 (K_4 - K_1) - H_1^2 (H_2 - H_1)$. At the next step, we will focus on the case of $\theta_1 > 0$. We will return to the case of $\theta_1 < 0$ at Step 2$''$ and that of $\theta_1 = 0$ at the end of the next secton.

Step 2 (case $\theta_1 > 0$): Next, we will erase m_1. Solve the second and third conditions of (4.1) for n_1 and t_1, and substitute them into the first condition. By solving the resulting equation for m_1 and using $m_1 > 0$, we obtain

$$\begin{array}{c} (m_1 =)\ \frac{1}{\theta_1}[H_1 H_2 (H_2 - H_1) m_2 + \rho^2 (K_4 - K_1) \\ \{(H_2 - H_1) K_6 + K_1\} n_6 \\ +\rho^2 K_1 (K_4 - K_1) n_7 - \rho^3 K_1 K_4 (K_4 - K_1) m_4 \\ -\rho (H_2 - H_1) \{K_4 - K_1 + H_1 K_6\} t_6 - \\ \rho H_1 (H_2 - H_1) t_5 \\ -\rho H_1 (H_2 - H_1) K_5 + \rho^2 K_1 (K_4 - K_1) H_2] > 0. \end{array} \tag{4.2}$$

By using (4.2), we will erase m_1 from the second and third conditions of (4.1). That is, by $H_2 - H_1 > 0$,[6] the second and third conditions

[6]From (3.6), (3.9), $c - b > 0$ and $z > y > x$

can be transformed as follows.[7]

$$n_7 - \left(\rho^2 K_1/H_1\right) t_5 = A^{n_7}\left(n_1; m_2, m_4, n_6\right); \tag{4.3}$$

$$n_7 - \left(\rho^2 K_1/H_1\right) t_5 = B^{n_7}\left(t_1; m_2, m_4, t_6\right). \tag{4.4}$$

In order to erase m_1 from the fourth condition of (4.1), again, use (4.2) to obtain

$$\begin{gathered}(1 - m_1 - m_2 - m_4 =)\ \tfrac{1}{\theta_1}[-\{H_1\left(H_2 - H_1\right)^2 + \\ \rho^3 K_1^2\left(K_4 - K_1\right)\}m_2 - \rho^2\left(K_4 - K_1\right)\{\left(H_2 - H_1\right)K_6 + K_1\} \\ n_6 - \rho^2 K_1\left(K_4 - K_1\right)n_7 \\ +\{H_1^2\left(H_2 - H_1\right) + \rho^3 K_1\left(K_4 - K_1\right)^2\}m_4 \\ +\rho\left(H_2 - H_1\right)\{\left(K_4 - K_1\right) + H_1 K_6\}t_6 + \rho H_1\left(H_2 - H_1\right)t_5 \\ +\rho H_1\left(H_2 - H_1\right)K_4 - \rho^2 K_1\left(K_4 - K_1\right)H_2 \\ +\rho^3 K_1^2\left(K_4 - K_1\right) - H_1^2\left(H_2 - H_1\right)] > 0.\end{gathered} \tag{4.5}$$

In what follows, we will first focus on the case of $\theta_1 > 0$. Note that $K_4 > K_1$ by (3.6) and (3.13).[8] Therefore, by $\theta_1 > 0$, (4.2)

$$H_1 = \frac{c - b - K_1(z - y)}{z - x} < H_2 = \frac{c - b}{z - x}$$

hold.

[7] $A^{n_7}(n_1; m_2, m_4, n_6)$ and $B^{n_7}(t_1; m_2, m_4, n_6)$ are defined as follows:

$$\begin{aligned}A^{n_7}(n_1; m_2, m_4, n_6) = \frac{1}{H_1^2}[&-\theta_1 n_1 - \rho K_1 H_1 H_2 m_2 + \rho H_1^2 K_4 m_4 \\ &-\{H_1^2 + \rho^3 K_1(K_4 - K_1)K_6\}n_6 \\ &+\rho^2 K_1\{H_1 K_6 + K_4 - K_1\}t_6 \\ &+H_1(\rho^2 K_1 K_4 - H_1 H_2)];\end{aligned}$$

$$\begin{aligned}B^{n_7}(t_1; m_2, m_4, t_6) = \frac{1}{\rho K_1 H_1}[&\rho\theta_1 t_1 - \rho^2 K_1^2 H_2 m_2 + \rho^2 H_1 K_1 K_4 m_4 \\ &-\rho H_1\{K_1 + (H_2 - H_1)K_6\}n_6 \\ &+\{\rho^3 K_1^2 K_6 + H_1(H_2 - H_1)\}t_6 \\ &+\rho K_1(\rho^2 K_1 K_4 - H_1 H_2)].\end{aligned}$$

[8] From (3.6), (3.13), $c - a > 0$ and $z > y > x$

and (4.5) can be rewritten as follows:[9]

$$M_1^{n_7}(m_2, m_4, t_6, n_6) < n_7 - \frac{H_1\,(H_2 - H_1)}{\rho K_1\,(K_4 - K_1)} t_5 < M_3^{n_7}(m_2, m_4, t_6, n_6). \tag{4.6}$$

The next lemma follows from the above construction.

Sublemma 4.2: *Let $\theta_1 > 0$. Condition C.ii if Sublemma 4.1 is satisfied if and only if the following condition is satisfied:*

C.iii: *There is $\pi'' = (t_1, n_1, t_5, n_7, m_2, m_4, t_6, n_6) >> 0$ satisfying (4.3) through (4.6) together with $t_1 < 1$ and $n_1 < 1$.*

Step 3 (case $\theta_1 > 0$): Next, we will eliminate t_1 and n_1 from C.i. In what follows, we use $A^{n_7}(n_1)$ and $B^{n_7}(t_1)$ as shorthands for $A^{n_7}(n_1; m_2, m_4, n_6)$ and $B^{n_7}(t_1; m_2, m_4, t_6)$ whenever it is not necessary to emphasize that they depend parameters (m_2, m_4, n_6). Notice that condition (4.6) does not depend on either t_1 or n_1. Condition (4.3) depends on n_1 but not on t_1. Condition (4.4) depends

---

$$K_1 = \frac{c - a - H_1(y - x)}{z - x} < K_4 = \frac{c - a}{z - x}$$

hold.

[9] $M_1^{n_7}(m_2, m_4, t_6, n_6)$ and $M_3^{n_7}(m_2, m_4, t_6, n_6)$ are defined as follows.

$$M_1^{n_7}(m_2, m_4, t_6, n_6) =$$

$$\begin{aligned}
[-H_1H_2(H_2 - H_1)m_2 - \rho^2(K_4 - K_1)\{(H_2 - H_1)K_6 + K_1\}n_6 \\
+\rho K_1K_4(K_4 - K_1)m_4 + \rho(H_2 - H_1)\{K_4 - K_1 + H_1K_6\}t_6 \\
+\rho H_1(H_2 - H_1)K_4 - \rho^2 K_1(K_4 - K_1)H_2]/(\rho^2 K_1(K_4 - K_1));
\end{aligned}$$

$$M_3^{n_7}(m_2, m_4, t_6, n_6) =$$

$$\begin{aligned}
[-\{H_1(H_2 - H_1)^2 + \rho^3 K_1^2(K_4 - K_1)\}m_2 \\
-\rho^2(K_4 - K_1)\{(H_2 - H_1)K_6 + K_1\}n_6 + \{H_1^2(H_2 - H_1) \\
+\rho^3 K_1(K_4 - K_1)^2\}m_4 + \rho(H_2 - H_1)\{K_4 - K_1 + H_1K_6\}t_6 \\
-H_1^2(H_2 - H_1)]/(\rho^2 K_1(K_4 - K_1)).
\end{aligned}$$

on t_1 but not on n_1. By $\theta_1 > 0$, $A^{n_7}(n_1)$ and $B^{n_7}(t_1)$ are decreasing in n_1 and t_1. Moreover, $0 < t_1 < 1$ and $0 < n_1 < 1$. By these facts, conditions (4.3) and (4.4) are equivalent to the following:

$$\begin{aligned} A^{n_7}(1; m_2, m_4, n_6) &< n_7 - (\rho^2 K_1/H_1)t_5 \\ &< A^{n_7}(0; m_2, m_4, n_6) \end{aligned} \tag{4.7}$$

$$\begin{aligned} B^{n_7}(1; m_2, m_4, t_6) &< n_7 - \left(\rho^2 K_1/H_1\right) t_5 \\ &< B^{n_7}(0; m_2, m_4, t_6) \end{aligned} \tag{4.8}$$

Sublemma 4.3: *Let $\theta_1 > 0$. Condition C.iii if Sublemma 4.2 is satisfied if and only if the following condition is satisfied:*

C.iv: *There is $\pi''' = (t_5, n_7, m_2, m_4, t_6, n_6) >> 0$ satisfying* (4.6), (4.7) *and* (4.8).

At the next step, we will eliminate t_5 and n_7 from C.iv. For this purpose, denote by A the region in the $t_5 - n_7$ space defined by condition (4.7). Denote by B that defined by (4.8). In order for condition C.iv of Sublemma 4.3 to hold (i.e., for π''' to satisfy conditions (4.6), (4.7) and (4.8)), it is necessary that $A \cap B \neq \phi$. Since the boundaries of regions A and B are parallel to one another, by (4.7) and (4.8), $A \cap B \neq \phi$ if and only if the following condition are satisfied.

$$A^{n_7}(0) - B^{n_7}(1) > 0; \tag{4.9}$$

$$B^{n_7}(0) - A^{n_7}(1) > 0. \tag{4.10}$$

As explained below, (4.9) and (4.10) are equivalent to

$$-H_1 < t_6 - \rho K_6 n_6 < \rho K_1; \tag{4.11}$$

this follows from $\theta_1 > 0$ and the fact that, by definition, the following holds:

$$A^{n_7}(0) - B^{n_7}(1) = \theta_1(t_6 - \rho K_6 n_6 + H_1)/(\rho K_1 H_1^2); \tag{4.12}$$

$$B^{n_7}(0) - A^{n_7}(1) = -\theta_1(t_6 - \rho K_6 n_6 - \rho K_1)/(\rho K_1 H_1^2). \quad (4.13)$$

Note that the lower boundary of region $A \cap B$, which plays a critical role at the next step, can be characterized as follows:

$$n_7 - (\rho^2 K_1/H_1)t_5 = A^{n_7}(1; m_2, m_4, n_6)$$
$$if\ t_6 - \rho K_6 n_6 \geq \rho K_1 - H_1; \quad (4.14)$$

$$n_7 - (\rho^2 K_1/H_1)t_5 = B^{n_7}(1; m_2, m_4, t_6)$$
$$if\ t_6 - \rho K_6 n_6 < \rho K_1 - H_1. \quad (4.15)$$

Define $\theta_2 = t_6 - \rho K_6 n_6 - (\rho K_1 - H_1)$. At the next step, we will focus on the case of $\theta_2 \geq 0$; we will return to the case of $\theta_2 < 0$ at Step 4′.

Step 4 (case $\theta_1 > 0$; subcase $\theta_2 \geq 0$): We now eliminate t_5 and n_7 from condition C.iv. of Sublemma 4.3. Notice that condition (4.6), like (4.7) and (4.8), determines a region between two parallel lines in the $t_5 - n_7$ space. Denote by M the region defined by (4.6). Condition C.iv is equivalent to the condition that $A \cap B$ intersects region M in the strictly positive orthant of the $t_5 - n_7$ space.

See Figures 2 and 3. Note the following facts. By $\theta_1 > 0$, the slope of the boundaries of region M, i.e., $\frac{H_1(H_2-H_1)}{\rho K_1(K_4-K_1)}$, is smaller than or equal to that of the boundaries of regions A and B, $\rho^2 K_1/H_1$. $A \cap B$ intersects with region M in the strictly positive orthant of the $t_5 - n_7$ space. By $\theta_2 \geq 0$, the lower boundary of $A \cap B$ is given by (4.14). Suppose the vertical intercept of the lower boundary of $A \cap B$ is non-negative ($A^{n_7}(1) \geq 0$). Then, if and only if that intercept lies below the vertical intercept of the upper boundary of M, $A \cap B$ intersects M in the strictly positive orthant of the $t_5 - n_7$ space. Suppose, instead, the vertical intercept of the lower boundary of $A \cap B$ is negative ($A^{n_7}(1) < 0$). Then, if and only if the horizontal intercept of the lower boundary lies to the right of that of the upper boundary of M, $A \cap B$ intersects M.

Denoting by $M_1^{t_5} = M_1^{t_5}(m_2, m_4, t_6, n_6,)$ and $M_3^{t_5} = M_3^{t_5}(m_2, m_4, t_6, n_6,)$ the horizontal intercepts of the lower and upper boundaries of region M, we may summarize this as follows. Regions $A \cap B$ and M intersect each other in the strictly positive orthant of the $t_5 - n_7$ space if and only if one of the following two cases holds.

$$M_3^{n_7}(m_2, m_4, t_6, n_6,) > A^{n_7}(1; m_2, m_4, n_6) \geq 0; \tag{4.16}$$

$$A^{n_7}(1; m_2, m_4, n_6) < 0 \ , \ A^{t_5}(1; m_2, m_4, n_6) > M_3^{t_5}(m_2, m_4, t_6, n_6,). \tag{4.17}$$

Note that by $K_1 > 0$ and $t_6 - \rho K_6 n_6 \geq \rho K_1 - H_1$, (4.11) is equivalent to

$$\rho K_1 - H_1 \leq t_6 - \rho K_6 n_6 < \rho K_1 \tag{4.18}$$

The next lemma follows.

Sublemma 4.4: *Let* $\theta_1 > 0$ *and* $t_6 - \rho K_6 n_6 \geq \rho K_1 - H_1$. *Condition C.iv of Sublemma 4.3 is satisfied if and only if the following condition is satisfied:*

C.v: *There is* $\pi'''' = (m_2, m_4, t_6, n_6) >> 0$ *satisfying* (4.18) *and either* (4.16) or (4.17).

Step 5 (case $\theta_1 > 0$; *subcase* $\theta_2 \geq 0$*):* Next, we will eliminate m_2 and m_4 from Condition (v). Note that condition (4.18) is independent of m_2 and m_4. We may demonstrate that condition (4.16) is equivalent to the following:

$$\begin{array}{c} -\rho K_1 H_1 H_2 m_2 + \rho H_1^2 K_4 m_4 \geq \\ \{H_1^2 - \rho^3 K_1 (K_4 - K_1) K_6\} n_6 - \\ \rho^2 K_1 \{H_1 K_6 + K_4 - K_1\} t_6 \\ +\rho^3 K_1^2 (K_4 - K_1) + H_1^3 - \rho^2 K_1 H_1 K_4; \end{array} \tag{4.19}$$

$$\begin{array}{c} -H_1 (H_2 - H_1) m_2 + H_1^2 m_4 < \rho^2 (K_4 - K_1) K_6 n_6 \\ -\rho (H_1 K_6 + K_4 - K_1) t_6 - \rho H_1 K_4 \\ +H_1^2 + \rho^2 K_1 (K_4 - K_1) . \end{array} \tag{4.20}$$

Given $\pi''''' = (t_6, n_6)$, conditions (4.19) and (4.20) determine a region between two lines in the $m_2 - m_4$ space. Call this region Γ;

see Figure 4. Then, condition (4.16) is satisfied for $(m_2, m_4) >> 0$ if and only if region Γ intersects the strictly positive orthant of the $m_2 - m_4$ space. Since $K_1H_2 + K_4H_1 - K_4H_2 > 0$,[10] we may demonstrate that the slope of the boundary of (4.19) is larger than that of (4.20) in the $m_2 - m_4$ space. Therefore, region Γ intersects the strictly positive orthant if and only if the boundaries of (4.19) and (4.20) intersect each other in the positive orthant. This condition can be expressed as follows:[11]

$$\begin{aligned} &-\{\rho^3 (K_4 - K_1)^2 K_6 - H_1^2\} n_6 \\ &+\rho^2 (K_4 - K_1)(H_1K_6 + K_4 - K_1)(t_6 - C^{t_6}) < 0; \end{aligned} \tag{4.21}$$

$$\begin{aligned} &-\{\rho^3 K_1 (K_4 - K_1) K_6 - H_1 (H_2 - H_1)\} n_6 \\ &+\rho^2 K_1 (H_1K_6 + K_4 - K_1)(t_6 - D^{t_6}) < 0. \end{aligned} \tag{4.22}$$

Similarly, we may demonstrate that (4.17) is equivalent to the following.

$$\begin{aligned} &-\rho K_1H_1H_2m_2 + \rho H_1^2K_4m_4 < \\ &\{H_1^2 + \rho^3K_1 (K_4 - K_1) K_6\} n_6 - \\ &\rho^2 K_1 \{H_1K_6 + K_4 - K_1\} t_6 \\ &+\rho^3 K_1^2 (K_4 - K_1) + H_1^3 - \rho^2 K_1H_1K_4; \end{aligned} \tag{4.23}$$

$$-\rho K_1 m_2 + \rho (K_4 - K_1) m_4 > n_6 - \rho K_1 + H_1. \tag{4.24}$$

---

[10]This may be shown from (3.6), (3.10) and (3.13) and by using

$$[(z-y)(y-x)/(z-x)^2] < 1$$

[11]C^{t_6} and D^{t_6} are defined as follows.

$$C^{t_6} = \frac{-\rho^3K_1(K_4 - K_1)^2 + \rho^2H_1K_4(K_4 - K_1) - \rho H_1^2K_4 + H_1^3}{\rho^2(K_4 - K_1)(H_1K_6 + K_4 - K_1;};$$

$$D^{t_6} = \frac{-\rho^3K_1^2(K_4 - K_1) + \rho^2H_1K_1K_4 - \rho H_1H_2K_1 + H_1^2(H_2 - H_1)}{\rho^2K_1(H_1K_6 + K_4 - K_1)}.$$

In much the same way as (4.19) and (4.20), we may demonstrate that Conditions (4.23) and (4.24) as well are equivalent to (4.21) and (4.22).

Sublemma 4.5: *Let $\theta_1 > 0$ and $t_6 - \rho K_6 n_6 \geq \rho K_1 - H_1$. Condition C.v of Sublemma 4.4 is satisfied if and only if the following condition is satisfied.*

C.vi: *There is $\pi''''' = (t_6, n_6) >> 0$ satisfying* (4.18), (4.21) *and* (4.22).

Step 6 (case $\theta_1 > 0$; subcase $\theta_2 \geq 0$): This is the last step, at which we eliminate t_6 and n_6 from C.vi. Conditions (4.18), (4.21) and (4.22) each determine a region in the $n_6 - t_6$ space. Call those regions Θ, C and D, respectively. Then, it suffices to derive conditions under which these regions have a non-empty intersection in the $n_6 - t_6$ space. It can be demonstrated that, among the slopes of the boundaries of regions Θ, C and D, the slope of the boundaries of region Θ is the largest and that the slope of the boundary of region C is the smallest. Denote by S^C and S^D the slopes of the boundaries of regions C and D. Moreover, denote by Θ^{n_6} and Θ^{t_6} the horizontal and vertical intercepts of the lower boundary of region Θ. For the moment, focus on the case in which $S^C \neq 0$ and $S^D \neq 0$; we will deal with the case of $S^C = 0$ and/or $S^D = 0$, separately. Then, C.vi of Sublemma 4.5 is equivalent to the following conditions (see Figure 5 and 6)

Condition 4.1
Case I: *$\rho K_1 \geq H_1$ and*

$$(I.a)\ C^{t_6} > \Theta^{t_6} \text{ and } D^{t_6} > \Theta^{t_6}$$

Case II: *$\rho K_1 < H_1$ and one of the following three conditions holds:*

$$(II.b)\ S^C > 0, S^D > 0, \Theta^{n_6} > C^{n_6} \text{ and } \Theta^{n_6} > D^{n_6};$$

$$(II.c)\ S^C > 0, S^D < 0, C^{n_6} > D^{n_6} \text{ and } \Theta^{n_6} > D^{n_6};$$

(*II.d*) $S^C < 0, S^D < 0, C^{n_6} > 0 and\ D^{n_6} > 0$.

The next lemma follows from Sublemmas 1 through 5 and the above discussion.

Lemma 4.1: *Suppose* $S^C \neq 0$, $S^D \neq 0$, $\theta_1 > 0$ *and* $t_6 - \rho K_6 n_6 \geq \rho K_1 - H_1$. *Condition C.i of Theorem 3.1 is satisfied if and only if Condition 4.1 is satisfied.*

We now return to the end of Step 3 and focus on the case of $\theta_2 < 0$.

Step 4′ (case $\theta_1 > 0$; *subcase* $\theta_2 < 0$*)*: In this case, the lower boundary of region $A \cap B$ is given by (4.15). By using this fact, we obtain the counterparts of (4.18), (4.16) and (4.17) as follows.

$$-H_1 < t_6 - \rho K_6 n_6 < \rho K_1 - H_1 \tag{4.25}$$

$$M_3^{n_7}(m_2, m_4, t_6, n_6,) > B^{n_7}(1; m_2, m_4, t_6) \geq 0; \tag{4.26}$$

$$\begin{aligned} B^{n_7}(1; m_2, m_4, t_6) < 0 \ , \ & B^{t_5}(1; m_2, m_4, t_6) \\ & > M_3^{t_5}(m_2, m_4, t_6, n_6,). \end{aligned} \tag{4.27}$$

The counterpart of Sublemma 4.4 is as follows:

Sublemma 4.6: *Let* $\theta_1 > 0$ *and* $t_6 - \rho K_6 n_6 < \rho K_1 - H_1$. *Condition C.iv of Sublemma 4.3 is satisfied if and only if the following condition is satisfied:*
C.v′: *There is* $\pi'''''= (m_2, m_4, t_6, n_6) >> 0$ *satisfying* (4.25) *and either* (4.26) *or* (4.27).

Step 5′ (case $\theta_1 > 0$; *subcase* $\theta_2 < 0$*)*: Following the line of discussion at Step 5, we may obtain the counterparts of (4.23) and (4.24) as follows:[12]

$$\begin{aligned} & \rho H_1 \left[(H_2 - H_1) K_6 + K_1 \right] n_6 \\ & + \left[\rho^3 K_1 K_6 (K_4 - K_1) - H_1 (H_2 - H_1) \right] \\ & (t_6 - E^{t_6}) < 0; \end{aligned} \tag{4.28}$$

[12] E^{t_6} and F^{t_6} are defined as follows.

$$\rho\left(H_2-H_1\right)\left[\left(H_2-H_1\right)K_6+K_1\right]n_6 \\ +\left[\rho^3K_1^2K_6-\left(H_2-H_1\right)^2\right]\left(t_6-F^{t_6}\right)<0. \tag{4.29}$$

Sublemma 4.7: *Let* $\theta_1 > 0$ *and* $t_6 - \rho K_6 n_6 < \rho K_1 - H_1$. *Condition C.v′ of Sublemma 4.6 is satisfied if and only if the following condition is satisfied:*

C.vi′: *There is* $\pi''''' = (t_6, n_6) >> 0$ *satisfying* (4.25), (4.28) *and* (4.29).

Step 6′ (case $\theta_1 > 0$*; subcase* $\theta_2 < 0$*)*: Conditions (4.25), (4.28) and (4.29) each determine a region in the $n_6 - t_6$ space. Call those regions Θ', E and F, respectively. Denote by S^E and S^F the slopes of the boundaries of regions E and F. The equation of the upper boundary of Θ' is the same as that of the lower boundary of Θ. So, the horizontal and the vertical intercepts of the upper boundary of Θ' are, respectively, Θ^{n_6} and Θ^{t_6}. Following the line of discussion at Step 6, we may prove that C.vi' is equivalent to the following condition, given $S^E \neq 0$ and $S^F \neq 0$.

Condition 4.2

Case I: $\rho K_1 \geq H_1$ *and one of the following three conditions:*

$$(I.b)\ S^E > 0, S^F > 0, \Theta^{t_6} > E^{t_6} \, and \ \Theta^{t_6} > F^{t_6};$$

$$(I.c)\ S^E > 0, S^F < 0, \Theta^{t_6} > E^{t_6} \, and \ F^{t_6} > E^{t_6};$$

$$(I.d)\ S^E < 0, S^F > 0, E^{t_6} > 0 and \ F^{t_6} > 0.$$

Case II: $\rho K_1 < H_1$ *and*

$$(II.a)\ E^{n_6} > \Theta^{n_6} and \ F^{n_6} > \Theta^{n_6}$$

$$E^{t_6} = \frac{-\rho^3K_1^2(K_4-K_1)+\rho^2H_1K_1K_4-\rho H_1H_2K_1+H_1^2(H_2-H_1)}{\rho^3K_1(K_4-K_1)K_6-H_1(H_2-H_1)};$$

$$F^{t_6} = \frac{-\rho^3K_1^3+\rho^2H_2K_1^2-\rho H_2(H_2-H_1)K_1+H_1(H_2-H_{1111})^2}{\rho^3K_1^2K_6-(H_2-H_1)^2}$$

The counterpart of Lemma 4.1 is as follows:

Lemma 4.2: *Suppose* $S^E \neq 0$, $S^F \neq 0$, $\theta_1 > 0$ *and* $t_6 - \rho K_6 n_6 < \rho K_1 - H_1$. *Condition C.i of Theorem 3.1 is satisfied if and only if Condition 4.2 is satisfied.*

We now return to the end of Step 1 and focus on the case of $\theta_1 < 0$.

Step 2″/3″ (case $\theta_1 < 0$*)*: Return to the position at Step 2 at which we focused on the case of $\theta_1 > 0$. Instead, we now focus on the case of $\theta_1 < 0$. For the sake of simplicity, here, we now eliminate m_1, t_1 and n_1 by one step, corresponding to Steps 2 and 3. Since $\theta_1 < 0$, the counterparts of (4.6), (4.7) and (4.8) are as follows:

$$\begin{aligned} M_3^{n_7}(m_2, m_4, t_6, n_6) &< n_7 - \frac{H_1 (H_2 - H_1)}{\rho K_1 (K_4 - K_1)} t_5 \\ &< M_1^{n_7}(m_2, m_4, t_6, n_6); \end{aligned} \tag{4.30}$$

$$\begin{aligned} A^{n_7}(0; m_2, m_4, n_6) &< n_7 - (\rho^2 K_1 / H_1) t_5 \\ &< A^{n_7}(1; m_2, m_4, n_6); \end{aligned} \tag{4.31}$$

$$\begin{aligned} B^{n_7}(0; m_2, m_4, t_6) &< n_7 - \left(\rho^2 K_1 / H_1\right) t_5 \\ &< B^{n_7}(1; m_2, m_4, t_6); \end{aligned} \tag{4.32}$$

conditions (4.31) and (4.32) follow since $A^{n_7}(n_1; m_2, m_4, n_6)$ and $B^{n_7}(t_1; m_2, m_4, n_6)$ are increasing in n_1 and t_1, respectively, in the case of $\theta_1 < 0$.

Sublemma 4.8: *Let* $\theta_1 < 0$. *Condition C.iii of Sublemma 4.2 is satisfied if and only if the following condition is satisfied:*

C.iv″: *There is* $\pi''' = (t_5, n_7, m_2, m_4, t_6, n_6) >> 0$ *satisfying* (4.30), (4.31) *and* (4.31).

At the next step, we will eliminate t_5, n_7, m_2, m_4, t_6 and n_6 by one step. Before moving to this step, as in Step 3, denote by A and

B, respectively, the regions defined by (4.31) and (4.32). We may demonstrate that the condition for $A \cap B \neq \phi$ is identical to that derived at Step 4; i.e., $A \cap B \neq \phi$ if and only if (4.11) is satisfied Recall that at Step 4, we used the lower boundary of $A \cap B$. In the case of $\theta_1 < 0$, instead, we will use the upper boundary of $A \cap B$, which is given as follows:

$$n_7 - (\rho^2 K_1/H_1)t_5 = A^{n_7}(1; m_2, m_4, n_6) \quad if\ t_6 - \rho K_6 n_6 \geq \rho K_1 - H_1; \tag{4.33}$$

$$n_7 - (\rho^2 K_1/H_1)t_5 = B^{n_7}(1; m_2, m_4, t_6) \quad if\ t_6 - \rho K_6 n_6 < \rho K_1 - H_1. \tag{4.34}$$

We will first focus on the case of $\theta_2 \geq 0$, in which (4.33) holds. At Step $4'''/5'''/6'''$, we will return to the case of $\theta_2 < 0$.

Step 4″/5″/6″ (case $\theta_1 < 0$; subcase $\theta_2 \geq 0$): Denote by M the region defined by (4.30). In order to eliminate t_5 and n_7, as at Step 4, we derive conditions under which region M intersects region $A \cap B$ in the strictly positive orthant of the $t_5 - n_7$ space.

By $\theta_1 < 0$, the slope of the boundaries of region M, i.e., $\frac{H_1(H_2-H_1)}{\rho K_1(K_4-K_1)}$, is larger than or equal to that of the boundaries of regions A and B, $\rho^2 K_1/H_1$. Therefore, by $\theta_2 < 0$, in much the same way as Step 4, we may prove that M intersects region $A \cap B$ in the strictly positive orthant of the $t_5 - n_7$ space if and only if one of the following two condition is satisfied.

$$A^{n_7}(1; m_2, m_4, n_6) \geq 0\ ,\ A^{n_7}(1; m_2, m_4, n_6) > M_3^{n_7}(m_2, m_4, t_6, n_6) \tag{4.35}$$

$$A^{n_7}(1; m_2, m_4, n_6) < 0\ ,\ M_3^{t_5}(m_2, m_4, t_6, n_6) > A^{t_5}(1; m_2, m_4, n_6) \tag{4.36}$$

In order to erase m_2 and m_4, as at Step 5, we may demonstrate that (4.35) is equivalent to (4.19) and (4.20) and that (4.36) is equivalent of (4.23) and (4.24). As is seen at Step 5, these conditions boil down to conditions (4.21) and (4.22). Finally, by $t_6 - \rho K_6 n_6 \geq \rho K_1 - H_1$ and (4.11), (4.18) holds.

This implies that the discussion at Step 6 holds without any change. Therefore, the same result as at Step 6 holds for the case of $\theta_1 < 0$ as well.

Lemma 4.3: *Suppose $S^C \neq 0$, $S^D \neq 0$, $\theta_1 > 0$ and $t_6 - \rho K_6 n_6 \geq \rho K_1 - H_1$. Condition C.i of Theorem 3.1 is satisfied if and only if Condition 4.1 is satisfied.*

Next, we will focus on the case of $\theta_2 < 0$.

Step 4'''/5'''/6''' (case $\theta_1 < 0$; subcase $\theta_2 < 0$): In the case of $t_6 - \rho K_6 n_6 < \rho K_1 - H_1$, we may follow the above discussions to demonstrate that a lemma similar to Lemma 4.3 holds for the case of $\theta_1 < 0$.

Lemma 4.4: *Suppose $S^E \neq 0$, $S^F \neq 0$, $\theta_1 < 0$ and $t_6 - \rho K_6 n_6 < \rho K_1 - H_1$. Condition C.i of Theorem 3.1 is satisfied if and only if Condition 4.2 is satisfied.*

5. The Proof of Theorem 3.2

We will demonstrate Theorem 3.2.

5.1 Case of $\theta_1 \neq 0$

Lemmas 4.2 through 4.4 imply that in the cases of $t_6 - \rho K_6 n_6 \geq \rho K_1 - H_1$ and $t_6 - \rho K_6 n_6 < \rho K_1 - H_1$, respectively, Conditions 4.1 and 4.2 can characterize Condition C.i of Theorem 3.1 regardless of the sign of $\theta_1 \neq 0$, . Therefore, by expressing Conditions 4.1 and 4.2 in terms of H_1, H_2, K_1, K_4, K_6 and ρ, we may summarize these lemmas as follows.

Lemma 5.1: *Let $\theta_1 \neq 0$, $S^C \neq 0$, $S^D \neq 0$, $S^E \neq 0$ and $S^E \neq 0$. Condition C.i of Theorem 3.1 is satisfied if and only if the following condition is satisfied.*

C.vii: Case I: $\rho K_1 \geq H_1$ *and one of the following conditions is satisfied.*

$$(I.a') \quad \begin{array}{c} \rho^2 (K_4 - K_1) K_6 + \rho (K_4 - K_1) - H_1 < 0, \text{ and} \\ \rho^2 K_1 K_6 + \rho K_1 - (H_2 - H_1) < 0. \end{array}$$

$$(I.b') \quad \begin{array}{c} \rho^3 K_1 (K_4 - K_1) K_6 - H_1 (K_4 - K_1) < 0, \\ \rho^3 K_1^2 K_6 - (H_2 - H_1)^2 < 0, \\ \rho^2 (K_4 - K_1) K_6 + \rho (K_4 - K_1) - H_1 < 0 \text{ and} \\ \rho^2 K_1 K_6 + \rho K_1 - (H_2 - H_1) < 0 \end{array}$$

$$(I.c') \quad \begin{array}{c} \rho^3 K_1 (K_4 - K_1) K_6 - H_1 (H_2 - H_1) < 0 \\ \rho^3 K_1^2 K_6 - (H_2 - H_1)^2 > 0 \\ \rho^2 (K_4 - K_1) K_6 - \rho (K_4 - K_1) - H_1 < 0 \\ \rho^2 K_1 K_6 - \rho (H_2 - H_1) K_6 - (H_2 - H_1) > 0 \end{array}$$

$$(I.d') \quad \begin{array}{c} \rho^3 K_1 (K_4 - K_1) K_6 - H_1 (H_2 - H_1) > 0, \\ \rho^3 K_1^2 K_6 - (H_2 - H_1)^2 > 0, \\ \rho^2 K_1 (K_4 - K_1) - \rho K_1 H_1 + H_1 (H_2 - H_1) < 0, \text{ and} \\ \rho^2 K_1^2 - \rho K_1 (H_2 - H_1) + (H_2 - H_1) > 0 \end{array}$$

Case II: $\rho K_1 < H_1$ *and one of the following conditions is satisfied:*

$$(II.a') \quad \begin{array}{c} \rho^2 (K_4 - K_1) K_6 - \rho H_1 K_6 - H_1 > 0, \text{ and} \\ \rho^2 K_1 K_6 - \rho (H_2 - H_1) K_6 - (H_2 - H_1) > 0 \end{array}$$

$$(II.b') \quad \begin{array}{c} \rho^3 (K_4 - K_1)^2 K_6 - H_1^2 > 0, \\ \rho^3 K_1 (K_4 - K_1) K_6 - H_1 (H_2 - H_1) > 0, \\ \rho^2 (K_4 - K_1) K_6 - \rho H_1 K_6 - H_1 > 0, \text{ and} \\ \rho^2 K_1 K_6 - \rho (H_2 - H_1) K_6 - (H_2 - H_1) > 0 \end{array}$$

$$(II.c')\quad \begin{array}{c} \rho^3 K_1 (K_4 - K_1) K_6 - H_1 (H_2 - H_1) > 0, \\ \rho^3 (K_4 - K_1)^2 K_6 - H_1^2 < 0, \\ \rho^2 K_1 K_6 - \rho (H_2 - H_1) K_6 - (H_2 - H_1) > 0, \ and \\ \rho^2 (K_4 - K_1) K_6 + \rho (K_4 - K_1) - H_1 < 0 \end{array}$$

$$(II.d')\quad \begin{array}{c} \rho^3 K_1 (K_4 - K_1) K_6 - H_1 (H_2 - H_1) < 0, \\ \rho^3 (K_4 - K_1)^2 K_6 - H_1^2 < 0, \\ \rho^2 (K_4 - K_1)^2 - \rho (K_4 - K_1) H_1 + H_1^2 < 0, \ and \\ \rho^2 K_1 (K_4 - K_1) - \rho K_1 H_1 + H_1 (H_2 - H_1) < 0 \end{array}$$

We will rewrite condition C.vii of Lemma 5.1 by introducing parameters $\eta, = \frac{H_2 - H_1}{K_1}$, $\kappa = \frac{K_4 - K_1}{H_1}$ and $B = \frac{c-b}{c-a}$. Then, by (3.7), (3.10) and (3.14), the following relationship holds among the parameters introduced above.

$$\eta = \frac{H_2 - H_1}{K_1} = \frac{z - y}{z - x}, \ \kappa = \frac{K_4 - K_1}{H_1} = \frac{y - x}{z - x}, \ B = \frac{c - b}{c - a}. \tag{5.1}$$

Moreover, by (3.7), (3.10), (3.14) and (3.19), we can express K_6 and H_1/K_1 as follows

$$K_6 = \frac{\eta}{\kappa} \ and \ \frac{H_1}{K_1} = \frac{\eta - B}{B\kappa - 1}. \tag{5.2}$$

By (3.1) and (3.3), parameters κ, η and B satisfy the following:

$$\eta + \kappa = 1;$$

$$\kappa > 0, \quad \eta > 0 \quad and \quad B > 0. \tag{5.4}$$

The next lemma expresses condition C.vii of Lemma 5.1 in terms of η, κ and B.

Lemma 5.2: *Let $\theta_1 \neq 0$. Condition C.i of Theorem 3.1 is satisfied if and only if the following condition is satisfied.*

C.viii: Case I: $\frac{\eta-B}{B\kappa-1} \leq \rho$ *and one of the following three conditions are satisfied*

$$(I.a'')\rho^2\eta + \rho\kappa - 1 < 0 \textit{ and } (\eta - \rho)(\kappa - \rho^2) > \rho^3;$$

$$(I.c'')\ \rho^3 < 1, \quad \rho^3 > \eta\kappa, \quad \rho^2\eta + \rho\kappa < 1 \quad \textit{and} \quad \rho\eta + \kappa < \rho^2;$$

$$(I.d'')\ \rho^3 > 1, \quad \rho^3 > \eta\kappa, \quad \rho^2\kappa + \eta < \rho \quad \textit{and} \quad \rho^2 - \rho\eta + \rho\eta < 0.$$

Case II: $\frac{\eta-B}{B\kappa-1} > \rho$ *and one of the following three conditions are satisfied.*

$$(II.a'')\ (\kappa - \rho^{-1})(\eta - \rho^{-2}) > \rho^{-3} \quad \textit{and} \quad \rho\eta + \kappa < \rho^2;$$

$$(II.c'')\ \rho^3 > 1, \quad \rho^3\eta\kappa < 1, \quad \rho\eta + \kappa < \rho^2 \quad \textit{and} \quad \rho^2\eta + \rho\kappa < 1;$$

$$(II.d'')\ \rho^3 \quad < \quad 1, \quad \rho^3\eta\kappa < 1, \quad (\rho\kappa)^2 - \rho\kappa + 1 < 0 \quad \textit{and}$$
$$\rho^2\kappa + \eta \quad < \quad \rho.$$

Proof: In order to transform Lemma 5.1 into this lemma, we need to demonstrate that $S^C \neq 0$, $S^D \neq 0$, $S^E \neq 0$ and $S^E \neq 0$. If $S^C = 0$, by 4.21), $\rho^3(K_4 - K_1)^2 K_6 - H_1^2 = \dot{0}$. This implies, by the definition of η and κ, $\eta\kappa = \rho^{-3} > \dot{1}$, which contradicts (5.3); i.e., $S^C \neq 0$. If $S^D = 0$, by (4.22), $\rho^3 K_1(K_4 - K_1)K_6 - H_1(H_2 - H_1) = \dot{0}$. This means $\rho^3 = 1$ by (5.1) and (5.2), which contradicts $\rho < 1$; i.e., $S^D \neq 0$. By (4.28), $S^E = 0$ and $S^F = 0$, respectively, imply $\rho H_1[(H_2 - H_1)K_6 + K_1] = 0$ and $\rho(H_2 - H_1)[(H_2 - H_1)K_6 + K_1] = 0$, both of which contradict $\rho > 0$, $H_2 > H_1$, $K_6 > 0$ and $K_1 > 0$; i.e., $S^E \neq 0$ and $S^F \neq 0$.

Cases (I.b′) and (II.b′) are special cases of (I.a′) and (II.a′), respectively. Therefore, by using (5.1) and (5.2), we may rewrite C.vii of Lemma 5.1 as C.viii of this lemma. Q.E.D.

We will demonstrate the following lemma.

Lemma 5.3: *Let $\theta_1 \neq 0$. Condition C.i of Theorem 3.1 is satisfied if and only if* $0 < \rho < \hat{\rho} = (3 - \sqrt{5})/2$.

Proof: As the above discussion demonstrates, it suffices to demonstrate that there is (κ, η) that satisfies Condition 4 if and only if $0 < \rho < \hat{\rho}$. By the above discussion. First, we will demonstrate that all the cases but (I.a″) of Condition 5 are void. To this end, note the following: The fourth condition of (I.c″) is not compatible with $\eta + \kappa = 1$ for $\rho < 1$; the first condition of (I.d″) contradicts our hypothesis $\rho < 1$; the second condition of (II.a″) is not compatible with $\eta + \kappa = 1$ $\rho < 1$; the first condition of (II.b″) contradicts $\rho < 1$; the third condition of (II.d″) is never satisfied. Thus, it suffices to focus on case (I.a″).

Given (5.3) and (5.4), $\frac{\eta - B}{B\kappa - 1} > 0$ holds if and only if $\kappa < 1/B$ and $\eta < B$. Therefore, $0 < \frac{\eta - B}{B\kappa - 1} \leq \rho$ if and only if

$$B > \eta \geq \rho(B\kappa - 1) + B. \tag{5.5}$$

See Figure 7. The intersection of line (5.3) and the lower boundary of the region defined by (5.5) is given by

$$(\kappa(\rho), \eta(\rho)) = ((1 + \rho - B)/(\rho B + 1), (B - \rho(1 - B))/(\rho B + 1)). \tag{5.6}$$

Therefore, given (5.3), (5.5) are satisfied if and only if $\eta(\rho) \leq \eta < B$, which is equivalent to

$$\eta < B \leq \frac{\rho + \eta}{1 + \rho - \rho\eta}. \tag{5.7}$$

In short, conditions on κ, η and B are summarized by (5.3), (5.4) and (5.7). Thus, (I.a″) must be satisfied under conditions (5.3), (5.4) and (5.7). The first condition of (I.a″) is satisfied for any (κ, η) on line $\eta + \kappa = 1$. In other words, it is satisfied, given (5.3). Note that the region defined by the second condition of (I.a″) intersects line $\eta + \kappa = 1$ if and only if $0 < \rho < \hat{\rho} = \left(3 - \sqrt{5}\right)/2$. In order to prove Lemma 10, therefore, it suffices to demonstrates that we may choose (κ, η) satisfying the second condition of (I.a″) at the same time as (5.3), (5.4) and (5.7) for any ρ satisfying $0 < \rho < \hat{\rho}$.

Let $0 < \rho < \hat{\rho}$. Think of the boundary of the region defined by the second condition of (I.a″); i.e.,

$$(\eta - \rho)(\kappa - \rho^2) = \rho^3. \tag{5.8}$$

This boundary intersects $\eta + \kappa = 1$ at two points in the positive orthant of the $\kappa - \eta$ space. The value of η at the upper intersection is

$$\eta^* = \frac{1 + \rho - \rho^2 + \sqrt{(1 + \rho - \rho^2)^2 - 4\rho}}{2}.$$

Choose $B = \eta^*$, and choose $\eta < B$ sufficiently close to B. Then, it is easy to demonstrate that condition (5.7) is satisfied. Moreover, choose κ by condition (5.3). Then, the second condition of (I.a″) as well as (5.7), (5.3) and (5.4) are satisfied. Q.E.D.

5.2 Proof of Theorem 3.1

Lemma 5.2 imply that in order to complete the proof of Theorem 3.1, it suffices to demonstrate that in the case of $\theta_1 = 0$, there is no π satisfying condition (3.27) for any ρ, $\hat{\rho} \leq \rho < 1$. Suppose that $\theta_1 = 0$ and that there is π^* satisfying condition (3.27) for some ρ, $\hat{\rho} \leq \rho < 1$. Then we may choose parameters x, y, x, a, b and c in such a way that (3.27) is satisfied with π^*. Then, we may demonstrate that it is possible to perturb the value of these parameters in such a way that $\theta_1 \neq 0$ and (3.27) remains to hold with π^*. This contradicts Lemma 5.2. Theorem 3.1 is proved.

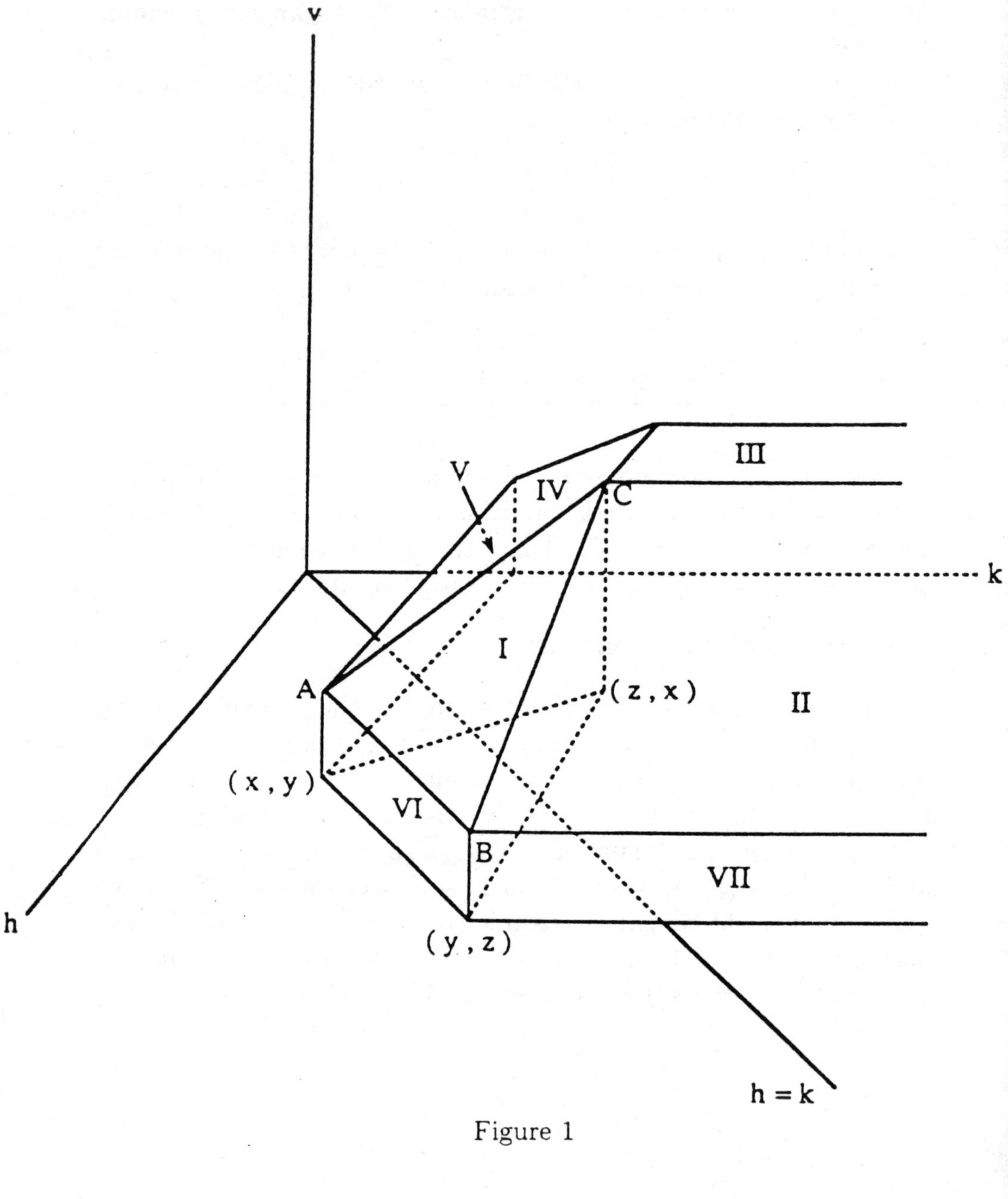

Figure 1

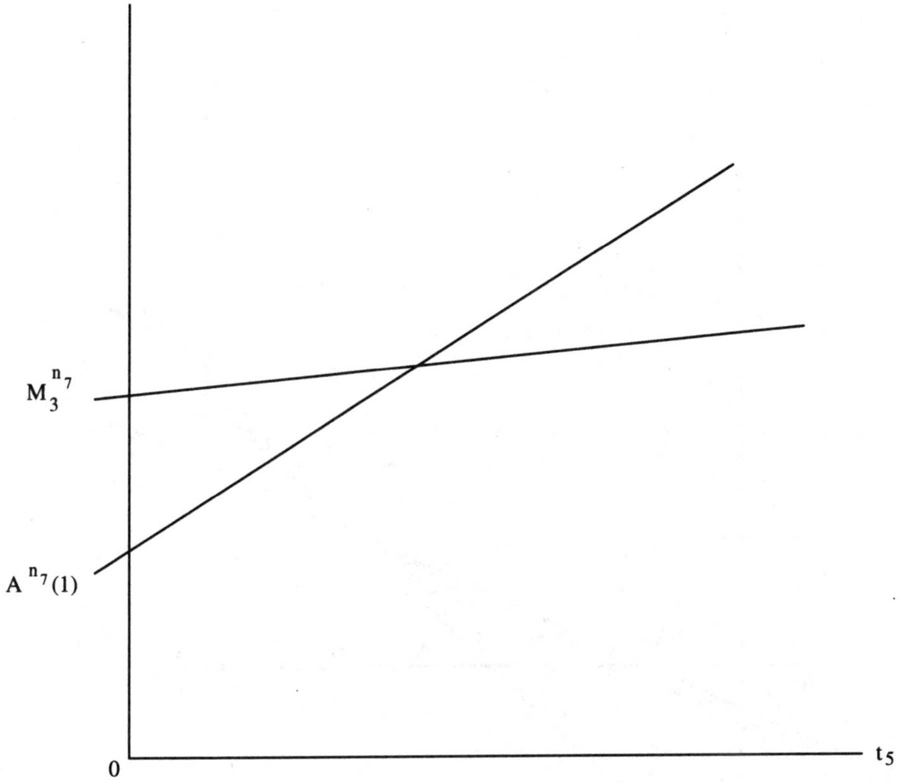

Figure 2. $t_6 - \rho K_6 n_6 \geqslant 0$, $A^{n_7}(1) \geqslant 0$

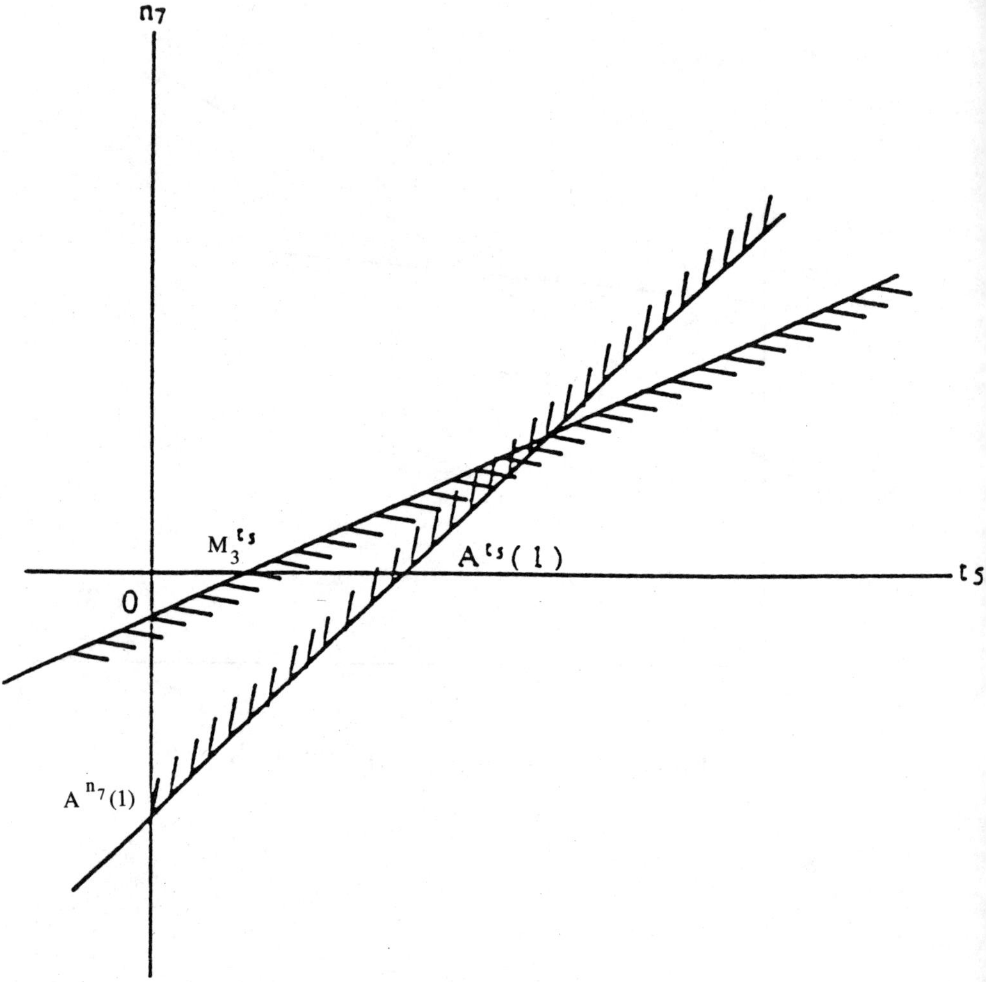

Figure 3 $t_6 - \rho K_6 n_6 > 0$. $A^{n_7}(1) < 0$

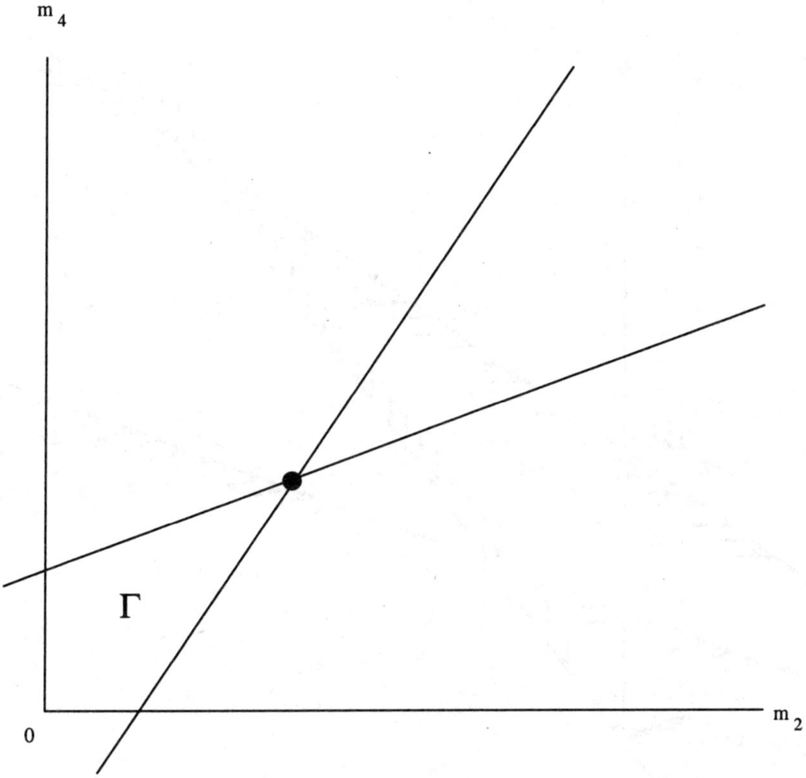

Figure 4

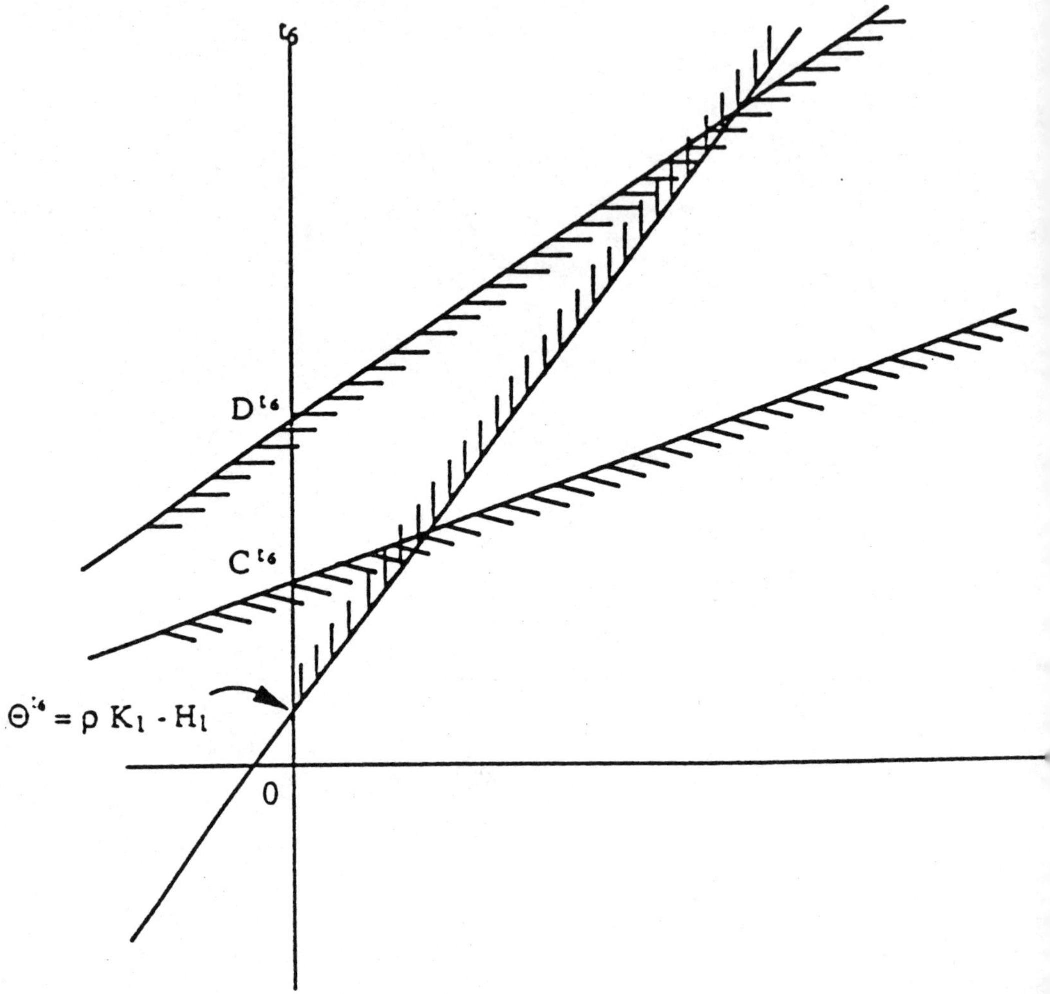

Figure 5 (Ia) $\Theta^{t_6} \geqslant 0$

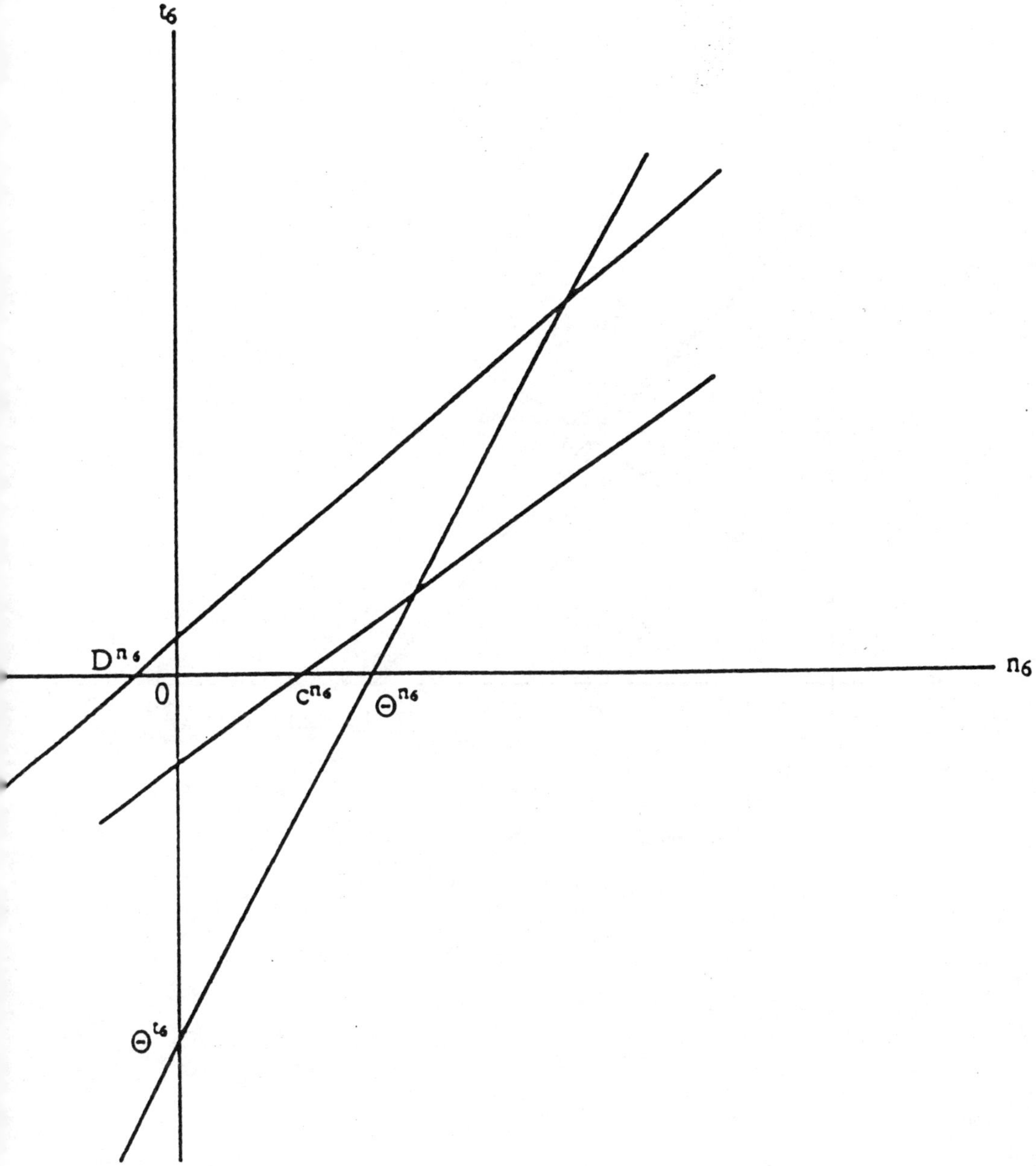

Figure 6 (IIb) $\Theta^{t_6} < 0$

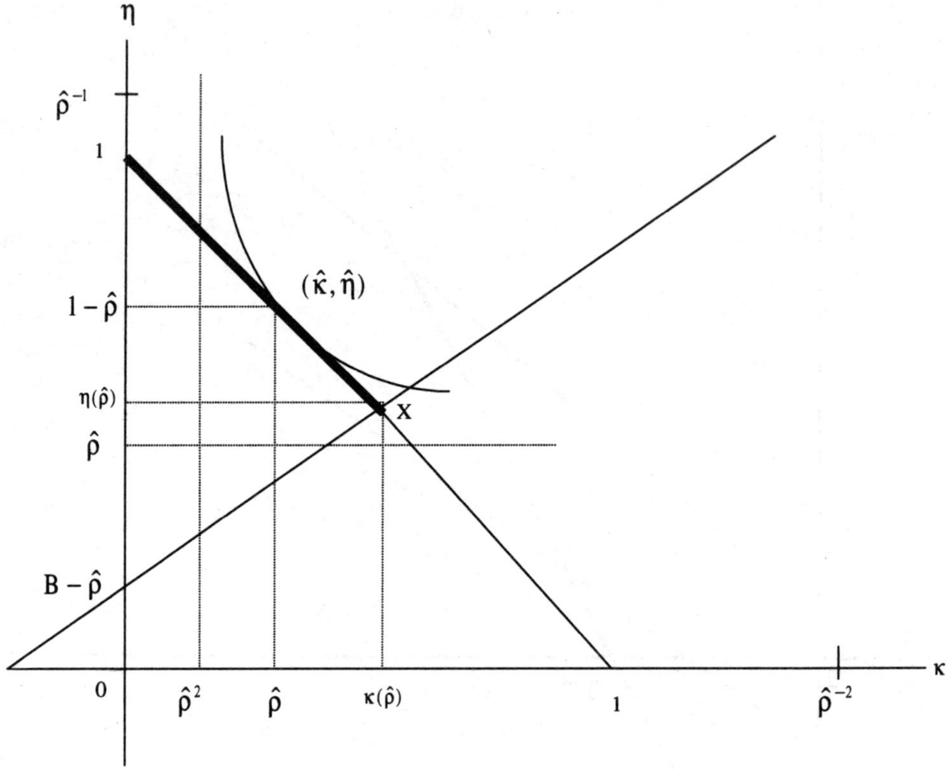

Figure 7 (Ia) with $\widehat{\rho} = (3 - \sqrt{5})/2$

13

On the Relationship between Discounting and Entropy of Dynamic Optimization Models

Tapan Mitra

1. Introduction

In a standard aggregative dynamic optimization framework (Ω, u, δ), where Ω is the transition possibility (technology) set, u is a (reduced form) utility function defined on this set, and $0 < \delta < 1$ is a discount factor, the relation between the magnitude of the discount factor and the extent of complicated behavior generated by the corresponding (optimal) policy function has been a topic of extensive study.

Boldrin and Montrucchio (1986) showed that any twice continuously differentiable function can be a policy function of an appropriately defined dynamic optimization model. However, when the C^2 function was taken to be the logistic map ($h(x) = 4x(1-x)$ for $x \in [0,1]$), which is well-known to exhibit complicated behavior, the dynamic optimization model (obtained by using their constructive method) for which the logistic map was the policy function was seen to have an extremely small discount factor (about 0.01). Subsequently, Sorger (1992a) showed that if the policy function of *any* dynamic optimization model is the logistic map, then its associated discount factor must be smaller than 0.5.

These results suggested that substantial discounting might be *necessary* to obtain complicated (or "chaotic") optimal behavior. This was confirmed by Sorger (1992b) when he showed, using the theory of stochastic dominance, that if any dynamic optimization model (Ω, u, δ) exhibits a period-three cycle, then the discount factor, δ, must satisfy

$$\delta < (\sqrt{5} - 1)/2 \approx 0.618 \tag{1.1}$$

In subsequent work, Sorger (1994) refined the above bound to

$$\delta < 0.5479 \tag{1.2}$$

The bound was further refined to

$$\delta < [(\sqrt{5} - 1/2]^2 \approx 0.3819 \tag{1.3}$$

in Mitra (Chapter 11) and Nishimura and Yano (Chapter 12). Furthermore, the bound in (1.3) was shown to be "exact" in the sense that whenever $\delta < [(\sqrt{5} - 1)/2]^2$, one can construct a transition possibility set, Ω, and a reduced form utility function, u, such that the dynamic optimization model (Ω, u, δ) has an optimal program exhibiting a period-three cycle.

A period-three cycle is a special case of what is known as "topological chaos", which occurs whenever there is a periodic cycle of a period not equal to a power of 2. [Topological chaos implies the existence of an infinite number of periodic cycles of distinct periodicities, as well as the existence of an uncountable "scrambled set"; see Section 2 for details]. Thus, if we focus on the existence of any periodic point of period $q = np$, where $n > 1$ is an odd integer and $p = 2^k$, with k a non-negative integer, it can be shown [see Mitra (Chapter 11)] that, combining (1.3) with the "Sarkovskii order" (see Proposition 2.1 below), the following bound on the discount factor can be obtained:

$$\delta < [(\sqrt{5} - 1)/2]^{1/2^k} \tag{1.4}$$

In studying the closely related phenomenon of "turbulence" (see Section 2 for a definition), Mitra (1994) established that if any dynamic optimization model (Ω, u, δ) exhibits turbulence, then the discount factor, δ, must satisfy

$$\delta < (1/4) \tag{1.5}$$

Furthermore, the bound in (1.5) is "exact" in the sense that whenever $\delta < 0.25$, one can construct a transition possibility set, Ω, and a utility function, u, such that the dynamic optimization model (Ω, u, δ) exhibits turbulence. Using the "Sarkovskii order" and (1.5), the bound obtained in (1.4) can be refined to

$$\delta < [1/2]^{1/2^k} \tag{1.6}$$

as shown in Mitra (1994).

All these results certainly indicate a close relationship between discounting and complex optimal behavior. However, this relationship could be made more precise if we had a convenient numerical measure of complicated behavior. From the topological point of view, such a numerical measure is the "topological entropy" of a dynamical system. Thus, one could proceed to examine whether the upper bound on the discount factor goes on decreasing as the extent of complicated optimal behavior goes on increasing. This approach was pioneered by Montrucchio (1994), who established (under a strong concavity assumption on the utility function) that

$$\delta \leq 1/e^{\psi(h,A)} \tag{1.7}$$

where A is a compact, invariant set contained in the interior of the "state space" of the dynamic optimization model, and $\psi(h, A)$ is the topological entropy on the set A of the policy function, h, when the discount factor is δ. Montrucchio and Sorger (1994) have shown that the inequality holds even when the strong concavity assumption on the utility function is replaced by a relatively mild strict concavity assumption on the utility function [of the type used in Section 3 of this chapter].

In this chapter, we do three things. First, we establish the Montrucchio-Sorger result (1.7) by using what I have called the "value-loss approach" to minimum impatience problems. Focusing on "value-losses" (that one suffers by deviating from certain price-supported activities) has been the cornerstone of the approach of McKenzie [see McKenzie (1986) for a comprehensive survey] to problems of "turnpike" theory. It appears to be a natural concept to focus on for the study of complicated optimal behavior as well, and the inequality (1.7) is seen to rest firmly on a version of the well-known "value-loss lemma", which was introduced to (final-state) turnpike theory by Radner (1961), and which has occupied center-stage in this [and in consumption-oriented turnpike theory] literature thereafter.

Second, under an additional assumption of "bounded steepness" of the utility function (see Section 5 for details) we extend the Montrucchio-Sorger result to establish that

$$\delta \leq 1/e^{\psi(h,X)} \tag{1.8}$$

where X is the compact state space itself. A problem in applying the Montrucchio-Sorger result to obtain discount factor restrictions for topological chaos is that, in general, given a policy function it is difficult to identify compact invariant sets in the *interior* of the state space. There is, of course, no such difficulty in applying formula (1.8).

Noting that topological chaos may not be "observable", we consider a natural alternate measure of chaos, metric entropy, and show (using a mathematical result comparing the magnitude of the topological entropy with the metric entropy of a dynamical system) that

$$\delta \leq 1/e^{\phi_\mu(h,X)} \tag{1.9}$$

where μ is a probability measure on the Borel σ-field of X, invariant under h, and $\phi_\mu(h, X)$ the metric entropy.

Results (1.8) and (1.9) [Theorems 5.2 and 5.3 of this chapter] summarize the relationship between discounting and complicated behavior in dynamic optimization models, where "complicated behavior" is measured in the topological sense in (1.8) and in the measure-theoretic sense in (1.9).

Third, I study the implications of result (1.8) for periodic and turbulent optimal programs. The concept of topological entropy has been thoroughly studied in the theory of dynamical systems, and powerful methods have been developed for computing the topological entropy of dynamical systems, exhibiting periodic and turbulent behavior, by Block, Guckenheimer, Misiurewicz and

Young (1980). We use these results to show that the inequality (1.8) yields the discount factor restriction

$$\delta \leq (\sqrt{5} - 1)/2 \tag{1.10}$$

for period-three cycles, and

$$\delta \leq (1/2) \tag{1.11}$$

for turbulence. These are fairly strong bounds, but turn out to be the square-roots of the "exact" bounds obtained in (1.3) and (1.5) respectively. This naturally prompts us to ask whether there is a more refined relationship than (1.8) between the discount factor and the topological entropy, which would yield the exact discount factor restrictions for period-three cycles and turbulence as special cases. We leave this as an open question.

2. Chaos

2.1 Periodic Orbits

Let I be a compact interval in $\Re$, the set of reals. Let $f : I \to I$ be a continuous map of the interval I into itself. The pair (I, f) is called a *dynamical system*; I is called the *state space* and f the *law of motion* of the dynamical system.

We write $f^0(x) = x$ and for any integer $n \geq 1$, $f^n(x) = f[f^{n-1}(x)]$. If $x \in I$, the sequence $\tau(x) = \{f^n(x)\}_0^\infty$ is called the *trajectory* from (the initial condition) x. The *orbit* from x is the set $\gamma(x) = \{y : y = f^n(x) \text{ for some } n \geq 0\}$.

A point $x \in I$ is a *fixed point* of f if $f(x) = x$. A point $x \in I$ is called a *periodic point* of f if there is $k \geq 1$ such that $f^k(x) = x$. The smallest such k is called the *period* of x. [In particular, if $x \in I$ is a fixed point of f, it is periodic with period 1]. If $x \in I$ is a periodic point with period k, we also say that the orbit of x (or trajectory from x) is periodic with period k.

The following fundamental result on the existence of periodic orbits is due to Sarkovskii (1964). A good discussion of this result is contained in Block and Coppel (1992).

Proposition 2.1: *Let the positive integers be totally ordered in the following way:*

$$3 \prec 5 \prec 7 \prec 9 \prec \ldots \prec 2.3 \prec 2.5 \prec \ldots \prec 2^2.3 \prec 2^2.5$$
$$\prec \ldots \prec 2^3 \prec 2^2 \prec 2 \prec 1.$$

If f has a periodic orbit of period n and if $n \prec m$, then f also has a periodic orbit of period m.

2.2 Aperiodic Orbits

In order to study the nature of trajectories which are not periodic, it is useful to define a "scrambled" set. A set $S \subset I$ is called a *scrambled set* if it possesses the following two properties:

(i) If x, $y \in S$ with $x \neq y$, then

$$\limsup_{n \to \infty} |f^n(x) - f^n(y)| > 0$$

and

$$\liminf_{n \to \infty} |f^n(x) - f^n(y)| = 0$$

(ii) If $x \in S$ and y is any periodic point of f,

$$\limsup_{n \to \infty} |f^n(x) - f^n(y)| > 0$$

Thus trajectories starting from points in a scrambled set are not even "asymptotically periodic". Furthermore, for any pair of initial states in the scrambled set, the trajectories move apart and return close to each other infinitely often.

The following theorem, due to Li and Yorke (1975), is fundamental in establishing a connection between the existence of period-three cycles and the existence of an uncountable scrambled set.

Proposition 2.2: *Assume that there is some point x^* in I such that:*

$$f^3(x^*) \leq x^* < f(x^*) < f^2(x^*)$$
$$(\text{or } f^3(x^*) \geq x^* > f(x^*) > f^2(x^*)) \tag{2.1}$$

Then (i) for every positive integer $k = 1, 2, ...,$ *there is a periodic point of period* k.

(ii) there is an uncountable scrambled set $S \subset I$.

We will say that the dynamical system (I, f) exhibits *Li-Yorke chaos* if conditions (i) and (ii) of Proposition 2.2 are satisfied. It is easy to check that (I, f) exhibits Li-Yorke chaos if and only if (I, f)has a periodic point of period three.

Generalizing the Li-Yorke result, and using Sarkovskii's theorem, Li, Misiurewicz, Pianigiani, and Yorke (1982) established the following result.

Proposition 2.3: *Suppose* f *has a periodic point of period which is not a power of 2. Then (i)* f *has infinitely many periodic points of different periods; and (ii)*I *has an uncountable scrambled set.*

2.3. Turbulence[1]

The map $f : I \to I$ will be called *turbulent* if there exist compact (non-degenerate) subintervals J, K with at most one common point such that

$$J \cup K \subset f(J) \cap f(K)$$

The concept of turbulence is related to the occurrence of periodic points in the following result.

Proposition 2.4: *(i) If* $f : I \to I$ *is turbulent then* f *has periodic point of all periods.*

(ii) If $f : I \to I$ *has a periodic point of odd period* $n > 1$, *then* f^2 *is turbulent.*

---

[1]The definition of turbulence used here follows Block and Coppel (1992). The phenomenon is also referred to as a "2-horseshoe" by Alseda, Llibre and Misiurewicz (1993). The concept has evolved from the earlier dicussions of Ruelle and Takens (1971) and Lasota and Yorke (1977).

2.4. Topological Entropy[2]

There is a well-known measure of the extent of complicated behavior that a dynamical system can exhibit, and it is called "topological entropy". Let A be a compact subset of I, which is invariant under f (that is, $f(A) \subset A$). Let $J = (J_1, ..., J_p)$ be a finite open cover of A (that is, $\bigcup_{i=1}^{p} J_i \supset A$). Denote by $f^{-1}J$ the open cover $(f^{-1}J_1, ..., f^{-1}J_p)$. If $J = (J_1, ..., J_p)$ and $J' = (J'_1, ..., J'_q)$ are open covers of A, then their join JVJ' is the open cover consisting of all sets of the form $J_i \cap J'_j$ for all $i \in \{1, ..., p\}$, $j \in \{1, ..., q\}$. Let $N(J)$ be the minimum number of sets in the cover J which still covers A. Then, we can define

$$\psi(f, A, J) = \lim_{n\to\infty} (1/n) \log N(JVf^{-1}JV...Vf^{-n+1}J)$$

The *topological entropy* of f on A is defined as

$$\psi(f, A) = \sup_J \psi(f, A, J)$$

where the supremum is taken over all open covers J. For $A = I$, we often write $\psi(f)$ instead of $\psi(f, I)$.

We will use a more "operational" definition of topological entropy, which we now provide. A finite set $E \subset A$ is called (n, ε)-*separated* ($n = 1, 2, ...$ and $\varepsilon > 0$) if for every x, $y \in E$, $x \neq y$, there is $0 \leq k < n$ such that $|f^k(x) - f^k(y)| \geq \varepsilon$. Let $s(n, \varepsilon)$ denote the maximal cardinality of an (n, ε)-separated set. We define

$$\psi_\varepsilon(f, A) = \limsup_{n\to\infty} (1/n) \log s(n, \varepsilon)$$

and the *topological entropy* of f as

$$\Psi(f, A) = \lim_{\varepsilon\to 0} \psi_\varepsilon(f, A).$$

[2]The formal definition of topological entropy was given by Adler, Konheim and McAndrew (1965). Bowen (1971a) provided the more "operational" definition (which we use here). In our context, the two definitions are equivalent; for a proof, see Bowen (1971b).

The following result, due to Misiurewicz (1979), relates topological entropy to the existence of periodic points with period which is not equal to a power of 2.

Proposition 2.5: *The following two statements are equivalent: (i) f has a periodic point of period which is not a power of 2;(ii) f has positive topological entropy.*

We will say that a dynamical system (I, f) exhibits *topological chaos* if f has a periodic point whose period is not a power of 2. This definition follows Block and Coppel (1992).

A justification for adopting the above definition of topological chaos is that, while it is weaker than the period-three condition of Li and Yorke it is strong enough to ensure the existence of an uncountable scrambled set. On the other hand, the definition is weak enough so that a dynamical system (I, f) exhibits topological chaos if and only if it has positive topological entropy.

2.5. Metric Entropy

A difficulty with the notion of "topological chaos" is that it may not be "observed", and therefore might be unsuitable for the purpose of signaling "unpredictability" of outcomes of a dynamical system. If by chaos we mean unpredictability, the natural measure of it is metric entropy (also known as "measure-theoretic entropy" or "Kolmogorov-Sinai invariant"), since it measures the "uncertainty" in an experiment about the outcome, or equivalently the "information" gained by conducting the experiment.

Let $\wp$ be the Borel σ-field of I, and v a probability measure on $\wp$. Thus, $(I, \wp, v)$ is a probability space. If f is $\wp$-measurable, then v is called *invariant* under h if $v(E) = v(f^{-1}(E))$ for all E in $\wp$.

Suppose the dynamical system (I, f) has an invariant measure v. We call $J = \{J_1, ..., J_p\}$ a (finite) *measurable partition* of I if the J_i are disjoint measurable subsets of I, and $\bigcup_i J_i \supset I$. If $J = \{J_1, ..., J_p\}$ is a finite measurable partition of I, then the *entropy of the partition* J is

$$H(f, I, v, J) = -\sum_{i=1}^{p} v(J_i) \log v(J_i)$$

If $J = \{J_1, ..., J_p\}$ is a measurable partition, we denote by $f^{-1}J$ the measurable partition $\{f^{-1}J_1, ..., f^{-1}J_p\}$. If $J = \{J_1, ..., J_p\}$ and $J' = \{J'_1, ..., J'_q\}$ are measurable partitions of X, then JVJ' is the measurable partition consisting of sets of the form $J_i \cap J'_j$ for all $i \in \{1, ..., p\}$, $j \in \{1, ..., q\}$. We can then define

$$\Phi(f, I, v, J) = \lim_{n\to\infty} (1/n) H(f, I, v, JVf^{-1}JV...Vf^{-n+1}J)$$

Finally, the *metric entropy* is defined as

$$\Phi_v(f, I) = \sup_J \Phi(f, I, v, J)$$

where the supremum is taken over all partitions J of I for which $\Phi(f, I, v, J) < \infty$.

3. Dynamic Optimization

3.1 The Model

The framework is described by a triplet (Ω, u, δ), where Ω, a subset of $\Re_+ \times \Re_+$, is a *transition possibility set*, $u : \Omega \to \Re$ is a *utility function* defined on this set, and δ is the *discount factor* satisfying $0 < \delta < 1$.

The transition possibility set describes the states $z \in \Re_+$ that it is possible to go to tomorrow, if one is in state $x \in \Re_+$ today. We define a correspondence $\Gamma : \Re_+ \to \Re_+$ by $\Gamma(x) = \{y \in \Re_+ : (x, y) \in \Omega\}$ for each $x \in \Re_+$.

A program $\{x_t\}_0^\infty$ from $\mathbf{x} \in \Re_+$ is a sequence satisfying

$$x_0 = \mathbf{x} \text{ and } (x_t, x_{t+1}) \in \Omega \text{ for } t \geq 0$$

If one is in state x today and one moves to state z tomorrow (with $(x, z) \in \Omega$) then there is an immediate utility (or "reward" or "return") generated, measured by the utility function, u.

The discount factor, δ, is the weight assigned to tomorrow's utility (compared to today's) in the objective function. The *discount rate* (associated with the discount factor, δ) is given by $\rho = (1/\delta) - 1$.

The following assumptions are imposed on the transition possibility set, Ω, and the utility function, u:

A1: *(i)* $(0,0) \in \Omega$, *(ii)* $(0,z) \in \Omega$ *implies* $z = 0$.
A2: Ω *is (i) closed, and (ii) convex.*
A3: *There is* $\xi > 0$ *such that* $(x,z) \in \Omega$ *and* $x \geq \xi$ *implies* $z < x$.
A4: *If* $(x,z) \in \Omega$ *and* $x' \geq x$, $0 \leq z' \leq z$ *then* $(x',z') \in \Omega$.

Clearly, we can pick $0 < \zeta < \xi$, such that if $x > \zeta$ and $(x,z) \in \Omega$, then $z < x$. It is straightforward to verify that if $(x,z) \in \Omega$, then $z \leq \max(\zeta, x)$. It follows from this that if $\{x_t\}_0^\infty$ is a program from $\mathbf{x} \in \Re_+$, then $x_t \leq \max(\zeta, \mathbf{x})$ for $t \geq 0$. In particular, if $\mathbf{x} \leq \zeta$, then $x_t \leq \zeta$ for $t \geq 0$. This leads us to choose the closed interval, $[0,\zeta]$ as the natural *state space* of our model, which we will denote by X. We denote the interval $[0,\xi]$ by Y.

The following assumptions are imposed on the utility function, u:

A5: u *is concave on* Ω*; further if* (x,z) *and* (x',z') *are in* Ω, *and* $x \neq x'$, *then for every* $0 < \lambda < 1$,
$u(\lambda(x,z)+$
$(1-\lambda)(x',z')) > \lambda u(x,z) + (1-\lambda)u(x',z')$.
A6: u *is upper semi-continuous on* Ω.
A7: *If* x, $x' \in Y$, $(x,z) \in \Omega$, $x' \geq x$ *and* $0 \leq z' \leq z$, *then* $u(x',z') \geq u(x,z)$.

We will refer to a triplet (Ω, u, δ) satisfying (A.1) - (A.7) as a *dynamic optimization model.*

A program $\{\hat{x}_t\}_0^\infty$ from $\mathbf{x} \geq 0$ is an *optimal program* if

$$\sum_0^\infty \delta^t u(x_t, x_{t+1}) \leq \sum_0^\infty \delta^t u(\hat{x}_t, \hat{x}_{t+1})$$

for every program $\{x_t\}_0^\infty$ from $\mathbf{x}$.

Under (A1)-(A7), there is a unique optimal program from every $\mathbf{x} \in \Re_+$.

3.2 Value and Policy Functions

The *value function* $V : \Re_+ \to \Re$ is defined by

$$V(x) = \sum_0^\infty \delta^t u(\hat{x}_t, \hat{x}_{t+1})$$

where $\{\hat{x}_t\}_0^\infty$ is the optimal program from $x \in \Re_+$.

The *policy function* $h : \Re_+ \to \Re_+$ is defined by

$$h(x) = \hat{x}_1$$

where $\{\hat{x}_t\}_0^\infty$ is the optimal program from $x \in \Re_+$.

The properties of the value and policy functions can be summarized in the following result. This is based on Dutta and Mitra (1989) and Stokey, Lucas and Prescott (1989).

Proposition 3.1: *(i) The value function V is strictly concave and continuous on $\Re_+$ and non-decreasing on Y. Further, V is the unique continuous function on $Y \equiv [0, \xi]$ which satisfies the functional equation of dynamic programming:*

$$V(x) = \max_{y \in \Gamma(x)} [u(x, y) + \delta V(y)]$$

(ii) The policy function h satisfies the following property: for each $x \in \Re_+$, $h(x)$ is the unique solution to the constrained maximization problem:

$$\begin{array}{ll} \text{Maximize} & u(x, y) + \delta V(y) \\ \text{Subject to} & y \in \Gamma(x) \end{array}$$

Further, h is continuous on $\Re_+$.

In view of the definition of the policy function h, the optimal program from $x \in X$ is the trajectory $\{h^t(x)\}_0^\infty$ generated by the policy function. Thus, an optimal program from $x \in X$ can be called *periodic* (with period k) if x is a periodic point of h (with period k).

4. Duality Theory

4.1 Price Characterization of Optimality

Optimality can be characterized in terms of dual variables or shadow prices. The basic result of the theory, describing this characterization, can be stated as follows. [A full discussion can be found in Weitzman (1973) and McKenzie (1986)].

Proposition 4.1: (a) *If $\{x_t\}_0^\infty$ is an optimal program from $\mathbf{x} \in X$ and $\mathbf{x} > 0$, and there is some $(\bar{x}, \bar{y}) \in \Omega$ with $\bar{y} > 0$, then there is a sequence $\{p_t\}_0^\infty$ of non-negative prices such that for $t \geq 0$,*

(i) $\delta^t V(x_t) - p_t x_t \geq \delta^t V(x) - p_t x$ for all $x \geq 0$

(ii) $\delta^t u(x_t, x_{t+1}) + p_{t+1} x_{t+1} - p_t x_t \geq \delta^t u(x, y) + p_{t+1} y - p_t x$ for all $(x, y) \in \Omega$

(iii) $\lim_{t \to \infty} p_t x_t = 0$

(b) *If $\{x_t\}_0^\infty$ is a program from $\mathbf{x} \geq 0$, and there is a sequence $\{p_t\}_0^\infty$ of non-negative prices such that for $t \geq 0$,* (ii) *and* (iii) *above are satisfied, then $\{x_t\}_0^\infty$ is an optimal program from $\mathbf{x}$.*

If $\{x_t\}_0^\infty$ is a program from $\mathbf{x} \geq 0$, and $\{p_t\}_0^\infty$ is a non-negative sequence of prices satisfying (i), (ii) and (iii) of Proposition 4.1(a), we will say that the program $\{x_t\}_0^\infty$ is *price supported* by $\{p_t\}_0^\infty$. When $\{x_t\}_0^\infty$ is price supported by $\{p_t\}_0^\infty$, we refer to $\{p_t\}_0^\infty$ as a sequence of *present-value prices.* Associated with $\{p_t\}_0^\infty$ is a sequence $\{P_t\}_0^\infty$ of *current value prices* defined by

$$P_t = (p_t / \delta^t) \text{ for } t \geq 0$$

If the value function has finite steepness at zero, it is possible to choose the current value prices associated with any optimal program to be uniformly bounded above. To establish this, we proceed formally as follows. Normalize $u(0, 0) = 0$, and note then that $V(0) = 0$. Then for all x in X, $(x, 0)$ is in Ω (by (A4)) and

$u(\zeta, 0) \geq u(x, 0) \geq 0$ by (A7). Further, for all x in X with $x > 0$, $V(x) \geq 0$ and $[V(x)/x]$ is decreasing in x by Proposition 3.1. We define

$$\rho \equiv \lim_{x \to 0+} [V(x)/x]$$

In the above definition, ρ can be infinite. If ρ happens to be finite, we say that V has "finite steepness".

Proposition 4.2: *Let (Ω, u, δ) be a dynamic optimization model. Suppose there is $(\hat{x}, \hat{y}) \in \Omega$ such that $\hat{x} \in X$ and $\hat{y} > 0$. Further, suppose that*

$$\rho \equiv \lim_{x \to 0+} [V(x)/x] < \infty \tag{4.1}$$

If $\{x_t\}_0^\infty$ is an optimal program from any $x \in X$, there is a price support $\{p_t\}_0^\infty$ of $\{x_t\}_0^\infty$ such that

$$P_t \equiv (p_t/\delta^t) \leq \beta \text{ for } t \geq 0 \tag{4.2}$$

where β is given by

$$\beta \equiv \max\ (\rho, [u(\zeta, 0) - u(\hat{x}, \hat{y}) + \rho\hat{x}]/\delta\hat{y}) \tag{4.3}$$

Proof: Consider, first, any optimal program $\{x_t\}_0^\infty$ from x in X, with $x > 0$. There are two cases to consider (i) $x_t > 0$ for all $t \geq 0$; (ii) $x_t = 0$ for some t.

In case (i), using Proposition 4.1 (i),

$$V(x_t) - P_t x_t \geq 0 \text{ for } t \geq 0$$

so that $P_t \leq [V(x_t)/x_t] \leq \rho \leq \beta$ for $t \geq 0$.

In case (ii), let T be the first period for which $x_t = 0$. Then $T \geq 1$, $x_t > 0$ for $t = 0, ..., T-1$, and $x_t = 0$ for $t \geq T$. By the argument used in case (i), $P_t \leq \rho$ for $t = 0, ..., T-1$. For $t = T$, using Proposition 4.1 (ii), we get

$$u(x_{T-1}, 0) - P_{T-1}x_{T-1} \geq u(\hat{x}, \hat{y}) + \delta P_T \hat{y} - P_{T-1}\hat{x}$$

so that $P_T \leq [u(\zeta, 0) - u(\hat{x}, \hat{y}) + \rho\hat{x}]/\delta\hat{y} \leq \beta$ by (4.3).

If $P_t \leq P_T$ for all $t > T$, then we are done. Otherwise, let τ be the first period ($> T$) for which $P_t > P_T$. Then $P_{\tau-1} \leq P_T < P_\tau$, and using Proposition 4.1 (ii), we get for all $(x, z) \in \Omega$,

$$0 \geq u(x, z) + \delta P_\tau z - P_{\tau-1}x \geq u(x, z) + \delta P_{\tau-1}z - P_{\tau-1}x.$$

Furthermore, since $x_{\tau-1} = 0$, we have

$$0 \geq V(x) - P_{\tau-1}x \text{ for all } x \geq 0$$

Thus, defining $P'_t = P_t$ for $t = 0, ..., \tau - 1$ and $P'_t = P_{\tau-1}$ for $t \geq \tau$, and $p'_t = \delta^t P'_t$ for $t \geq 0$, we see that $\{p'_t\}_0^\infty$ provides a price support to $\{x_t\}_0^\infty$, and $P'_t \leq \beta$ for $t \geq 0$.

The final step is to obtain a price support for the optimal program $\{0\}_0^\infty$ from 0, which satisfies (4.2). Since $\rho < \infty$, we have $V'_+(0) = \rho < \infty$. Thus, by concavity of V, we have

$$V(0) - P0 \geq V(x) - Px \text{ for all } x \geq 0 \tag{4.4}$$

where $P = V'_+(0) = \rho$. Then, by the induction argument of Weitzman (1973) [see also McKenzie (1986)], we can get a price support $\{p_t\}_0^\infty$ of the "zero program" $\{0\}_0^\infty$, with $P_0 = p_0 = P$, obtained in (4.4). Now, following the analysis of case (ii) above (identifying period T with period 0), we get a price support $\{p'_t\}_0^\infty$ of $\{0\}_0^\infty$, such that $P'_t \leq \beta$ for $t \geq 0$.

4.2 The Value-Loss Method

The value-loss method is based on the observation that at the prices supporting an optimal program, there is no activity which yields a higher "generalized profit" at any date (value of utility plus value of terminal stocks minus value of initial stocks at that date) than the activity chosen along the optimal program at that date. In other words, there are no arbitrage possibilities available at the supporting prices.

This observation leads to a basic tool for analyzing minimum impatience results (see Mitra (Chapter 11) for a proof) which we state in the following proposition.[3]

Proposition 4.3: *Let (Ω, u, δ) be a dynamic optimization model. Suppose $\{x_t\}_0^\infty$ is an optimal program with price support $\{p_t\}_0^\infty$, and $\{y_t\}_0^\infty$ is an optimal program with price support $\{q_t\}_0^\infty$. Denoting (p_t/δ^t) by P_t and (q_t/δ^t) by Q_t for $t \geq 0$, we have*
(i) $\delta(P_{t+1} - Q_{t+1})(y_{t+1} - x_{t+1}) \leq (P_t - Q_t)(y_t - x_t)$ *for* $t \geq 0$
(ii) $(P_t - Q_t)(y_t - x_t) \geq 0$ *for* $t \geq 0$
Furthermore, if $y_t \neq x_t$ for some t, then the inequalities in (i) and (ii) are strict for that t.

5. On a Relationship between Discounting and Complexity

5.1 The Montrucchio-Sorger Result

If we consider topological entropy to be an appropriate measure of "complexity" of a dynamical system, then a natural way to study the relationship between discounting and complicated optimal behavior is to find the relationship between the discount factor of a dynamic optimization model and the topological entropy of its policy function. This is the approach taken in Montrucchio (1994), where he establishes (under strong concavity assumptions on the utility function) that if (Ω, u, δ) is a dynamic optimization model with policy function h, A is a compact, invariant set contained in the interior of X, and $\psi(h, A)$ is the topological entropy of h on A, then the discount factor, δ, is related to the topological entropy by the inequality

$$\delta \leq e^{-\psi(h,A)} \tag{5.1}$$

Subsequently, it has been shown in Montrucchio and Sorger (1994) that the strong concavity assumption on the utility func-

[3] The inequalities of Proposition 4.3 must be very familiar objects from the point of view of the turnpike theory literature, where they suggest a natural choice of a Lyapunov function for the study of global asymptotic stability of optimal growth paths; see especially Brock and Scheinkman (1976), Cass and Shell (1976) and McKenzie (1986). The same inequalities have figured prominently in the literature on the intertemporal decentralization of the transversality condition; see especially Brock and Majumdar (1988) and Dasgupta and Mitra (1988).

tion can be dispensed with in deriving the inequality (5.1). In particular, it follows from their result that inequality (5.1) holds under the standard assumptions used in Section 3 of this chapter.

In this section, we will show how the relationship (5.1) (which I refer to henceforth as the Montrucchio-Sorger result) can be derived using the value-loss approach, thereby providing a unified view of the minimum impatience results obtained in the literature on chaotic optimal behavior.

In order to establish (5.1), we need two preliminary results. To describe the results, define $Z \equiv X - \{0\}$, and let A be a compact invariant set contained in Z. Given any x and y in A, let $\{p_t\}_0^\infty$ and $\{q_t\}_0^\infty$ be the price sequences supporting the optimal programs from x and y respectively. Denote by $\{P_t\}_0^\infty$ and $\{Q_t\}_0^\infty$ the "current" price sequences corresponding to $\{p_t\}_0^\infty$ and $\{q_t\}_0^\infty$ respectively, and to simplify notation denote P_0 by P and Q_0 by Q.

We know from Proposition 4.3 that

$$(P - Q)(y - x) \geq 0 \tag{5.2}$$

and the inequality in (5.2) is strict when $y \neq x$.

We need to establish, first, that there is $\beta > 0$, such that for all x, y in A,

$$(P - Q)(y - x) \leq \beta|x - y| \tag{5.3}$$

This amounts to establishing a uniform bound on the initial period supporting prices associated with optimal programs starting from initial stocks in A.

We also need to establish that given any $\varepsilon > 0$, there is $\alpha(\varepsilon) > 0$ such that

$$x, y \in A \text{ and } |x - y| \geq \varepsilon \text{ imply } (P - Q)(y - x) \geq \alpha(\varepsilon) \tag{5.4}$$

One recognizes this, of course, as a version of the well-known "value-loss lemma" appearing prominently in the turnpike literature since Radner (1961).

We now proceed to establish these two results formally.

Lemma 5.1: *Let A be a compact invariant set contained in Z. There is $\beta > 0$ such that for all x, y in A,*

$$(P - Q)(y - x) \leq \beta \, |x - y| \tag{5.5}$$

Proof: Given A, we can find $a \in Z$ such that $A \subset [a, \zeta]$. Denote $V'_-(a)$ by β.

Without loss of generality, suppose $y \geq x$. Then

$$(P - Q)(y - x) \leq P(y - x) \leq V'_-(x)(y - x) \tag{5.6}$$

Since $x \in A$, we have $V'_-(x) \leq V'_-(a) = \beta$. So (5.6) yields (5.5) immediately.

Lemma 5.2: *Let A be a compact invariant set contained in Z. For every $\varepsilon > 0$, there exists $\alpha(\varepsilon) > 0$ such that*

$$x, y \in A \text{and } |x - y| \geq \varepsilon \text{imply } (P - Q)(y - x) \geq \alpha(\varepsilon) \tag{5.7}$$

Proof: Given A, we can find $a \in Z$ such that $A \subset [a, \zeta]$. Denote $V'_-(a)$ by β.

If the Lemma were not true, there would exist a sequence (x^s, y^s) $s = 1, 2, \ldots$, with x^s, $y^s \in A$ and $|x^s - y^s|$ for all s, such that $(P^s - Q^s)(y^s - x^s) \to 0$ as $s \to \infty$.

Now, using (i) of Proposition 4.1, $0 \leq P^s \leq V'_-(x^s) \leq V'_-(a) = \beta$, $0 \leq Q^s \leq V'_-(y^s) \leq V'_-(a) = \beta$, so we can find a subsequence s' of s, such that $|P^{s'}| \to \hat{P}$, $|Q^{s'}| \to \hat{Q}$, $x^{s'} \to \bar{x}$ and $y^{s'} \to \bar{y}$ as $s' \to \infty$.

Denoting $(\bar{y} + \bar{x})/2$ by $\bar{z}$, and using (i) of Proposition 4.1 again, $V(x^s) - P^s x^s \geq V(\bar{z}) - P^s \bar{z}$ for all $s \geq 1$. Thus, by the continuity of V,

$$V(\bar{x}) - \hat{P}\bar{x} \geq V(\bar{z}) - \hat{P}\bar{z} \tag{5.8}$$

Since $|x^{s'} - y^{s'}| \geq \varepsilon$ for all s', we have $|\bar{x} - \bar{y}| \geq \varepsilon$. Thus, using the strict concavity of V, we get

$$V(\bar{z}) > \frac{1}{2}V(\bar{x}) + \frac{1}{2}V(\bar{y}) \tag{5.9}$$

Using (5.8) and (5.9), we obtain

$$V(\bar{x}) - \hat{P}\bar{x} > V(\bar{y}) - \hat{P}\bar{y} \tag{5.10}$$

Using y^s and Q^s in place of x^s and P^s respectively in the above steps, we can obtain similarly

$$V(\bar{y}) - \hat{Q}\bar{y} > V(\bar{x}) - \hat{Q}\bar{x} \tag{5.11}$$

Combining (5.10) and (5.11) yields

$$(\hat{P} - \hat{Q})(\bar{y} - \bar{x}) > 0 \tag{5.12}$$

But since $(P^{s'} - Q^{s'})(y^{s'} - x^{s'}) \to 0$ as $s' \to \infty$, and $P^{s'} \to \hat{P}$, $Q^{s'} \to \hat{Q}$, $y^{s'} \to \bar{y}$, $x^{s'} \to \bar{x}$ as $s' \to \infty$, we have

$$(\hat{P} - \hat{Q})(\bar{y} - \bar{x}) = 0 \tag{5.13}$$

which contradicts (5.12) and establishes the result.

We are now in a position to state and prove (a version of) the Montrucchio-Sorger result.

Theorem 5.1: *Let (Ω, u, δ) be a dynamic optimization model and let $h : X \to X$ be its policy function. Assume that A is a compact subset of X which is contained in Z, and which is invariant under h. Then,*

$$\delta \leq e^{-\psi(h,A)}$$

Proof: Let n be a positive integer, ε a positive real number, and B an (n, ε)-separated subset of A. For every x, y in B with $x \neq y$, there exists $t \in \{0, 1, ..., n-1\}$ such that $|h^t(x) - h^t(y)| > \varepsilon$. Using Proposition 4.3, we have

$$\delta^t (P_t - Q_t)(y_t - x_t) \leq (P - Q)(y - x) \tag{5.15}$$

where $x_t = h^t(x)$, $y^t = h^t(y)$, $P = P_0$ and $Q = Q_0$.

Using Lemma 5.1, we get

$$(P - Q)(y - x) \leq \beta |x - y| \tag{5.16}$$

Using Lemma 5.2 and $|x_t - y_t| > \varepsilon$, we get

$$(P_t - Q_t)(y_t - x_t) \geq \alpha(\varepsilon) \tag{5.17}$$

since $(x_t, x_{t+1}, ...)$ is an optimal program from x_t with current value supporting prices $(P_t, P_{t+1}, ...)$, and a similar remark applies to y_t and Q_t.

Combining (5.15), (5.16), (5.17), we get

$$\delta^t \alpha(\varepsilon) \leq \beta|x - y| \tag{5.18}$$

and since $t \leq n - 1$, and $0 < \delta < 1$,

$$\delta^n \alpha(\varepsilon)/\beta < |x - y| \tag{5.19}$$

Since $B \subset A \subset X$, where $X = [0, \zeta]$, and B is an (n, ε)-separated set, the number of elements of B must satisfy the inequality:

$$\text{Card B } < [\beta\zeta/\delta^n \alpha(\varepsilon)] + 1 < [2\beta\zeta/\delta^n \alpha(\varepsilon)] \tag{5.20}$$

Using (5.20), we obtain

$$\begin{aligned}
\psi_{\in}(h, A) &= \limsup_{n\to\infty}(1/n)\log[2\beta\zeta/\delta^n\alpha(\varepsilon)] \\
&= \limsup_{n\to\infty}[(1/n)\log[2\beta\zeta/\alpha(\varepsilon)] + (1/n)\log(1/\delta)^n] \\
&\leq \limsup_{n\to\infty}(1/n)\log(1/\delta)^n = \log(1/\delta)
\end{aligned}$$

This implies that $\psi(h, A) = \lim_{\in\to 0}\psi_{\in}(h, A) \leq \log(1/\delta)$. This yields

$$e^{\psi(h,A)} \leq (1/\delta)$$

which is equivalent to (5.14).

Remark 5.1: Notice that if A is a compact subset of X which is contained in the interior of X, and which is invariant under h, then, by Theorem 5.1, we have

$$\delta \leq e^{-\psi(h,A)}$$

which is the Montrucchio-Sorger result for our aggregative framework.

5.2 Discounting and Topological Entropy

A possible difficulty in applying Theorem 5.1 of the previous section is in identifying suitable compact subsets of Z which are invariant under h. Notice that while X itself is clearly a compact set which is invariant under h, the Montrucchio-Sorger result does not imply that

$$\delta \leq e^{-\psi(h,X)} \tag{5.21}$$

and it is clear from our method of proof (and also from the proofs of Montrucchio (1994) and Montrucchio and Sorger (1994)) that this is not an easy extension of Theorem 1.

If, for example, h is the well-known logistic map ($h(x) = 4x(1-x)$ for $x \in X \equiv [0,1]$), then h is *topologically transitive*; that is, for any pair of open sets U_1, U_2 in X, there exists a positive integer k such that $h^k(U_1) \cap U_2$ is non-empty. For a proof of this fact, see Devaney (1989, p. 51). Then, by Lemma 37 of Block and Coppel (1992, p. 155), every proper closed subset of $[0,1]$, which is invariant under h, has empty interior. That is, these sets are "thin", and *would not* include, for instance, any open intervals. They *would* include finite sets, for instance those consisting of the points of a periodic cycle (of periodicity exceeding 1), but clearly the topological entropy of h on such sets is zero. Thus, compact subsets of Z which are invariant under h, might not be the sets that we would necessarily want to focus on.

This difficulty seems especially acute in the one-dimensional case, and less so in the multi-dimensional setting, for which the results of Montrucchio (1994) and Montrucchio-Sorger (1994) were developed. For instance, if $h : \Re^2 \to \Re^2$ is the Henon map ($h_1(x_1,x_2) = 1 + x_2 - 1.4\, x_1^2$, $h_2(x_1,x_2) = 0.3\, x_1$) then, as Henon (1976, p. 76) demonstrates, there is a rectangular region in the interior of $\Re^2$ which is mapped into itself by h.

In this section, we show under an additional assumption, how the formula (5.21) can be obtained. The extra assumption we use for this purpose involves "bounded steepness" of the reduced-form utility function, a concept introduced to the optimal growth literature in Gale (1967).

We now proceed formally as follows. Recall from Section 4 that we normalized $u(0,0) = 0$, so $V(0) = 0$. Then for all $x \in X$, $(x, 0)$ is in Ω (by (A4)) and $u(x, 0) \geq 0$ by (A7). Further, $[u(x, 0)/x]$ is decreasing in x on X, by (A5). Our additional assumption is:

A8: $\sigma \equiv \lim_{x \to 0+} [u(x, 0)/x] < \infty$

From this point onwards, we refer to a triplet (Ω, u, δ) as a dynamic optimization model if it satisfies assumptions (A1)-(A8).

We now proceed to establish the basic relationship between the discount factor and the topological entropy of the corresponding policy function.

Theorem 5.2: *Let* (Ω, u, δ) *be a dynamic optimization model and let* $h : X \to X$ *be its policy function. Then*

$$\delta \leq e^{-\psi(h,X)} \tag{5.22}$$

Proof: Notice, first, that if $h(x) = 0$ for all $x \in X$, then clearly $\psi(h, X) = 0$ and (5.22) holds trivially. Thus we concentrate on the situation where there is some $\hat{x} \in X$ such that $h(\hat{x}) > 0$.

Recall from Section 4 that we defined:

$$\rho \equiv \lim_{x \to 0+} [V(x)/x]$$

We subdivide our proof into two cases: (i) $\rho < \infty$ (ii) $\rho = \infty$.

Case I: $[\rho < \infty]$

In this case, defining β by (4.3), and applying Proposition 4.2, if $\{x_t\}_0^\infty$ is any optimal program from $x \in X$, then there is a price support $\{p_t\}_0^\infty$ of $\{x_t\}_0^\infty$ such that

$$P_t \equiv (p_t/\delta^t) \leq \beta \text{ for } t \geq 0$$

Then, we can follow the proofs of Lemma 5.1, Lemma 5.2 and Theorem 5.1 to obtain (5.22), replacing A by X in the appropriate steps of the proofs.

Case II: $[\rho = \infty]$

We can choose $0 < a < \zeta$, such that

$$[V(x)/x] > [\sigma/(1-\delta)] \text{ for } 0 < x \leq a \tag{5.23}$$

where σ is given by (A8).

We claim, first, that

$$h(x) > x \text{ for } 0 < x \leq a \tag{5.24}$$

For, suppose, on the contrary that there is some $0 < x \leq a$, for which $h(x) \leq x$. Then, using Proposition 3.1, $V(x) = u(x, h(x)) + \delta V(h(x)) \leq u(x, h(x)) + \delta V(x)$, so that

$$V(x) \leq u(x, h(x))/(1-\delta)$$

Using (A7), $u(x, h(x)) \leq u(x, 0)$, and we get

$$[V(x)/x] \leq \frac{[u(x,0)/x]}{(1-\delta)} \leq \frac{\sigma}{(1-\delta)}$$

which contradicts (5.23) and establishes (5.24).

Second, we claim that

$$h(x) > 0 \text{ for } 0 < x \leq \zeta \tag{5.25}$$

Suppose, on the contrary, there is some $0 < x \leq \zeta$ such that $h(x) = 0$. Then $\{x_t\}_0^\infty$ given by $(x, 0, 0, ...)$is the optimal program from x. Given (5.24), Proposition 4.1 can be used to get a price support $\{p_t\}_0^\infty$ of $\{x_t\}_0^\infty$. Denoting (p_t/δ^t) by P_t for $t \geq 0$, we have

$$V(x_1) - P_1 x_1 \geq V(y) - P_1 y \text{ for all } y \geq 0$$

Since $x_1 = 0$, and $V(0) = 0$, we get

$$V(y) \leq P_1 y \text{ for all } y \geq 0 \tag{5.26}$$

But, by letting $y \to 0$ in (5.26), we then contradict the fact that $\rho = \infty$. This establishes (5.25).

Since h is continuous on $[a, \zeta]$, (5.25) implies that there is $b' > 0$ such that

$$h(x) \geq b' \text{ for all } x \in [a, \zeta] \tag{5.27}$$

Define $b = min[a, b']$ and $A = [b, \zeta]$. Then, we claim that A is a compact, invariant set. Since compactness is clear, we proceed to check the invariance property. We have either (I) $b = a$, or (II) $b \neq a$. If $b = a$, then $b' \geq a = b$, and (5.27) implies that $h(x) \geq b$ for $x \in [b, \zeta]$. If $b \neq a$, then $b = b' < a$, and (5.27) implies that $h(x) \geq b$ for all $x \in [a, \zeta]$. For $x \in [b, a)$, we have $h(x) > x$ by (5.24), and so $h(x) \geq b$. Thus for all $x \in [b, \zeta]$, we get $h(x) \geq b$. Since $h(x) \leq \zeta$ for all $x \in [b, \zeta]$, we have established that $A = [b, \zeta]$ is an invariant set.

If $0 < x < b$, then there is some T large enough such that $h^T(x) \in A$ and so $h^t(x) \in A$ for $t \geq T$. If $x \in A$, then $h^t(x) \in A$ for $t \geq 0$. If $x = 0$, then $h^t(x) = 0$ for $t \geq 0$.

Define $C = \bigcap\limits_{t=0}^{\infty} h^t(X)$. Then C is compact and invariant, and $C = C_1 \cup C_2$ where $C_1 = \bigcap\limits_{t=0}^{\infty} h^t(A)$ and $C_2 = \{0\}$. Clearly, C_1 and C_2 are compact, invariant sets.

Now, $\psi(h, X) = \psi(h, C)$ by Corollary 4.1.8, p. 196 of Alseda, Llibre and Misiurewicz (1993). Further, $\psi(h, C_2) = 0$, since from the definition of topological entropy, if W is any finite set, then the topological entropy of any map $g : W \to W$ is zero. Thus, using Lemma 4.1.10, p. 197 of Alseda, Llibre and Misiurewicz (1993), $\psi(h, C) = \psi(h, C_1)$. Summarizing, we have $\psi(h, X) = \psi(h, C_1)$.

Since $C_1 \subset A$, we note that C_1 is a compact, invariant set in Z, and so by Theorem 5.1,

$$\delta \leq e^{-\psi(h, C_1)}$$

Since $\psi(h, C_1) = \psi(h, X)$, we have established (5.22).

Remark 5.2: We have, so far, not been able to construct a proof of Theorem 5.2, which dispenses with assumption (A8). However, this assumption establishes the strong claim in (5.24), and that does not seem to be necessary for the rest of the proof in case (ii) to be valid.

5.3 Discounting and Metric Entropy

We have already indicated in Section 2.6 that topological chaos may not be observable, and so topological entropy might not be an appropriate measure of the complexity of a dynamical system. In this context, a natural alternative measure to consider is the metric (or measure-theoretic) entropy of a dynamical system, and to conclude that the system exhibits complicated behavior when the metric entropy is positive.

Interestingly, if we adopt this point of view, our analysis so far is still seen to be extremely useful in studying the relation between discounting and complicated behavior of a dynamical system. This is clear by noting the basic relationship between topological and metric entropy, due to Goodwyn (1968).[4]

Proposition 5.1: *Let (f, I) be a dynamical system (as in Section 2.1), and $M(f, I)$ be the set of all f-invariant probability measures on the Borel sets of I. If $\mu \in M(f, I)$, then*

$$\Phi_\mu(f, I) \leq \psi(f, I) \tag{5.28}$$

In view of Proposition 5.1 and Theorem 5.2, we have the following result relating discounting and metric entropy.

Theorem 5.3: *Let (Ω, u, δ) be a dynamic optimization model and let $h : X \to X$ be its policy function. If μ is an h-invariant probability measure on the Borel sets of X, then*

$$\delta \leq e^{-\Phi_\mu(h, X)} \tag{5.29}$$

[4]The result of Goodwyn (1968) is applicable to more general dynamical systems than ours. It turns out that the topological entropy, $\psi(f, I)$, is the *supremum* over all $\mu \in M(f, I)$ of the metric entropies, $\Phi_\mu(f, I)$. This was first established by Dinaburg (1970). For a discussion of these results in the most general setting, see Goodman (1971).

Remark 5.3: Nishimura, Sorger and Yano (Chapter 9) have provided an example in which for every $0 < \delta < 1$, there is a dynamic optimization model (Ω, u, δ) such that the policy function, h_δ, exhibits ergodic chaos. However, as they have noted, the metric entropy of h_δ converges to zero as δ converges to 1. Theorem 5.3 shows that this is true not only for that example but in general; that is, if $(\Omega_s, u_s, \delta_s)$ is a sequence of dynamic optimization models with policy functions $h_s (s = 1, 2, ...)$, and $\delta_s \to 1$ as $s \to \infty$, then the metric entropy of h_s must converge to zero at $s \to \infty$.

6. Discount Factor Restrictions for Turbulence and Topological Chaos

In this section, we demonstrate the usefulness of the Montrucchio-Sorger result, by deriving a number of implications of it. Specifically, we show how the result can be used to derive discount factor restrictions for policies exhibiting turbulence and topological chaos.

The basic technical background that we need to study discount factor restrictions for periodic programs is a result due to Block, Guckenheimer, Misiurewicz and Young (1980) which provides a formula for the topological entropy of any continuous function exhibiting a periodic program with period not equal to a power of 2. We state this result here for ready reference.

Proposition 6.1: *Let $f : I \to I$ be a continuous map with an orbit of period $q = np$ where $n > 1$ is odd and $p = 2^k$ with $k \geq 0$. Then*

$$\psi(f) \geq (\log\ \lambda_n)/p \tag{6.1}$$

where λ_n is the unique positive root of the equation

$$z^n - 2z^{n-2} - 1 = 0 \tag{6.2}$$

If we combine Proposition 6.1 with Theorem 5.2, we obtain the following result immediately.

Theorem 6.1: *Let (Ω, u, δ) be a dynamic optimization model which exhibits a periodic program of period $q = np$ where $n > 1$ is* *odd and* $p = 2^k$ *with* $k \geq 0$. *Then*

$$\delta \leq 1/\lambda_n^{1/p} \tag{6.3}$$

where λ_n *is the unique positive root of the equation*

$$z^n - 2z^{n-2} - 1 = 0 \tag{6.4}$$

Given Theorem 6.1, all one needs to obtain suitable discount factor restrictions for policy functions which exhibit periodic programs of positive topological entropy (and, therefore, which exhibit topological chaos) is an accurate calculation of the (unique) positive root of the polynomial in (6.4).

We illustrate this point with the simplest case, where the dynamic optimization model (Ω, u, δ) exhibits a period-three cycle. Here, of course, $n = 3$, $p = 1$ (so $k = 0$) and $q = np = 3$.

The polynomial in (6.4) reduces to

$$z^3 - 2z - 1 = 0 \tag{6.5}$$

It is easy to verify that

$$\lambda_3 = [\sqrt{5} + 1]/2 \tag{6.6}$$

is the unique positive root of (6.5). Thus, the discount factor restriction for a period-three cycle, by applying Theorem 6.1, is

$$\delta \leq 1/\lambda_3 \tag{6.7}$$

Now, the magnitude $(1/\lambda_3)$ can be written as

$$\begin{aligned} 2/(\sqrt{5}+1) &= 2(\sqrt{5}-1)/(\sqrt{5}+1)(\sqrt{5}-1) \\ &= 2(\sqrt{5}-1)/(5-1) \\ &= (\sqrt{5}-1)/2 \end{aligned}$$

This calculation leads to the following Corollary, which was first established by Sorger (1992b) by using entirely different methods.

Corollary 6.1: *Let (Ω, u, δ) be a dynamic optimization model, which exhibits a period-three cycle. Then*

$$\delta \leq (\sqrt{5} - 1)/2 \tag{6.8}$$

In order to obtain a discount factor restriction for a turbulent policy function, we need the following mathematical result.

Proposition 6.2: *Let $f : I \to I$ be a continuous map. Let $J_1, ..., J_p$ be closed intervals with pairwise disjoint interiors and let $A = (a_{ik})$ be the $p \times p$ matrix defined by*

$$a_{ik} = \begin{cases} 1 & \text{if } J_k \subset f(J_i) \\ 0 & \text{otherwise} \end{cases}$$

Then $\psi(f) \geq \log \lambda$, where λ is the maximal eigenvalue of A.

If we combine Proposition 6.2 with Theorem 5.2, we obtain the following result immediately.

Theorem 6.2: *Let (Ω, u, δ) be a dynamic optimization model, with policy function, $h : X \to X$. If h exhibits turbulence, then*

$$\delta \leq (1/2) \tag{6.9}$$

Proof: If h exhibits turbulence, then we can find closed subintervals J_1 and J_2 of X, with at most one point in common, such that

$$J_1 \cup J_2 \subset f(J_1) \cap f(J_2)$$

Then the interiors of J_1 and J_2 are disjoint, and the matrix $A = (a_{ik})$ of Proposition 6.1 is

$$A = \begin{bmatrix} 1 & 1 \\ 1 & 1 \end{bmatrix}$$

The maximal eigenvalue of A is 2. So, applying Proposition 6.1, we get

$$\psi(f) \geq \log\ 2 \tag{6.10}$$

Using (6.10) and Theorem 5.2, we get (6.9).

We can use Theorem 6.1 to obtain an upper bound on the discount factor, δ, that must be satisfied in order that a dynamic optimization model (Ω, u, δ) yield a periodic optimal program of odd period greater than one.

Corollary 6.2: *Suppose (Ω, u, δ) is a dynamic optimization model with policy function, h. Let $n > 1$ be any odd integer. If h has a periodic orbit of period n, then*

$$\delta < (1/\sqrt{2}) \tag{6.11}$$

Proof: Using Theorem 6.1, we know that

$$\delta \leq (1/\lambda_n) \tag{6.12}$$

where λ_n is the unique positive root of the equation

$$z^n - 2z^{n-2} - 1 = 0 \tag{6.13}$$

Define $g(z) = z^n - 2z^{n-2} - 1$ for $z \geq 0$. For $z = \sqrt{2}$, $\{g(z)/z^{n-2}\} = z^2 - 2 - (1/z^{n-2}) < z^2 - 2 = 0$. And if $z \geq \sqrt{2 + \{1/(\sqrt{2})^{n-2}\}}$, then $z^2 - 2 - (1/z^{n-2}) > z^2 - 2 - \{1/(\sqrt{2})^{n-2}\} \geq 2 + \{1/(\sqrt{2})^{n-2}\} - 2 - \{1/(\sqrt{2})^{n-2}\} = 0$. Thus we know that

$$\sqrt{2} < \lambda_n < \sqrt{2 + \{1(\sqrt{2})^{n-2}\}} \tag{6.14}$$

Combining (6.12) and (6.14), we get (6.11).

More generally, Proposition 6.1, Theorem 5.2, and (6.14) can be used to obtain upper bounds on discount factors that must hold in order that optimal programs be periodic with period $q = np$ where $n > 1$ is odd and $p = 2^k$ with $k \geq 0$.

Corollary 6.3: *Suppose (Ω, u, δ) is a dynamic optimization model with policy function, h. Let $n > 1$ be an odd integer, k be a non-negative integer and $q = n2^k$. If h has a periodic orbit with period q, then*

$$\delta < (1/\sqrt{2})^{(1/2^k)} \tag{6.15}$$

7. An Open Question

There are two results which provide "exact" discount factor restrictions for chaotic behavior in aggregate dynamic optimization models, the first dealing with the case of period-three cycles (see Mitra (Chapter 11) and Nishimura and Yano (Chapter 12)), and the second dealing with turbulence (see Mitra (1994b)). We state them here in order to provide a self-contained discussion.[5]

Proposition 7.1: *Let (Ω, u, δ) be a dynamic optimization model, with a policy function h. If h exhibits a period-three cycle, then*

$$\delta < [(\sqrt{5}-1)/2]^2 \tag{7.1}$$

Conversely, if δ satisfies (7.1), then one can construct a transition possibility set, Ω, and a reduced form utility function u, such that the dynamic optimization model (Ω, u, δ) has an optimal program exhibiting a period-three cycle.

Proposition 7.2: *Let (Ω, u, δ) be a dynamic optimization model, with a policy function h. If h exhibits turbulence, then*

$$\delta < (1/4) \tag{7.2}$$

Conversely, if δ satisfies (7.2), then one can construct a transition possibility set, Ω, and a reduced form utility function, u, such that the dynamic optimization model (Ω, u, δ) has a policy function which exhibits turbulence.

The observation we would like to make is that neither the discount factor restriction (7.1) for period-three cycles nor the discount factor restriction (7.2) for turbulence can be obtained from Theorem 5.2 which provides the basic relationship between the discount factor and the topological entropy of the corresponding policy function. In fact, as we have seen in Section 6, Theorem 5.2

[5]These results were proved for dynamic optimization models satisfying (A.1)-(A.7). Clearly, the "necessity part" of the results remain valid for dynamic optimization models satisfying (A.1)-(A.8). Furthermore, it can be checked that the examples used to demonstrate the "sufficiency part" satisfy (A.8), besides (A.1)-(A.7).

yields a discount factor restriction of $(\sqrt{5}-1)/2$ for period-three cycles (Theorem 6.1) and a discount factor restriction of $(1/2)$ for turbulence (Theorem 6.2). Interestingly, these restrictions are exactly the square roots of the exact restrictions appearing in (7.1) and (7.2) respectively.

The open question which this observation naturally raises is whether there is a more refined relationship (than is provided in Theorem 5.2) between the discount factor and the topological entropy, which would yield the exact discount factor restrictions for period-three cycles and turbulence [in (7.1) and (7.2) respectively] as special cases.

References

[1] Abraham, R., Marsden, J.E., Ratiu, T. (1988): Manifolds, Tensor Analysis and Applications. 2nd Edition. Springer-Verlag, New York

[2] Adler, R.L., Konheim, A.G., McAndrew, M.H. (1965): Topological Entropy. Trans. Amer. Math. Soc. 114, 309-319

[3] Alseda, L., Llibre, J., Misiurewicz, M. (1993): Combinatorial Dynamics and Entropy in Dimension One. World Scientific, Singapore

[4] Amir, R., Mirman, L.J., Perkins, W.R. (1991): One Sector Neo-Classical Optimal Growth: Optimality Conditions and Comparative Dynamics. Intern. Econ. Review 32, 625-644

[5] Araujo, A. (1991): The Once but not Twice Differentiability of the Policy Function. Econometrica 59, 1383-1393

[6] Araujo, A., Scheinkman, J.A. (1977): Smoothness, Comparative Dynamics, and the Turnpike Property. Econometrica 45, 601-620

[7] Arrow, K.J., Debreu, G. (1954): Existence of an Equilibrium for a Competitive Economy. Econometrica. 22, 265-290

[8] Arrow, K.J., Kurz, M. (1970): Pubic Investment, the Rate of Return and Optimal Fiscal Policy. Johns Hopkins Press, Baltimore

[9] Atsumi, H. (1965): Neoclassical Growth and the Efficient Program of Capital Accumulation. Rev. Econ. Studies 32, 127-136

[10] Aubin, J.P. (1977): Applied Abstract Analysis. John Wiley and Sons, New York

[11] Azariadis, C. (1981): Self-Fulfilling Prophecies. J. Econ. Theory 25, 380-396

[12] Bala, V., Majumdar, M. (1992): Chaotic Tatonnement. Econ. Theory 2, 437-446.

[13] Bank, B. et. al (1983): Non-Linear Parametric Optimization. Birkhauser, Basel, Boston, Massachusetts

[14] Banks, J., Brooks, J., Cairns, G., Davis, G., Stacey, P. (1992): On Devaney's Definition of Chaos. Amer. Math Monthly 99, 332-334.

[15] Baumol, W.J., Benhabib, J. (1989): Chaos: Significance, Mechanism and Economic Applications. J. Econ. Perspectives 3, 77-105

[16] Becker, R.A. (1980): On the Long-Run Steady State in a Simple Model of Equilibrium with Heterogeneous Households. Q. J. Econ. 95, 375-382

[17] Becker, R.A., Foias, C. (1987): A Characterization of Ramsey Equilibrium. J. Econ. Theory 41, 173-184

[18] Becker, R.A., Foias, C. (1990): Convergent Ramsey Equilibria. (Special issue dedicated to Nicholas GeorgescuRoegen). Lib. Math. 10, 41-52

[19] Becker, R.A., Foias, C. (1994): The Local Bifurcation of Ramsey Equilibrium. Econ. Theory 4, 719-744 [Chapter 5 in this volume]

[20] Becker, R.A., Boyd III, J.H., Foias, C. (1991): The Existence of Ramsey Equilibrium. Econometrica 59, 441-460

[21] Bellman, R. (1957): Dynamic Programming. Princeton University Press, Princeton

[22] Benedicks, M., Carleson, L. (1985): On Iterations of 1-ax^2 on (-1,1). Ann. Math. 122, 1-25

[23] Benhabib, J., Day, R. (1982): A Characterization of Erratic Dynamics in the Overlapping Generations Model. J. Econ. Dynam. Control 4, 37-55

[24] Benhabib, J., Nishimura, K. (1979): On the Uniqueness of Steady States in an Economy with Heterogeneous Capital Goods. Int. Econ. Review 20, 59-82

[25] Benhabib, J., Nishimura, K. (1985): Competitive Equilibrium Cycles. J. Econ. Theory 35, 284-306

[26] Benhabib, J., Jafarey, S., Nishimura, K. (1988): The Dynamics of Efficient Intertemporal Allocations with Many Agents. J. Econ. Theory 44, 301-320

[27] Benhabib, J., Rustichini, A. (1990): Equilibrium Cycling with Small Discounting. J. Econ. Theory 52, 423-432

[28] Benhabib, J. Rustichini, A. (1993): A Vintage Capital Model of Investment and Growth: Theory and Evidence. In: Becker, R. et. al. (Eds): General Equilibrium, Growth and Trade II. Academic Press, New York

[29] Benveniste, L.M., Scheinkman, J.A.(1979): On the Differentiability of the Value Function in Dynamic Models of Economics. Econometrica 47, 727-732

[30] Berge, C. (1963): Topological Spaces. Macmillan, New York.

[31] Berndt, E. (1976): Reconciling Alternative Estimates of the Elasticity of Substitution. Rev. Econ. Stat. 58, 59-68

[32] Bewley, T.F. (1982): An Integration of Equilibrium Theory and Turnpike Theory. J. Mathematical Economics 10, 233-268

[33] Bewley, T.F. (1986): Dynamic Implications of the Form of the Budget Constraint. In: Sonnenschein, H.F. (Ed.): Models of Economics Dynamics. Springer-Verlag, New York

[34] Blackwell, D. (1965): Discounted Dynamic Programming. Ann. Math. Stat. 36, 226-235

[35] Block, L.S., Guckenheimer, J., Misiurewicz, M., Young, L.S. (1980): Periodic Points and Topological Entropy of One Dimensional Maps. In: Nitecki, Z., Robinson, C. (Eds.): Global Theory of Dynamical Systems. Lecture Notes in Mathematics, 819, Springer-Verlag, Berlin, 18-34

[36] Block, L.S., Coppel, W.A. (1992): Dynamics in One Dimension. Lecture Notes in Mathematics, Springer-Verlag, Berlin.

[37] Boldrin, M., Deneckere,R. (1990): Sources of Complex Dynamics in Two-Sector Growth Models. J. Econ. Dynam. Control 14, 627-653

[38] Boldrin, M., Montrucchio, L. (1986a): Cyclic and Chaotic Behavior in Intertemporal Optimization Models. Math. Modelling 8, 697-700

[39] Boldrin, M., Montrucchio, L. (1986b): On the Indeterminancy of Capital Accumulation Paths. J. Econ. Theory 40, 26-39

[40] Boldrin, M., Montrucchio, L. (1989): On the Differentiability of the Policy Function. Mimeo, UCLA

[41] Boldrin, M., Woodford, M. (1990): Equilibrium Models Displaying Endogeneous Fluctuations and Chaos: A Survey. J. Monetary Econ. 25, 189-222

[42] Bowen, R. (1971a): Entropy for Group Endomorphisms and Homogeneous Spaces. Trans. Amer. Math. Soc. 154, 401-414.

[43] Bowen, R. (1971b): Periodic Points and Measures for Axiom A Diffeomorphisms. Trans. Amer. Math. Soc. 154, 377-397

[44] Boyce, W.E., DiPrima, R.C. (1977): Elementary Differential Equations and Boundary Value Problems. 3rd Edition. John Wiley and Sons, New York

[45] Boyd III, J.H. (1990): Dynamic Tax Incidence with Heterogeneous Households. Working Paper, University of Rochester

[46] Brock, W.A. (1986): Distinguishing Random and Deterministic Systems. J. Econ. Theory 40, 168-195

[47] Brock, W.A., Dechert, W.D. (1991): Nonlinear Dynamical Systems: Instability and Chaos in Economics. In: Hildenbrand, W., Sonnenschein, H. (Eds.): Handbook of Mathematical Economics. North Holland, Amsterdam

[48] Brock, W.A., Majumdar, M. (1988): On Characterizing Optimal Competitive Programs in Terms of Decentralizable Conditions. J. Econ. Theory 45, 262-273

[49] Brock, W.A., Scheinkman, J. (1976): Global Asymptotic Stability of Optimal Control Systems with Applications to the Theory of Economic Growth. J. Econ. Theory 12, 164-190

[50] Carleson, L. (1991): Stochastic Behavior of Deterministic Systems. J. Econ. Behav. Organiz. 16, 85-92

[51] Carr, J. (1981): Applications of Centre Manifold Theory. Springer-Verlag, New York

[52] Cass, D. (1965): Optimum Growth in an Aggregative Model of Capital Accumulation. Rev. Econ. Stud. 32, 233-240

[53] Cass, D., Majumdar, M. (1979) Efficiency, Consumption Value Maximization and Capital Value Transversality: A Unified View. In: Green, J., Scheinkman, J.A. (Eds.): General Equilibrium, Growth and Trade. Essays in Honor of Lionel McKenzie. Academic Press, New York

[54] Cass, D., Shell, K. (1976): The Structure and Stability of Competitive Dynamical Systems. J. Econ. Theory 12, 31-70

[55] Cass, D., Shell, K. (1983) Do Sunspots Matter? J. Polit. Econ. 91, 193-227

[56] Clark, C.W. (1971): Economically Optimal Policies for the Utilization of Biologically Renewable Resources. Math. Biosciences 12, 245-260

[57] Clark, C.W. (1976): Mathematical Bioeconomics. Wiley, New York

[58] Clarke, F.H. (1983): Optimization and Nonsmooth Analysis. John Wiley and Sons, New York

[59] Collet, P., Eckmann, J.P.(1982): Iterated Maps on the Interval as Dynamical Systems. Birkhäuser, Boston 1980

[60] Dasgupta, P.A.(1982): The Control of Resources. Blackwell, Oxford

[61] Dasgupta, S., Mitra, T. (1988): Characterization of Intertemporal Optimality in Terms of Decentralizable Conditions: The Discounted Case. J. Econ. Theory 45, 274-287.

[62] Dasgupta, S., Mitra, T. (1990): On Price Characterization of Optimal Plans in a Multi-Sector Economy. In: Dutta, B., Gangopadhyay, S., Mookherjee, D., Ray, D. (Eds.): Economic Theory and Policy. Oxford University Press, Bombay

[63] Day, R.H. (1982): Irregular Growth Cycles. Amer. Econ. Rev. 72, 406-414.

[64] Day, R.H. (1994): Complex Economic Dynamics. MIT Press, Cambridge

[65] Day, R.H., Pianigiani, G. (1991): Statistical Dynamics and Economics. J. Econ. Behav. Organiz. 16, 37-84

[66] Day, R.H., Shafer, W. (1987): Ergodic Fluctuations in Economic Models. J. Econ. Behav. Organiz. 8, 339-361

[67] Debreu, G. (1976): Least Concave Utility Functions. J. Math. Economics 3, 121-129

[68] Dechert, W.D. (1984): Does Optimal Growth Preclude Chaos? Z. Nationalökon 44, 57-61

[69] Dechert, W.D., Nishimura, K. (1983): A Complete Characterization of Optimal Growth Paths in an Aggregated Model with a Non-Concave Production Function. J. Econ. Theory 31, 332-354

[70] Denardo, E. (1967): Contraction Mappings in the Theory Underlying Dynamic Programming. SIAM Review 9, 165-177

[71] Deneckere, R., Pelikan, S. (1986): Competitive Chaos. J. Econ. Theory 40, 13-25

[72] Devaney, R.L. (1986): Chaotic Dynamical Systems. Benjamin Cummings Publishing Co., Menlo Park

[73] Devaney, R.L. (1992): A First Course in Chaotic Dynamical Systems: Theory and Experiment. Addison-Wesley, Reading

[74] Diamond, P. (1976): Chaotic Behavior of Systems of Difference Equations. International Journal of Systems Science 7, 953-956.

[75] Dinaburg, E.I. (1970): The Relation between Topological Entropy and Metric Entropy. Soviet Math. Dok. 11, 13-16

[76] Dorfman, R., Samuelson, P., Solow, R. (1958): Linear Programming and Economic Analysis. McGraw-Hill, New York

[77] Dutta, P.K. (1987): Capital Deepening and Impatience Equivalence in Stochastic Aggregative Growth Models. J. Econ. Dyn. and Control 11, 519-530

[78] Dutta, P.K., Sundaram, R.K. (1998): The Equilibrium Existence Problem in General Markovian Games. In: Organizations with Incomplete Information. Majumdar, M. (Ed.), Cambridge University Press, Cambridge

[79] Dutta, P.K., Mitra, T. (1989a): On Continuity of the Utility Function in Intertemporal Allocation Models: An Example. Intern. Econ. Review 30, 527-536

[80] Dutta, P.K., Mitra, T. (1989b): Maximum Theorems for Convex Structures with an Application to the Theory of Optimal Intertemporal Allocation. J. Math. Econ. 18, 77-86

[81] Eckmann, J.P., Ruelle, D. (1985): Ergodic Theory of Chaos and Strange Attractors. Rev. Modern Physics 57, 617-656

[82] Faustmann, M. (1968): Berechnung des Werthes, Welchen Walboden Sowie Nach Nicht Haubare Holzbestrande fur die Welwirtschaft Besitzen. Allegemeine Forest und Jagd Zeitung 25. Translated into English as: Calculation of the Value which Forestry Land and Immature Stands Possess for Forestry. In: Gane, M. (Ed.): Martin Faustmann and the Evolution of Discounted Cash Flow. Institute Paper No. 42, Commonwealth Forestry Institute, University of Oxford, 27-55

[83] Flynn, J. (1980): The Existence of Optimal Invariant Stocks in a Multi-Sector Economy. Rev. Econ. Studies 47, 809-811

[84] Frank, M., Stengos, T. (1988): Chaotic Dynamics in Economic Time-Series. J. Econ. Surveys 2, 103-133

[85] Gale, D. (1955): The Law of Supply and Demand. Math. Scand. 30, 155-169

[86] Gale, D. (1967): On Optimal Development in a Multi-Sector Economy. Rev. Econ. Stud. 34, 1-18

[87] Goodman, T.N.T. (1971): Relating Topological Entropy and Measure Entropy. Bull. London Math. Soc. 3, 176-180

[88] Goodwyn, L.W. (1969): Topological Entropy Bounds Measure-Theoretic Entropy. Proc. Amer. Math. Soc. 23, 679-688

[89] Grandmont, J.M. (1985): On Endogenous Competitive Business Cycles. Econ. 53, 995-1045

[90] Grandmont, J.M. (1986): Periodic and Aperiodic Behavior in Discrete One Dimensional Dynamical Systems. In: Hildenbrand, W., Mas-Colell, A. (Eds.): Contributions to Mathematical Economics in Honor of Gerard Debreu. North Holland, New York

[91] Guckenheimer, J. (1979): Sensitive Dependence to Initial Conditions for One Dimensional Maps. Commun. Math. Phys. 70, 133-160

[92] Guckenheimer, J., Holmes, P. (1983): Nonlinear Oscillations, Dynamical Systems, and Bifurcations of Vector Fields. Springer-Verlag, New York

[93] Guesnerie, R., Woodford, M. (1992): Endogeneous Cycles. In: Laffont, J.J. (Ed.): Advances in Economic Theory. Cambridge University Press, Cambridge

[94] Halmos, P. (1956): Lectures on Ergodic Theory. The Mathematical Society of Japan, Tokyo.

[95] Henon, M. (1976): A Two-Dimensional Mapping with a Strange Attractor. Commun. Math. Phys. 50, 69-77

[96] Hernández, A. (1991): The Dynamics of Competitive Equilibrium Allocations with Borrowing Constraints. J. Econ. Theory 55, 180-191

[97] Jakobson, M.V. (1981): Absolutely Continuous Invariant Measures for One-Parameter Families of One-Dimensional Maps. Commun. Math. Phys. 81, 39-88

[98] Kelley, A. (1967): The Stable Center-Stable, Center, Center-Unstable, Unstable Manifolds. J. Diff. Eq. 3, 546-570

[99] Khan, M.A., Mitra, T. (1986): On the Existence of a Stationary Optimal Stock for a Multi-Sector Economy: A Primal Approach. J. Econ. Theory 40, 319-328

[100] Koopmans, T.C. (1957): Three Essays on the State of Economic Science. McGraw-Hill

[101] Koopmans, T.C. (1964): Economic Growth at a Maximal Rate. Quart. Journ. Econ. 78, 355-394

[102] Koopmans, T.C. (1965): The Concept of Optimal Economic Growth. In: The Econometric Approach to Development Planning, Pontificae Academiae Scientiarum Scripta Varia. No. 28, North-Holland, Amsterdam

[103] Koopmans, T.C. (1967a): Objectives, Constraints and Outcomes in Optimal Growth Models. Econometrica 35, 1-15

[104] Koopmans, T.C. (1967b): Intertemporal Distribution and 'Optimal' Economic Growth. In: Fellner, W., et al. (Eds.): Ten Economic Studies in the Tradition of Irving Fisher. Chapter 5, Wiley, New York, 95-126

[105] Kurz, M. (1968): Optimal Economic Growth and Wealth Effects. Int. Econ. Rev. 9, 155-174

[106] Kurz, M. (1969): On the Inverse Optimal Problem. In: Kuhn, H.W. and Szegö, G.P. (Eds.): Mathematical Systems Theory and Economics I. Springer-Verlag, Berlin, 189-201

[107] Laitner, J. (1989): Dynamic Determinacy and the Existence of Sunspot Equilibria. J. Econ. Theory 47, 39-50

[108] Lasota, A., Yorke, J.A. (1973): On the Existence of Invariant Measures for Piecewise Monotonic Transformations. Trans. Amer. Math. Soc. 186, 481-488

[109] Lasota, A., Yorke, J.A. (1977): On the Existence of Invariant Measures for Transformations with Strictly Turbulent Trajectories. Bull. Acad. Polon. Sci. Ser. Sci. Math. Astronom. Phys. 25, 233-238

[110] Li, T., Yorke, J. (1975): Period Three Implies Chaos. Am. Math. Monthly 82, 985-992

[111] Li, T., Yorke, J. (1978): Ergodic Transformations from an Interval into Itself. Trans Amer Math Soc 235, 183-192

[112] Li, T., Misiurewicz, M., Pianigiani, G., Yorke, J. (1982): No Division Implies Chaos. Trans. Amer. Math. Soc. 273, 191-199

[113] Maitra, A. (1968): Discounted Dynamic Programming on Compact Metric Spaces. Sankhyā, Ser. A 30, 211-216

[114] Malinvaud, E. (1953): Capital Accumulation and Efficient Allocation of Resources. Econometrica 21, 233-268

[115] Majumdar, M. (1992): Decentralization in Infinite Horizon Economies. Westview Press, Boulder, Colorado

[116] Majumdar, M., Mitra, T. (1983): Dynamic Optimization with a Non-Convex Technology: The Case of a Linear Objective Function. Rev. Econ. Studies 50, 143-151

[117] Majumdar, M., Mitra, T. (1994a): Periodic and Chaotic Programs of Optimal Intertemporal Allocation in an Aggregative Model with Wealth Effect. Econ. Theory 4, 649-676 [Chapter 3 of this volume]

[118] Majumdar, M., Mitra, T. (1994b): Robust Ergodic Chaos in Discounted Dynamic Optimization Models. Econ. Theory 4, 677-688 [Chapter 7 in this volume]

[119] Majumdar, M., Nermuth, M. (1982): Dynamic Optimization in Non-Convex Models with Irreversible Investment: Monotonicity and Turnpike Results. Z. Nationalökonomie 42, 339-362

[120] Marshall, A., Olkin, I. (1979): Inequalities: Theory of Majorization and its Applications. Academic Press, New York

[121] Matsuyama, K. (1990): Sunspot Equilibria (Rational Bubbles) in a Model of Money-in-the-Utility-Function. J. Monet. Econ. 25, 137-144

[122] McFadden, D. (1967): The Evaluatin of Development Programmes. Rev. Econ. Studies 34, 25-50

[123] McKenzie, L.W. (1954): On Equilibrium in Graham's Model of World Trade and other Competitive Systems. Econometrica 22, 147-161

[124] McKenzie, L.W. (1963a): The Turnpike Theorem of Morishima. Rev. Econ. Studies 30, 169-176

[125] McKenzie, L.W. (1963b): Turnpike Theorems for a Generalized Leontief Model. Econometrica 31, 165-180

[126] McKenzie, L.W. (1963c): Maximal Paths in the von Neumann Model. In: Malinvaud, E., Bacharach, M.O.L. (Eds.): Activity Analysis in the Theory of Growth and Planning. Macmillan, London

[127] McKenzie, L.W. (1968): Accumulation Programs of Maximum Utility and the von Neumann Facet. In: Wolfe, J.N. (Ed.): Value, Capital and Growth. University Press, Edinburgh, 353-383

[128] McKenzie, L.W. (1971): Capital Accumulation Optimal in the Final State. In: Bruckmann, G., Weber, W. (Eds.): Contributions to the von Neumann Growth Model. Springer-Verlag, Vienna

[129] McKenzie, L.W. (1976): Turnpike Theory. Econometrica 44, 841-865

[130] McKenzie, L.W. (1982): A Primal Route to the Turnpike and Liapounov Stability. J. Econ. Theory 27, 194-209

[131] McKenzie, L.W. (1983): Turnpike Theory, Discounted Utility and the von Neumann Facet. J. Econ. Theory 30, 330-352

[132] McKenzie, L.W. (1986): Optimal Economic Growth: Turnpike Theorems and Comparative Dynamics. In: Arrow, K.J., Intriligator, M.D. (Eds.): Handbook of Mathematical Economics. Vol. III. North-Holland, Amsterdam, 1281-1355

[133] Milgrom, P., Shannon, C. (1994): Monotone Comparative Statics. Econometrica 62, 157-180

[134] Milnor, J. (1965): Topology from the Differentiable Viewpoint. University Press of Virginia, Charlottesville

[135] Mirman, L.J., Zilcha, I. (1975): On Optimal Growth under Uncertainty. J. Econ. Theory 11, 329-339

[136] Misiurewicz, M. (1979): Horseshoes for Mappings of the Interval. Bull. Acad. Polon. Sci. Ser. Sci. Math. 27, 167-169

[137] Misiurewicz, M. (1983): Maps of an Interval. In: Iooss, G., Hellerman, R., and Stora, R (Eds.): Chaotic Behavior of Deterministic Systems. North-Holland, Amsterdam

[138] Misiurewicz, M. (1981): Absolutely Continuous Measures for Certain Maps of an Interval. Publications Mathematiques 53, 17-51

[139] Mitra, T. (1994a): An Exact Discount Factor Restriction for Period Three Cycles in Dynamic Optimization Models. CAE Working Paper #94-11, Cornell University, J. Econ. Theory forthcoming [Chapter 11 in this volume]

[140] Mitra, T. (1994b): On Discount Rate Restrictions for Turbulence and Topological Chaos in Dynamic Optimization Models. Mimeo

[141] Mitra, T. (1995): On the Relationship between Discounting and Complicated Behavior in Dynamic Optimization Models. CAE Working Paper 95-11, Cornell University [Chapter 13 in this volume]

[142] Mitra, T. (1996): An Exact Discount Factor Restriction for Period-Three Cycles in Dynamics Optimization Models. J. Econ. Theory 69, 281-305

[143] Mitra, T., Nishimura, K. (1998): Discounting and Long-Run Behavior: Global Bifurcation Analysis of a Family of Dynamical Systems. J. Econ. Theory, forthcoming

[144] Mitra, T., Nishimura, K. (1999): Intertemporal Complementarity and Optimality: A Study of a Two-Dimensional Dynamical System, mimeo

[145] Mitra, T., Sorger, G. (1999): On the Existence of Chaotic Policy Functions in Dynamic Optimization. Japanese Econ. Review, forthcoming

[146] Mitra, T., Ray, D. (1984): Dynamic Optimization on a Non-Convex Feasible Set: Some General Results for Non-Smooth Technologies. Z. Nationalökonomie 44, 151-175

[147] Mitra, T., Ray, D., Roy, R. (1991): The Economics of Orchards: An Exercise in Point-Input, Flow-Output Capital Theory. J. Econ. Theory 53, 12-50

[148] Mitra, T., Wan, H.Y., Jr. (1985): Some Theoretical Results on the Economics of Forestry. Rev. Econ. Studies 52, 263-282

[149] Mitra, T., Wan, H.Y., Jr. (1986): On the Faustmann Solution to the Forest Management Problem. J. Econ. Theory 40, 229-249

[150] Montrucchio, L. (1988): The Occurrence of Erratic Fluctuations in Models of Optimization over Infinite Horizon. In: Ricci, G., Velupillai, K. (Eds.): Growth Cycles and Multisectoral Economics: The Goodwin Tradition. Springer Verlag, Berlin, 83-92

[151] Montrucchio, L. (1989): Dynamical Systems that Solve Continuous-Time Concave Optimization Problems: Anything Goes. In: Benhabib, J. (Ed): Cycles and Chaos in Economic Equilibrium. Princeton Univ. Press, Princeton, 277-288

[152] Montrucchio, L. (1994): Dynamic Complexity of Optimal Paths and Discount Factors for Strongly Concave Problems. J. Optimization Theory Applic. 80, 385-406

[153] Montrucchio, L. (1998): Thompson Metric, Contraction Property and Differentiability of Policy Functions. J. Econ. Beh. and Organization 33, 449-466

[154] Montrucchio, L. Sorger, G. (1994): Topological Entropy of Policy Functions in Concave Dynamic Optimization Models. J. Math. Econ. 25, 181-194

[155] Morishima, M. (1961): Proof of a Turnpike Theorem: The No Joint Production Case. Rev. Econ. Studies 28, 89-97

[156] Nikaido, H. (1956): On the Classical Multilateral Exchange Problem. Metroeconomica 8, 135-145

[157] Nikaido, H. (1968): Convex Structures and Economic Theory. Acadmic Press, New York

[158] Nishimura, K., Sorger, G., Yano, M. (1994): Ergodic Chaos in Optimal Growth Models with Low Discount Rates. Econ. Theory 4, 705-717 [Chapter 9 of this volume]

[159] Nishimura, K., Yano, M. (1992): Non-Linear Dynamics and Chaos in Optimal Growth. Kyoto University Discussion Paper

[160] Nishimura, K., Yano, M. (1993a): Endogenous Real Business Cycles and International Specialization. In: Ethier, W., Helpman, E., Neary, P. (Eds.): Theory, Policy and Dynamics in International Trade: Essays in Honor of Ronald W. Jones. Cambridge University Press, Cambridge

[161] Nishimura, K., Yano, M. (1993b): Optimal Chaos when Future Utilities are Discounted Arbitrarily Weakly. Researches in Nonlinear Analysis and Mathematical Economics, Lecture Note Series. Institute for Mathematical Analysis

[162] Nishimura, K., Yano, M. (1993c): Interlinkage in the Endogenous Real Business Cycles of International Economies. Econ. Theory 3, 153-170

[163] Nishimura, K., Yano, M. (1994): Optimal Chaos, Nonlinearity and Feasibility Conditions. Econ. Theory 4, 689-704 [Chapter 4 of this volume]

[164] Nishimura, K., Yano, M. (1995a): Non-Linearity and Business Cycles in a Two-Sector Equilibrium Model: An Example with Cobb-Douglas Production Functions. In: Maruyama, T., Takahashi, W. (Eds.): Nonlinear and Convex Analysis in Economic Theory. Springer-Verlag, 231-245

[165] Nishimura, K., Yano, M. (1995b): Nonlinear Dynamics and Chaos in Optimal Growth: An Example. Econometrica 63, 981-1001

[166] Nishimura, K., Yano, M. (1996):On the Least Upper Bound of Discount Factors that are Compatible with Optimal Period-Three Cycles. J. Econ. Theory 66, 306-333 [Chapter 12 in this volume]

[167] Ortega, J.M., Rheinboldt, W.C. (1970): Iterative Solution of Nonlinear Equations in Several Variables. Academic Press, New York

[168] Parthasarathy, K. (1967): Probability Measures on Metric Spaces. Academic Press, New York

[169] Peck, J. (1988) On the Existence of Sunspot Equilibria in an Overlapping Generations Model. J Econ Theory 44, 19-42

[170] Peleg, B. (1970): Efficiency Prices for Optimal Consumption Plans, III. J. Math. Analy. and Applications 32, 630-638

[171] Peleg, B. (1973): A Weakly Maximal Golden-Rule Program for a Multi-Sector Economy. Int. Econ. Review 14, 574-579

[172] Peleg, B., Ryder, H.E. Jr. (1972): On Optimal Consumption Plans in a Multi-Sector Economy. Rev. Econ. Studies 39, 159-169

[173] Peleg, B., Ryder, H.E. Jr. (1974): The Modified Golden-Rule of a Multi-Sector Economy. J. Math. Econ. 1, 193-198

[174] Pontryagin, L.S., et al. (1962): The Mathematical Theory of Optimal Processes. Interscience, New York

[175] Radner, R. (1961): Paths of Economic Growth that are Optimal with Regard Only to Final States. Rev. Econ. Stud. 28, 98-104

[176] Radner, R. (1967a): Dynamic Programming of Economic Growth. In Malinvaud, E., Bacharach, M. (Eds.): Activity Analysis in the Theory of Planning and Growth. St. Martin's Press, London 150-169

[177] Radner, R. (1967b): Efficiency Prices for Infinite-Horizon Production Programmes. Rev. Econ. Studies 34, 51-66

[178] Ramsey, F.P. (1928): A Mathematical Theory of Saving. Econ. J. 38, 543-559

[179] Rasband, S.N. (1990): Chaotic Dynamics of Nonlinear Systems. John Wiley, New York

[180] Rockafellar, R.T. (1970): Convex Analysis. Princeton University Press, Princeton

[181] Rockafellar, R.T. (1976): Saddle Points of Hamiltonian Systems in Convex Lagrange Problems Having a Non-Zero Discount Rate. J. Econ. Theory 12, 71-113.

[182] Ross, S.M. (1983): Introduction to Stochastic Dynamic Programming. Academic Press, New York

[183] Royden, H.L. (1988): Real Analysis. Macmillan, New York

[184] Ruelle, D., Takens, F. (1971): On the Nature of Turbulence. Comm. Math. Phys. 20, 167-192

[185] Ruelle, D. (1989): Elements of Differentiable Dynamics and Bifurcation Theory. Academic Press, New York

[186] Saari, D.G. (1991): Erratic Behavior in Economic Models. J. Econ. Behav. and Organiz. 16, 3-35.

[187] Samuelson, P.A. (1947): Foundations of Economic Analysis. Harvard University Press, Cambridge

[188] Samuelson, P.A. (1949): Market Mechanism and Maximization, Memo, Rand Corp., Santa Monica, California

[189] Samuelson, P.A. (1965): A Catenary Turnpike Theorem Involving Consumption and the Golden Rule. Amer.Econ. Review 55, 4860496

[190] Samuelson, P.A. (1973): Optimality of Profit-Including Prices under Ideal Planning. Proc. Nat. Acad. Sci. USA 70, 2109-2111

[191] Santos, M. (1991): Smoothness of the Policy Function in Discrete Time Economic Models. Econometrica 59, 1365-1382

[192] Sarkovskii, A. (1964): Coexistence of Cycles of a Continuous Map of a Line into Itself. Ukr. Math. Zh. 16, 61-71

[193] Scheinkman, J. (1976): On Optimal Steady State of n-Sector Growth Models when Utility is Discounted. J. Econ. Theory 12, 11-30

[194] Scheinkman, J. (1990): Nonlinearities in Economic Dynamics. Econ. J. 100, 33-48

[195] Scheinkman, J., Lebaron, B. (1989): Non-Linear Dynamics and Stock Returns. J. Business 62, 311-377

[196] Shell, K. (1977): Monnaie et Allocation Intertemporelle. mimeo., Seminaire d'Econometrie Roy-Malinvaud, Centre National de la Recherche Scientifique, Paris, November 21

[197] Sorger, G. (1990a): On the Optimality of Given Feedback Controls. J. Optim. Theory Appl. 65, 321-329.

[198] Sorger, G. (1990b): On the Minimal Rate of Impatience for Complicated Optimal Growth Paths. Forschungsbericht 130, Technische Universität Wien

[199] Sorger, G. (1992a): On the Minimum Rate of Impatience for Complicated Optimal Growth Paths. J. Econ. Theory 56, 160-179 [Chapter 10 in this volume]

[200] Sorger, G. (1992b): Minimum Impatience Theorems for Recursive Economic Models. Springer-Verlag, Berlin

[201] Sorger, G. (1992c): On the Sensitivity of Optimal Growth Paths. Working Paper, University of Vienna

[202] Sorger, G. (1992d): Period Three Implies Heavy Discounting. Mimeo

[203] Sorger, G. (1994a): Period Three Implies Heavy Discounting. Math. Operations Res. 19, 1-16

[204] Sorger, G. (1994b): Chaotic Ramsey Equilibrium. Mimeo, University of Vienna.

[205] Spear, S.E. (1991): Growth, Externalities, and Sunspots. J Econ Theory 54, 215-223

[206] Srinivasan, T.N. (1964): Optimum Savings in a Two-Sector Model of Economic Growth. Econometrica 32, 358-373.

[207] Stokey, N., Lucas, Jr. R., Prescott, E. (1989): Recursive Methods in Economic Dynamics. Harvard Univ. Press, Cambridge

[208] Strauch, R.E. (1966): Negative Dynamic Programming. Ann. Math. Stat. 37, 871-890

[209] Sutherland, W. (1967): On Optimal Development Programs when Future Utility is Discounted. Ph.D. Thesis, Brown University

[210] Sutherland, W. (1970): On Optimal Development in a Multi-Sectoral Economy: The Discounted Case. Rev. Econ. Studies 37, 585-589

[211] Topkis, D. (1978): Minimizing a Submodular Function on a Lattice. Operations Research 26, 305-321

[212] Uzawa, H. (1964): Optimal Growth in a Two-Sector Model of Capital Accumulation. Rev. Econ. Studies 31, 1-24

[213] Vinter, R.B., Lewis, R.M. (1980): A Verification Theorem which Provides a Necessary and Sufficient Condition for Optimality. IEEE Trans. Automat. Control 25, 84-89.

[214] von Neumann, J. (1945): A Model of General Economic Equilibrium. Rev. Econ. Studies 13, 1-9

[215] von Weizsäcker, C.C. (1965): Existence of Optimal Programs of Accumulation for an Infinite Time Horizon. Rev. Econ. Studies 32, 85-104

[216] Wan, H.Y., Jr. (1989): Optimal Evolution of Tree-Age Distribution for a Tree Farm. In: Castillo-Chavez, C. et. al. (Eds.): Mathematical Approaches to Ecological and Environmental Problems. Lect. Notes Bio-Math. Springer-Verlag, Berlin

[217] Wan, H.Y., Jr. (1993): A Note on Boundary Optimal Paths. In: Becker, R. et. al. (Eds.): General Equilibrium, Growth, and Trade II. Academic Press, New York

[218] Weitzman, M. (1973): Duality Theory for Infinite Horizon Convex Models. Manag. Sci. 19, 783-789.

[219] Woodford, M. (1984): Indeterminacy of Equilibrium in the Overlapping Generations Model: A Survey. Mimeo, Columbia University, New York

[220] Woodford, M. (1986a): Stationary Sunspot Equilibria: The Case of Small Fluctuations around a Deterministic Steady State. Mimeo, University of Chicago, Ilinois

[221] Woodford, M. (1986b): Stationary Sunspot Equilibria in a Finance Constrained Economy. J Econ Theory 40, 128-137

[222] Woodford, M. (1988a): Imperfect Financial Intermediation and Complex Dynamics. In: Barnett, W.A., Geweke, J., Shell, K. (Eds.): Economic Complexity: Chaos, Bubbles, and Nonlinearity. Cambridge University Press, Cambridge

[223] Woodford, M. (1988b): Expectations, Finance, and Aggregate Instability. In: Kohn, M., Tsiang, S.-C. (Eds.): Finance Constraints, Expectations, and Macroeconomics. Clarendon Press, Oxford

[224] Yano, M. (1990): Teoria del Equilibrio Con Sendas Convertentes. Cuadernos Economicos 46, 27-59